Advanced Control of Wheeled Inverted Pendulum Systems

Zhijun Li · Chenguang Yang · Liping Fan

Advanced Control of Wheeled Inverted Pendulum Systems

Springer

Zhijun Li
College of Automation Science and Engineering
South China University of Technology
Guangzhou, People's Republic of China
and
Department of Automation
Shanghai Jiao Tong University
Shanghai, People's Republic of China

Chenguang Yang
School of Computing and Mathematics
University of Plymouth
Plymouth, Devon, UK

Liping Fan
Department of Automation
Shanghai Jiao Tong University
Shanghai, People's Republic of China

ISBN 978-1-4471-2962-2 ISBN 978-1-4471-2963-9 (eBook)
DOI 10.1007/978-1-4471-2963-9
Springer London Heidelberg New York Dordrecht

Library of Congress Control Number: 2012942916

Printed on acid-free paper

Springer is part of Springer Science+Business Media (www.springer.com)

Preface

Recently, many robotic systems designed with wheeled inverted pendulum (WIP), such as the Segway PT and Segway type system based Robonauts (NASA robot astronauts), have been quite popular in the robotic community. In fact, WIP based robots are able to provide effective physical assistance to humans in various activities such as delivery and touring. The design with WIP guarantees high work capabilities, e.g., high speed, convenience without sacrifice of safety. The simple structure of the robot on the WIP contributes to reducing the weight of the system compared to a wheeled mobile robot and it ensures high safety against overturn when unpredictable collision occurs, because it is always controlled to maintain dynamic stability. The two coaxial wheels configuration as well as the small foot-print makes it able to achieve high mobility, to do stationary U-turns on dime, to navigate on various terrains and traverse small steps or curbs. These properties would definitely give the operator greater maneuverability and thus access to places most able-body people take for granted.

However, the control of the self unstable WIP system is a challenge. First of all, the kinematics of the system have been proved to be uncontrollable and therefore balancing of the pendulum is only achieved by dynamic effects. Active control must be applied to drive the wheels in the direction the upper part of the robot is falling to maintain the balance. In addition, WIP systems are different from either the conventional cart and pendulum systems or the conventional non-holonomic systems, many previous developed control approaches for these two kinds of systems are not applicable directly to the WIP systems.

Compared with other non-holonomic systems, the WIP systems are subject to (i) only kinematic constraints which geometrically restrict the direction of mobility, i.e., wheeled mobile robot; (ii) only dynamic constraints due to dynamic balance at passive degrees of freedom where no force or torque is applied, i.e., the robotic manipulator with passive link. It belongs to (iii) not only kinematic constraints but also dynamic constraints. While compared with cart and pendulum systems, the cart of WIP is no longer constrained to the guide rail but moves in its terrain while balancing the pendulum and the motors driving the wheels are directly mounted on the body of the pendulum. Therefore, many available control design approaches are not applicable to the WIP systems. Moreover, another notable control challenge of WIP lies in multiple under-actuated configurations, i.e., the number of the control inputs is less than the number of the degrees of freedom to be stabilized.

Although modeling and control of WIP systems have attracted much research attention in last decade, many fundamental problems are still either unexplored or less well understood. In particular, there still lacks a comprehensive framework that can cope with all the core issues in a systematic way. This motivated us to write the current monograph.

The book presents theoretical explorations on several fundamental problems for the modeling, motion planning, control and identification of the WIP systems. By integrating fresh concepts and state-of-the-art results to form a systematic approach for the motion control and identification, a fundamental theoretical framework is formed towards WIP systems.

The book is primarily intended for researchers and engineers in the robotics and control community. It can also serve as complementary reading for nonlinear system theory at the post-graduate level.

The book starts with a brief summary of the useful mathematical concepts and tools from operator, matrix and sign function, stability theory, as well as other stability results in Chap. 2. They are the tools used for stability analysis of the controllers presented later. Then, a detailed overview of the kinematics and dynamics of the WIP systems is presented in Chap. 3. To make the book self-content, we start from the basic Newton–Euler and Lagrange–Euler equation to derive the dynamics equation governing the general robotic systems, and then we focus on the expression and analysis of the WIP system dynamics.

In Chap. 4, we investigate control design approaches based on the linearized WIP dynamical model. Self-tuning PID controller is first studied, and then the optimal LQR control is designed, finally H_∞ control and backstepping are investigated. The corresponding simulation results of the proposed controls have been well compared.

In Chap. 5, we further develop nonlinear control for the WIP systems. First, we investigate the nonlinear feedback linearization approach. Then, taking advantage of the physical model properties, the system is decomposed into three sub-systems and model-based control and stability analysis have been studied for each of the sub-system.

In Chap. 6, both adaptive and robust motion control designs have been studied. Then control design is proceeded with on-line parameters estimation strategy and introduced to compensate for dynamics uncertainties and external disturbances. Then, hybrid motion/force adaptive robust control has been thoroughly studied. The unmodeled dynamics have been compensated by both online learning of the uncertain parameters and robust approach.

In Chap. 7, a number of intelligent control approaches have been investigated. First, the modeling and control method using the Least Squares Support Vector Machine (LS-SVM) have been utilized to design efficient model free control. Then, we further study the universal functional approximation of fuzzy logic and neural networks. All these intelligent control methods employ a systematic online adaptation mechanism without prepared off line learning.

In Chap. 8, aiming at shaping the controlled vehicle angular motion dynamics to be of minimized motion tracking errors and accelerations, we employ the linear quadratic regulation (LQR) optimization technique to obtain an optimal reference model. Adaptive control has then been developed using variable structure to ensure the reference model can be matched within a prescribed finite time horizon, in the presence of various uncertainties. The property of the optimized reference model as well as the minimized yaw and tilt angle accelerations guarantee rider's comfort.

In Chap. 9, the model reference control proposed in Chap. 8 has been further developed using neural network (NN) for the fully actuated angular motion subsystem. Same as in Chap. 8, we exploit the fact that the forward velocity can be indirectly affected by the reference trajectory of tilt angle, and design an NN based reference trajectory generator that guarantee the desired forward velocity is achieved.

Acknowledgments

For the creation of the book, we are very fortunate to have many helpful suggestions from our colleagues, friends and coworkers, through many stimulating and fruitful discussions. First of all, we would like to express our sincere gratitude to the careful scrutiny. We would also like to express our sincere appreciation to our coworkers who have contributed to the collaborative studies of Robotics.

We are greatly indebted to our parents and families for their love, support and sacrifice over the years. It is our pleasure to dedicate the book to them for their invisible yet invaluable contribution.

This work is supported by the National Natural Science Foundation of China (Nos. 60804003, 61174045, and 60935001) and International Science & Technology Cooperation Program of China (No. 2011DFA10950).

Shanghai, China — Zhijun Li
Plymouth, UK — Chenguang Yang
Shanghai, China — Liping Fan

Contents

List of Symbols

x^T or A^T — the transpose of vector x or matrix A
A^{-1} — the inverse of matrix A
A^+ — the Moore–Penrose pseudo-inverse of matrix A
$P > 0$ ($P \geq 0$) — matrix P is real symmetric and (semi-)positive definite
$P < 0$ ($P \leq 0$) — matrix P is real symmetric and (semi-)negative definite
$\mathrm{diag}[\cdots]$ — diagonal matrix with given diagonal elements
$\det A$ — the determinant of matrix A
$\lambda(A)$ — the set of eigenvalues of A
$\lambda_{\max}(A)$ — the maximum eigenvalue of real symmetric matrix A
$\lambda_{\min}(A)$ — the minimum eigenvalue of real symmetric matrix A
x_i, $x(j)$ — the ith element of vector x
$A(i, j)$, A_{ij} — the ijth element of matrix A
$|a|$ — the absolute value of number a
$\|x\|$ — the Euclidean norm of vector x
$\|A\|$ — the induced Euclidean norm of matrix A
$\|x\|_p$ — the l_p norm of vector x
$\|A\|_p$ — the induced l_p-norm of matrix A
$\langle x, y\rangle$ — the inner product of x and y in $\mathbf{R}^n$
$\mathcal{X}/\mathcal{Y}$ or $\frac{\mathcal{X}}{\mathcal{Y}}$ — the factor (quotient) space
$\{x, y, \ldots\}$ — a set of quantities x, y, etc.
$\max S$ — the maximum element of set S
$\min S$ — the minimum element of set S
$\sup S$ — the smallest number that is larger than or equal to each element of set S
$\inf S$ — the largest number that is smaller than or equal to each element of set S
$\arg\max S$ — the index of maximum element of ordered set S
$\arg\min S$ — the index of minimum element of ordered set S
$(\hat{\cdot})$ — the estimated of $(\cdot)$
$(\tilde{\cdot})$ — $(\cdot) - (\hat{\cdot})$
$\mathrm{sgn}(\cdot)$ — the signum function

sat$(\cdot)$	the saturation function with unit limits
$[a, b)$	the real number set $\{t \in \mathbf{R}\colon a \le t < b\}$ or the integer set $\{t \in \mathbf{N}\colon a \le t < b\}$
$[a, b]$	the real number set $\{t \in \mathbf{R}\colon a \le t \le b\}$ or the integer set $\{t \in \mathbf{N}\colon a \le t \le b\}$
iff	if and only if

Chapter 1
Introduction

Robotics in the twenty century plays a key role in the emerging industries. In 21st century, robotic technology is ubiquitous in various fields and it brings enormous benefits and profit to our human society. In the last century, robots were generally used to aid or replace human labor in modern factories. While in the recent decade, robots have been more used in daily environments such as home or office, where they serve as guard robots, service robots, care robots, exploration robots, and entertainment robots. Intelligent robots need a specific mechanism suitable for the specific application of the robot. Accordingly, we need to investigate the mechanically stable implementation of each type of mechanism.

Although intelligent robots can be created with various kinds of mechanisms, most mechanisms can be categorized as humanoid or animal-type robots. Humanoid robots, such as ASIMO [3], TOPIO [6], and NAO [5], are burdened by a highly complex control, though because of the similar external appearance as human they are largely considered appropriately as intelligent robots. Mechanisms that resemble animals or insects are also frequently adopted because of their human-friendly characteristics, as can be seen in the case of Sony's AIBO [2] and BigDog [4]. These animal-type robots have been manufactured and commercialized and they have the advantage of the natural intelligence out of their shapes.

The type of intelligent robot investigated in this book is a human-like mobile robot with two driven wheels and an inverted pendulum, the so called wheeled inverted pendulum (WIP) systems. This design is chosen because its mechanism has an innately clumsy motion for stabilizing the robot's body posture. The robot has a body with two wheels for moving in a plane and is similar to a human walking/running posture. Two independent driving wheels for compact structure without casters are used for position control, and for some actions that help the robot appear to be human motion.

Any physical systems possessing less actuators than degrees of freedom are underactuated systems. WIP robot has two driving wheels while it has four degrees of freedom, i.e., three of planar motions and one of tilt-angular motion. In addition, its dynamics are also nonlinear such that the WIP robot is an underactuated nonlinear system. The research for such kind of systems has been well carried out in the

Z. Li et al., *Advanced Control of Wheeled Inverted Pendulum Systems*,
DOI 10.1007/978-1-4471-2963-9_1,

Fig. 1.1 Self-balancing robot from the University of Electro-Communications

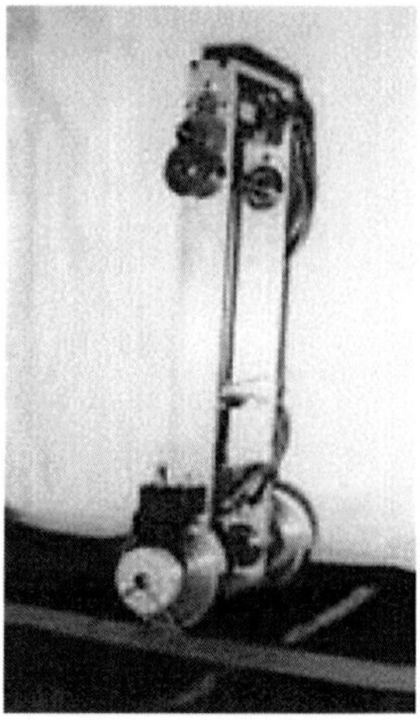

past years, however the applications were restricted to certain kinds of robots such as free-flying robots [120], snake-like robots [111] or crawling robots [69]. Readers may easily link WIP systems with the cart and pendulum systems [135], which seem quite similar at the first glance. However, there are major distinctions between these two kinds of systems, e.g., WIP system is drive by two wheels and its motion is not planar, and in addition, the motors driving the wheels are directly mounted on the pendulum body [103], while the actuator does not directly drive the pendulum in the cart and pendulum systems. Therefore, further study of dynamics modeling, control design as well as related analysis should be carried out for this special kind of mechanical system.

1.1 An Overview of WIP Robots

A number of studies specially focusing on the two-wheeled inverted pendulum robot have been undertaken. The idea that a robot can self-balance on wheel(s) has gained momentum over the last decade in various robot laboratories, for both academic research and practical application worldwide. In Fig. 1.1, the first two-wheel inverted pendulum robot was built by Kazuo Yamafuji, a Professor Emeritus at the University of Electro-Communications in Tokyo, in 1986, according to an article in the *Japan Times* [8]. The robot can simulate the behavior of an inverted pendulum, however, it is with different forms, the rolling cart is attached to a platform or two wheels.

The Equibot shown in Fig. 1.2 is balancing robot developed by Dani Piponi [7]. It uses a sharp infrared sensor to measure the distance to the floor and uses distance information to deduce its tilt angle.

The Segbot shown in Fig. 1.3 was developed in the spring of 2004 [1]. The goal was to design and build a two wheeled balancing robot based on the same principles of the Segway.

Fig. 1.2 Balancing Equibot

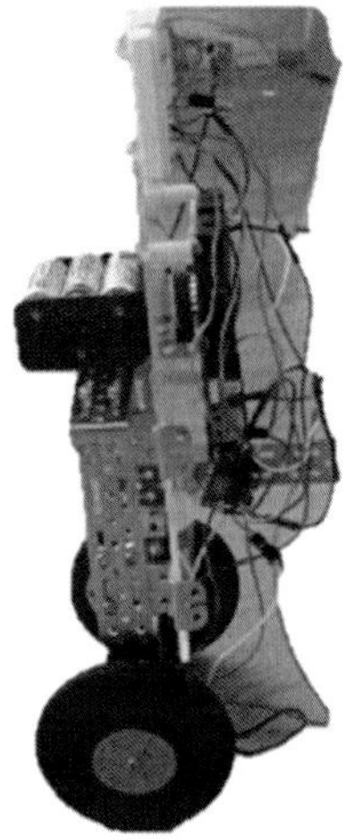

Fig. 1.3 The Segbot

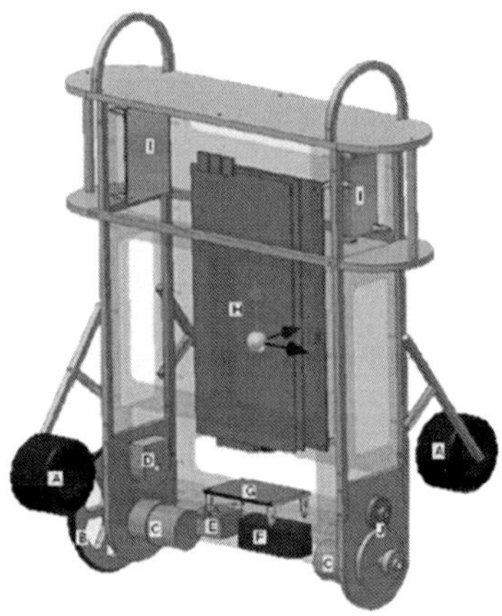

In Fig. 1.4, the UMASS uBot-5 [11] is another WIP robot, using two wheels in a differential drive configuration for mobility. Dynamically stable robots are well suited to environments designed for humans where both a high center of mass and a small footprint are often required. In the case of the uBot-5, which behaves much like an inverted pendulum, active stabilization becomes easier as the robot (and thus the center of mass) becomes taller. The uBot-5 can also employ whole body postural control afforded by its dynamically stable configuration to generate greater pushing and pulling forces.

Another example of an inverted pendulum that uses accelerometer and gyroscope is David P. Anderson's nBot [9], which is shown in Fig. 1.5. The data collected from both sensors is processed by the inertia measurement unit (IMU) that implements a filter to combine the two sensors into a single measurement.

The Industrial Electronics Laboratory at the Swiss Federal Institute of Technology has developed robot JOE [56] shown in Fig. 1.6, which is able to do stationary

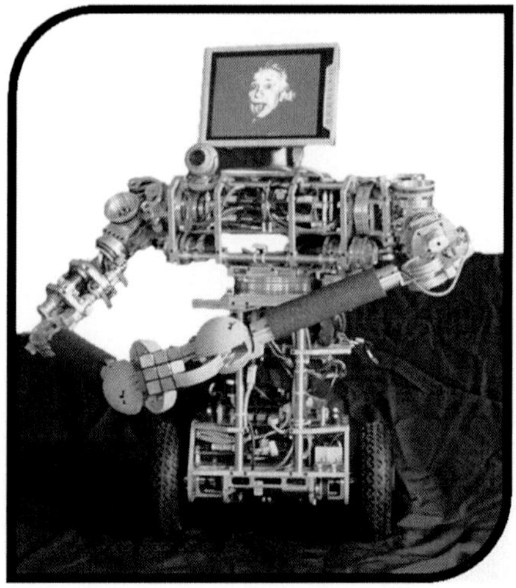

Fig. 1.4 The UMASS uBot-5

Fig. 1.5 The nBot

U-turns, because of its configuration of two coaxial wheels, each of which is coupled to a DC motor. A control system, made up of two decoupled state-space controllers, pilots the motors so as to keep the system in equilibrium.

As an attempt to solve the robotic unicycle problem, in Fig. 1.7, Ballbot [82] has been developed to balance itself on its single spherical wheel while traveling around. Ballbot balances with the aid of on-board sensors and computers and uses an actuator mechanism based on an inverse mouse-ball drive to move and change direction without needing to turn first.

The Rezero shown in Fig. 1.8, is another ball-driven balancing robot [10], which is composed of the coated ball, the three electrical motors, each driving an omni wheel and the main body, containing a microprocessor, an IMU, power electronics and batteries, are typical features of a Ballbot.

In 2000, a two-wheeled vehicle named B2 show in Fig. 1.9 for urban transformation was developed and demonstrated at INRIA [104]. The WIP robot vehicles in the framework of personal urban mobility and accessibility as an alternative to both bicycle and car transportation has been then rapidly developed. A specific feature of these wheeled vehicles is that they are dynamically stabilized systems using

Fig. 1.6 The Joe

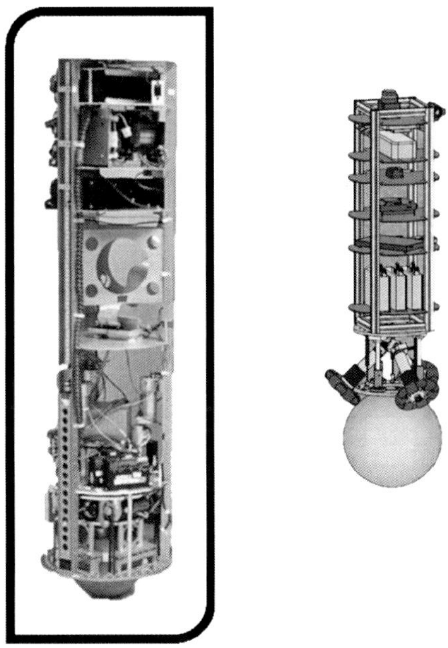

Fig. 1.7 The CMU Ballbot

their wheels for movement as well for balance. In Fig. 1.10, a commercialized WIP robot SEGWAY HT, invented by Dean Kamen [109], has been very popular in recent years. The SEGWAY HT is able to balance a human standing on its platform while the user traverses the terrain with it. This innovation uses five gyroscopes and a collection of other tilt sensors to keep itself upright. As a matter of fact, only three gyroscopes are necessary for the overall system, while the additional sensors are included as a safety precaution. The Segway RMP is created by combining a standard

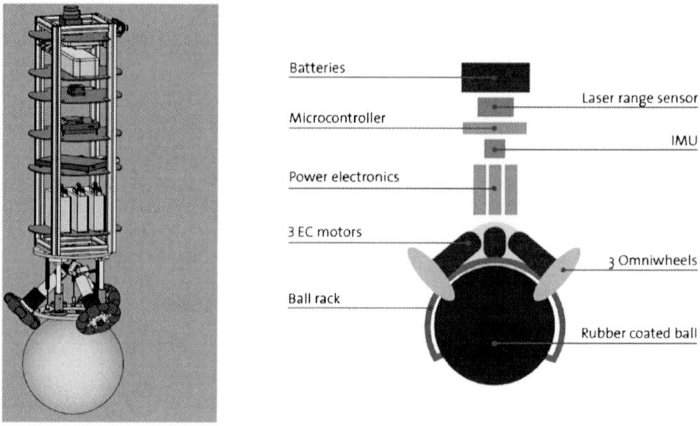

Fig. 1.8 The Ballbot Rezero

Fig. 1.9 The self-balancing vehicle B2 developed at INRIA

Segway Human Transporter (HT) with special purpose hardware and software (the RMP conversion kit). The result is a general-purpose robotic mobility platform.

1.2 Difficulties of Controlling WIP Systems

As above mentioned, underactuated mechanical systems are those systems that have less control inputs than generalized coordinates (degrees of freedom), i.e., they have generalized coordinates that are not actuated. In such systems, the unactuated generalized coordinates may indirectly be controlled by the actuated coordinates through dynamic coupling. The coupling is often nonlinear, hence the resulting dynamic constraints are generally nonintegrable and, therefore, second-order nonholonomic. Many examples of such systems exist; mainly in robotics, they include, among others, the underactuated robot manipulators, the gymnast robots, particularly the ac-

Fig. 1.10 Segway Human Transporter and Segway RMP

robot [130], the pendubot [135], planar vertical takeoff and landing aircraft [57], undersea vehicles [89] and other mobile robots [101].

Underactuation in these mechanical systems is generally introduced intentionally at the design level to reduce the manufacturing cost, weight and/or failure rate, such that the obtained systems may be able to perform complex tasks with a reduced number of actuators. However, such systems may require new approaches to design effective control strategies, and consequently they serve as a good platform for study of nonlinear control designing terms of both theoretical and practical interests. For these reasons, they are attracting more and more attention from researchers from the nonlinear control community and robotics.

WIP systems are not planar and the motors driving the wheels are directly mounted on the pendulum body [103], which are different from cart and pendulums [66, 135]. Recently, more researches have been done to investigate the dynamics and control of WIP systems [21, 40, 56, 103]. Since wheeled inverted pendulums belong to under-actuated dynamic system, where the number of control inputs is less than the number of degrees of freedom to be stabilized [68], it is difficult to apply the conventional control approach for under-actuated dynamic systems. Therefore, some control designs have been proposed to guarantee stability and robustness for wheeled inverted pendulums. Moreover, because of intrinsic under-actuated system, the dynamics of wheeled inverted pendulums systems are nonlinear and coupled, which can be described by coupled nonlinear differential equations. However, it is often possible to obtain an approximated linearized model around an operating point, where the signals involved are small enough. Several design of controllers and analysis techniques for linear systems were proposed, i.e., motion control using linear state-space model was proposed in [59], and the control was designed based on the dynamic equations linearized around an operating point [56]. In [106], dynamics involving the pitch of the inverted pendulum was studied and the rotation angles of the two wheels as the variables of interest were presented, and in [107], a linear controller was designed for stabilization for the robustness. In [19], only a planar model without yaw was considered, a linear stabilizing controller was derived based on this mode. In [75], although the exact dynamics of two-wheeled inverted pendulum were investigated, only the linear feedback control was developed on the dynamic model. In [103], based on partial feedback linearization, a two-level velocity controller and a stabilizing position controller were proposed.

Feedback linearization is an approach to nonlinear control design which has attracted a great deal of research interest in recent years. The central idea of the approach is to algebraically transform a nonlinear system dynamics into a (fully or partly) linear one, so that linear control techniques can be applied. In the transformation, the exact dynamics of system must be known beforehand, the model-based control can provide an effective solution to the problem. However, wheeled inverted pendulum is characterized by unstable balance and unmodeled nonlinear dynamics, and there are time varying external disturbances, in the form of parametric and functional uncertainties, which is not possible to model accurately. Modeling errors might undermine the control approach based on linearized model [56, 59], and the control proposed on the velocity level [103]. Therefore, model-based control may not be the ideal choice since the dynamics is not usually available. The presence of uncertainties and disturbances could disrupt the function of the model-based feedback control, and lead to the unstable balance. How to handle the parametric and functional uncertainties, unmodeled dynamics, and disturbances from the environment is one of the important issues in the control of wheeled inverted pendulum. The wheeled inverted pendulum is definitely different from other non-holonomic systems subject to (i) only kinematic constraints which geometrically restrict the direction of mobility, i.e., wheeled mobile robot [42, 49]; (ii) only dynamic constraints due to dynamic balance at passive degrees of freedom where no force or torque is applied, i.e., the manipulator with passive link [14, 30, 53]. It belongs to (iii) not only kinematic constraints but also dynamic constraints. Therefore, the wheeled inverted pendulum is more complex than the former two cases, so the previously proposed control approaches suitable for (i) and (ii) could not be applied directly to wheeled inverted pendulums.

A challenging problem is to control a wheeled inverted pendulum system whose cart is no longer constrained to the guide rail like cart-pendulum systems, but moves in its terrain while balancing the pendulum. Therefore, the wheeled inverted pendulum is subject to the nonholonomic constraints and belongs to nonholonomic system.

In physics and mathematics, a nonholonomic system is a system described by a set of parameters subject to differential constraints, such that when the system evolves along a path in its parameter space, where the parameters vary continuously (in the mathematical sense) and return to the identical values they held at the start of the path, the system itself may not have returned to its original state. A classical example of nonholonomic systems is a rigid disk rolling on a horizontal plane without slippage in [20], which is equivalent from the control perspective to a wheeled cart driven by two wheels. As a matter of fact, a car-like system in general is a nonholonomic system except a few examples of omnidirectional vehicles [22, 29, 125, 129]. Other examples of nonholonomic systems can be seen in underwater vehicle [38, 55], underactuated robotic manipulators [48, 117].

Due to Brockett's theorem, it is well known that wheeled mobile robot with restricted mobility cannot be stabilized to a desired configuration (or posture) via differentiable, or even continuous, pure-state feedback, although it is controllable. A number of approaches have been proposed for the problem, which can be classified as (i) discontinuous time-invariant stabilization [15], (ii) time-varying stabilization [108, 121] and (iii) hybrid stabilization [31, 114]. See the survey paper [77] for more details and references therein. One commonly used approach for controller design of mobile robot is to convert, with appropriate state and input transformations, the original systems into some canonical forms for which controller design can be carried out easier [65, 96, 99]. Using the special algebra structures of the canonical forms, various feedback strategies have been proposed to stabilize mobile robots in the literature [15, 49, 71, 99]. Recently, adaptive control strategies were proposed to stabilize the dynamic mobile robot systems with modeling or parametric uncertainties [49]. Hybrid control based on supervisory adaptive control was presented to globally asymptotically stabilize a wheeled mobile robot [62]. Adaptive state feedback control was considered in [34] using input-to-state scaling. Output feedback tracking and regulation were presented in [33] for practical wheeled mobile robots. In [70], robust exponential regulation for nonholonomic systems with input and state-driven disturbances was presented under the assumption that the bounds of the disturbances are known. However, these studies consider neither vehicles with inverted pendulum nor system dynamics.

Control of wheeled inverted pendulum with dynamics uncertainties is essential in many practical applications, especially for the case when the human transportation should be considered. To handle unknown dynamics of mechanical systems, robust and adaptive controls need to be extensive investigated for inverted pendulums and dynamic nonholonomic systems. Robust controls assume the known boundedness of unknown dynamics of the systems, nevertheless adaptive controls could learn the unknown parameters of interest through adaptive tuning laws.

Under the assumption of a good understanding of dynamics of the systems understudy, model based adaptive controls have been much investigated for dynamic nonholonomic systems. In [36], adaptive control was proposed for trajectory/force control of mobile manipulators subjected to holonomic and nonholonomic constraints with unknown inertia parameters, which ensures the motion of the system to asymptotically converge to the desired trajectory and force. In [50], adaptive state feedback and output feedback control strategies using state scaling and backstepping techniques were proposed for a class of nonholonomic systems in chained form with drift nonlinearity and parametric uncertainties. In [127], the nonholonomic kinematic subsystem was first transformed into a skew-symmetric form, then a virtual adaptive control designed at the actuator level was proposed to compensate for the parametric uncertainties of the kinematic and dynamic subsystems.

In [49], robust adaptive control was proposed for dynamic nonholonomic systems with unknown inertia parameters and disturbances, in which adaptive control techniques were used to compensate for the parametric uncertainties and sliding mode control was used to suppress the bounded disturbances. In [126], adaptive robust force/motion control was presented systematically for holonomic mechanical systems and a large class of nonholonomic mechanical systems in the presence of uncertainties and disturbances.

Because of the difficulty in dynamic modeling, adaptive neural network control, a non-model based approach, has been extensive studied for different classes of systems, such as robotic manipulators [44, 76, 80], and mobile robots [39]. In [43], adaptive neural network control for robot manipulator in the task space was proposed, which neither requires the inverse dynamical model nor the time-consuming off-line training process. In [87], a neuro-adaptive force/motion control based on the philosophy of the parallel approach had been developed for mobile manipulators under a class of the non-rigid uncertain constrained motions. In [63], adaptive neural fuzzy control for function approximation had been investigated for uncertain nonholonomic mobile robots in the presence of unknown disturbances. In [85], adaptive neural network controls had been developed for the motion control of mobile inverted pendulum subject to kinematic and dynamics constraints.

1.3 Outline of Book

The book contains nine chapters which exploit several independent yet related topics in detail.

Chapter 1 introduces the WIP system description, background and motivation of the study, and presents several general concepts and fundamental observations which provide a sound base for the book.

Chapter 2 concisely reviews the necessary mathematics including geometry, linear algebra, stability theory, as well as other useful tools such as Lie algebra and Lyapunov theory.

Chapter 3 describes the kinematics and dynamics of the WIP system. The system can be treated as two subsystems, namely the mobile platform and the pendulum, separately. First, since two-wheeled driven mobile platform has very unique control properties due to its nonholonomic nature, the modeling of the mobile platform should be addressed independently for the sake of clarity. Second, the inverted pendulum is also subject to second-order nonholonomic constraints. In this chapter, we derive the kinematics and dynamics equations for both the mobile platform and the inverted pendulum in order to investigate the dynamic interaction between the inverted pendulum and the platform. Deriving the entire equations of one mobile inverted pendulum in an explicit form such that we can conduct the simulations to verify the proposed controller in the other chapters.

Chapter 4 first presents a linearized model of the WIP system, then PD, LQR, H_∞ and backstepping controllers have been implemented. The pole placement method has been employed to design a PD controller, allowing the designer to specify the response of the closed loop system. Comparisons were made between the performance of an LQR controller and PD controller. The LQR controller is found to perform better when subject to the situation where the parameters vary with time. Finally, H_∞ and backstepping controllers are also developed.

In Chap. 5, we further develop nonlinear control for the WIP system. First, we investigate the nonlinear feedback linearization approach. Then, taking advantage of the physical model properties, the system is decomposed into three subsystems and model-based control and stability analysis have been studied for each of the subsystem.

In Chap. 6, both adaptive and robust motion control designs have been studied. Then control design is proceeded with on-line parameters estimation strategy and introduced to compensate for dynamics uncertainties and external disturbances. Then, hybrid motion/force adaptive robust control has been thoroughly studied. The unmodeled dynamics have been compensated by both online learning of the uncertain parameters and robust approach.

In Chap. 7, a number of intelligent control approaches have been investigated. First, the modeling and control method using the Least Squares Support Vector Machine (LS-SVM) have been utilized to design efficient model free control. Then, we further study the universal functional approximation of fuzzy logic and neural networks. All these intelligent control methods employ a systematic online adaptation mechanism without prepared off line learning.

In Chap. 8, aiming at shaping the tilt and yaw angular motion subsystem of the WIP system to minimize both motion tracking errors and angular accelerations, we employ linear quadratic regulation (LQR) optimization technique to obtain an optimal reference model, which is similar to an impedance model of spring and damper with an artificial force caused by the preset desired trajectories and it helps to enhance the driver's comfort. Adaptive control techniques combined with variable structure have been then exploited to compensate for the effect of uncertainties.

In Chap. 9, as an extension of Chap. 8, we further develop Neural Network (NN) based direct model reference control for the angular motion subsystem. The closed-loop angular motion subsystem is guaranteed to match the reference model in a prescribed finite time. As it is noted that the forward velocity can be indirectly affected by the reference of tilt angle, in this chapter we develop an NN based reference trajectory planner for tilt angle, which is called Adaptive Generator of Implicit Control Trajectory (AGICT), such that the forward velocity can be "controlled" to follow the desired velocity asymptotically.

Chapter 2
Mathematical Preliminaries

2.1 Introduction

In this chapter, the fundamental mathematical concepts and analysis tools in systems theory are summarized, which will be used in control design and stability analysis in the subsequent chapters. Much of the material is described in classical control theory textbooks and robotics books as standard form. Thus, some standard theorems, lemmas and corollaries, which are available in references, are sometimes given without a proof. This chapter serves as a short review and as a convenient reference when necessary. In addition, for the robotic control, stability analysis is the key core for all the closed-loop system, therefore, some metric or norms need to be defined such that system could be measured. Those norms that are defined to easily manipulate for the control design, also, all norms have some physical significance.

2.2 Matrix Algebra

Matrix Definition A vector $\mathbf{x}$ defined as a column $(n \times 1)$ vector and $m \times n$ matrix $\mathbf{X}$ that has m rows and n columns are given as follows:

$$\mathbf{x} = \begin{bmatrix} x_1 \\ x_2 \\ \vdots \\ x_n \end{bmatrix}, \quad \mathbf{x} = \begin{bmatrix} x_1 \ x_2 \ \dots \ x_n \end{bmatrix}^T, \quad \mathbf{X} = \begin{bmatrix} x_{11} & x_{12} & \cdots & x_{1n} \\ x_{21} & x_{22} & \cdots & x_{2n} \\ \vdots & \vdots & \ddots & \vdots \\ x_{m1} & x_{m2} & \cdots & x_{mn} \end{bmatrix} \tag{2.1}$$

Matrix $\mathbf{X}$ is defined as a square matrix if $m = n$, that is, $\mathbf{X}$ is an $m \times m$ matrix. Vectors may also be viewed as particular matrix $(n = 1)$, and x_{ij} is an entry from the ith row and jth column of the matrix $\mathbf{X}$. The transpose $\mathbf{X}^T$ of $m \times n$ matrix $\mathbf{X}$ is an $n \times m$ matrix, the entry x_{ij} is the element x_{ji} of the matrix $\mathbf{X}$. Moreover, the related properties are listed as follows:

$$\begin{aligned} \mathbf{X}^T &= [x_{ji}], & (\mathbf{X}^T)^T &= \mathbf{X}, \\ (\mathbf{XY})^T &= \mathbf{Y}^T\mathbf{X}^T, & [\mathbf{X}+\mathbf{Y}]^T &= \mathbf{X}^T + \mathbf{Y}^T \end{aligned} \tag{2.2}$$

Z. Li et al., *Advanced Control of Wheeled Inverted Pendulum Systems*,
DOI 10.1007/978-1-4471-2963-9_2,

Definition 2.1 The square matrix $\mathbf{X} \in \mathbb{R}^{n\times n}$ is said to be

(1) positive semi-definite (denoted by $\mathbf{X} \geq 0$) if $a^T\mathbf{X}a \geq 0$, $\forall a \in \mathbb{R}^n$;
(2) positive definite if $a^T\mathbf{X}a > 0$ for all nonzero $a \in \mathbb{R}^n$;
(3) negative semi-definite if $-\mathbf{X}$ is real positive semi-definite;
(4) negative definite if $-\mathbf{X}$ is positive definite;
(5) indefinite if $\mathbf{X}$ is positive for some $a \in \mathbb{R}^n$ and negative for other $a \in \mathbb{R}^n$;
(6) symmetric if $\mathbf{X}^T = \mathbf{X}$;
(7) skew-symmetric if $\mathbf{X}^T = -\mathbf{X}$;
(8) symmetric positive definite (semi-definite) if $\mathbf{X} > 0$ (≥ 0) and $\mathbf{X} = \mathbf{X}^T$; and
(9) a time-varying matrix $\mathbf{X}(t)$ is uniformly positive definite if there exists $a > 0$ such that $\mathbf{X}(t) \geq a\mathbf{I}$.

The product of $\mathbf{XY}$ of $m \times n$ matrix $\mathbf{X}$ by an $n \times r$ matrix $\mathbf{Y}$ is an $(m \times r)$ matrix $\mathbf{Z}$:

$$\mathbf{Z} = \mathbf{XY}, \quad z_{ij} = \sum_{p=1}^{n} x_{ip} y_{pj} \tag{2.3}$$

and satisfies

$$\mathbf{XY} \neq \mathbf{YX}, \qquad (\mathbf{XYZ}) = \mathbf{X}(\mathbf{YZ}), \qquad (\mathbf{X}+\mathbf{Y})\mathbf{Z} = \mathbf{XZ} + \mathbf{YZ} \tag{2.4}$$

If the matrices $\mathbf{X}$, $\mathbf{Y}$ and $\mathbf{Z}$ are symmetric, the following properties are satisfied:

$$\mathbf{XY} = \mathbf{YX}, \qquad \mathbf{XYZ} = \mathbf{YXZ} = \mathbf{ZXY} = \mathbf{ZYX} \tag{2.5}$$

Inner and Outer Product The inner product of two n-dimensional vector $\mathbf{x}$ and $\mathbf{y}$ is a scalar $\mathbf{z} = \mathbf{x}^T\mathbf{y} = \mathbf{y}^T\mathbf{x}$. The n-dimensional Euclidean space, denoted by E^n, is $\mathbb{R}^n$ with the inner product

$$\langle \mathbf{x}, \mathbf{y} \rangle = \mathbf{x}^T\mathbf{y} = \sum_{i=1}^{n} x_i y_i \tag{2.6}$$

The outer product of $\mathbf{x}$ and $\mathbf{y}$ is a matrix $\mathbf{Z}$ ($\mathbf{x} \in \mathbb{R}^m$, $\mathbf{y} \in \mathbb{R}^n$):

$$\mathbf{Z} = \mathbf{xy}^T = \begin{bmatrix} x_1 y_1 & x_2 y_2 & \cdots & x_1 y_n \\ x_2 y_1 & x_2 y_2 & \cdots & x_2 y_n \\ \vdots & \vdots & \ddots & \vdots \\ x_m y_1 & x_m y_2 & \cdots & x_m y_n \end{bmatrix} \tag{2.7}$$

Linear Independent of Vectors The vectors $\mathbf{x}$ are linearly independent if

$$\sum_{i=1}^{n} \alpha_i \mathbf{x} = 0 \quad \Rightarrow \quad \alpha_1 = \alpha_2 = \cdots = \alpha_n = 0 \tag{2.8}$$

The columns (rows) of $\mathbf{X} \in \mathbb{R}^{m\times n}$ are linearly independent if and only if $\mathbf{X}^T\mathbf{X}$ is a nonsingular matrix, i.e., $\det(\mathbf{X}^T\mathbf{X}) = |\mathbf{X}^T\mathbf{X}| \neq 0$.

Vector Norms Vector norms are positive scalars and are used as measures of length, size, distance, and so on, depending on context. An L_p norm is p-norm of an $(n \times 1)$ vector x defined as

$$\|\mathbf{x}\|_p = \left(\sum_{i=1}^{n} |x_i|^p\right)^{1/p}, \quad \text{for } 1 \le p < \infty \tag{2.9}$$

$$\|x\|_\infty = \max_{1 \le i \le n} |x_i|, \quad \text{for } p = \infty \tag{2.10}$$

The three most commonly used norms are $\|\mathbf{x}\|_1$, $\|\mathbf{x}\|_2$ and $\|\mathbf{x}\|_\infty$, which are defined as

$$\|\mathbf{x}\|_1 = \sum_{i=1}^{n} |x_i| \tag{2.11}$$

$$\|\mathbf{x}\|_2 = \sqrt{\mathbf{x}^T \mathbf{x}} = \sqrt{x_1^2 + x_2^2 + \cdots + x_n^2} = \left(\sum_{i=1}^{n} x_i^2\right)^{1/2} \tag{2.12}$$

$$\|\mathbf{x}\|_\infty = \max |x_i| \tag{2.13}$$

All p-norms are equivalent in the sense that if $\|\cdot\|_{p_1}$ and $\|\cdot\|_{p_2}$ are two different p-norms, then there exist positive constants c_1 and c_2 such that

$$c_1\|\mathbf{x}\|_{p_1} \le \|\mathbf{x}\|_{p_2} \le c_2\|\mathbf{x}\|_{p_1}, \quad \forall \mathbf{x} \in \mathbb{R}^n \tag{2.14}$$

Determinants of Matrix The determinant of an $n \times n$ square matrix $\mathbf{X}$ is defined as the sum of the signed products of all possible combinations of n elements, in which each element is taken from a different row and column. The determinant of $\mathbf{X}$ is denoted by $\det(\mathbf{X})$ as

$$\det(\mathbf{X}) = \sum^{n!} (-1)^p x_{1p_1} x_{1p_2} \ldots x_{1p_n} \tag{2.15}$$

where $p_1, p_2, \ldots, p_n$ is a permutation of $1, 2, \ldots, n$ and the sum is taken over all possible permutations. A permutation is a rearrangement of $1, 2, \ldots, n$ into some other order, such as $n, 1, \ldots, 2$, that is obtained by successive transpositions. A transposition is the interchange of places of two numbers in the list $1, 2, \ldots, n$. The exponent p of -1 is the number of transpositions it takes to go from the natural order to $p_1, p_2, \ldots, p_n$. There are $n!$ possible permutations of n numbers, so each determinant is the sum of $n!$ products.

Eigenvalues and Properties of Matrix For a given general matrix $\mathbf{X} \in \mathbb{R}^{n \times n}$, there can be up to n such special vectors satisfying $\det(\mathbf{X} - \lambda\mathbf{I}) = 0$. These vectors $\lambda_1, \ldots, \lambda_i, \ldots, \lambda_n$ are called the eigenvectors of $\mathbf{X}$, and the proportionality constants are called the eigenvalues.

The properties for the eigenvectors can be obtained as

- If $\lambda_i(\mathbf{X}) > 0$ (≥ 0), then $\mathbf{X} = \mathbf{X}^T \in \mathbb{R}^{n \times n}$ is positive (semi-)definite.
- If $\lambda_i(\mathbf{X}) < 0$ (≤ 0), $\mathbf{X} = \mathbf{X}^T \in \mathbb{R}^{n \times n}$ is negative (semi-)definite.

- If a matrix is symmetric, then its eigenvalues are all real.
- A necessary condition for a square matrix $\mathbf{X}$ to be positive definite is that its diagonal elements be strictly positive.
- A necessary and sufficient condition for a symmetric matrix $\mathbf{X}$ to be positive definite is that all its principal minors be strictly positive.
- If $\mathbf{X} = \mathbf{X}^T > 0$ (≥ 0) and $\mathbf{Y} = \mathbf{Y}^T > 0$ (≥ 0), then $\mathbf{X} + \mathbf{Y} > 0$ (≥ 0) and has all eigenvalues real positive but it is not true in general that $\mathbf{XY} > 0$ (≥ 0). If and only if $\mathbf{X}$ and $\mathbf{Y}$ are commutative, i.e., $\mathbf{XY} = \mathbf{YX}$, then $\mathbf{XY} > 0$ (≥ 0).
- If matrix $\mathbf{X}$ is symmetric positive semi-definite, then it can be decomposed as $\mathbf{X} = U^T \Lambda U$, where U is a unitary matrix and satisfies $U^T U = I$, and Λ is a diagonal matrix containing the eigenvalues of the matrix $\mathbf{X}$, moreover, the following inequality is satisfied $\lambda_{\min}(\mathbf{X})\|a\|^2 \leq a^T \mathbf{X} a \leq \lambda_{\max}(\mathbf{X})\|a\|^2$.
- If $\mathbf{X}$ is skew-symmetric, then $a^T \mathbf{X} a = 0$ for all $a \in \mathbb{R}^n$.
- If $\mathbf{X} \geq 0$, then $\mathbf{X} + \mathbf{X}^T$ is symmetric positive semi-definite.
- If matrix $\mathbf{X}$ is square, then it can be represented as the sum of a symmetric matrix and a skew-symmetric matrix as follows: $\mathbf{X} = \frac{\mathbf{X}+\mathbf{X}^T}{2} + \frac{\mathbf{X}-\mathbf{X}^T}{2}$.
- Since $a^T \mathbf{X} a = a^T \frac{\mathbf{X}+\mathbf{X}^T}{2} a$, the positive definiteness of a matrix can be determined by only the symmetric part of $\mathbf{X}$.
- If $\det(\mathbf{X}) \neq 0$, the square matrix $\mathbf{X}$ is non-singular. Otherwise, $\mathbf{X}$ is singular.

Diagonal Matrix Given $D = \mathrm{diag}[d_{ii}] \in \mathbb{R}^{n\times n}$, $d_{ii} > 0$, $i = 1, \ldots, n$, we have the following inequality

$$0 < \gamma_1 x^T x \leq x^T D x \leq \gamma_2 x^T x, \quad \forall x \in \mathbb{R}^n \tag{2.16}$$

where both γ_1 and γ_2 are positive constants, and $\gamma_1 \leq \min_{i=1,\ldots,n}\{d_{ij}\}$ and $\gamma_2 \geq \max_{i=1,\ldots,n}\{d_{ii}\}$.

Lemma 2.2 *For* $K = \mathrm{diag}[k_{ii}] \in \mathbb{R}^{n\times n}$, *and* $a = [a_1, a_2, \ldots, a_n]^T \in \mathbb{R}^n$, *if* $u = -K\,\mathrm{sgn}(x)$, *and* $k_{ii} \geq |d_i|$, *then* $x^T D(u - a) \leq 0$.

Lemma 2.3 *For* $K = \mathrm{diag}[k_{ii}] \in \mathbb{R}^{n\times n}$ *and* $\underline{k} = \min_{i=1,\ldots,n}\{k_{ii}\}$, *if* $u = -K\,\mathrm{sgn}_1(x)$, *and* $\underline{k} \geq \frac{\|D\|\|a\|}{\gamma_1}$ *(or* $\geq \frac{\gamma_2}{\gamma_1}\|a\|$*), then* $x^T D(u - a) \leq 0$.

Symmetric Positive Definite Matrix

Lemma 2.4 *Consider* $u = -k\,\mathrm{sgn}_1(x)$, *and* $0 < \gamma_1 x^T x \leq x^T D x \leq \gamma_2 x^T x$, *if* $k \geq \frac{\|D\|\|a\|}{\gamma_1}$ *(or* $\geq \frac{\gamma_2}{\gamma_1\|a\|}$*), then* $x^T D(u - a) \leq 0$.

Although $D(t)$ is time variant and usually unknown, the lower and upper bounds of D can be obtained from the view point of physical model for practical systems, therefore, it can be chosen such that $\gamma_1 \leq \lambda_{\min}(D)$ and $\gamma_2 \geq \lambda_{\max}(D)$. If D is symmetric negative definite, we have

$$-\gamma_2 x^T x \leq x^T D x \leq -\gamma_1 x^T x < 0, \quad \forall x \in \mathbb{R}^n \tag{2.17}$$

where γ_1 and γ_2 are positive constants.

Induced Norm of Matrices For an $m \times n$ matrix $\mathbf{X}$, the induced p-norm of $\mathbf{X}$ is defined by

$$\|\mathbf{X}\|_p = \sup_{x \neq 0} \frac{\|\mathbf{X}y\|_p}{\|y\|_p} = \sup_{\|y\|_p = 1} \|\mathbf{X}y\|_p \tag{2.18}$$

As similar as the vector norms, for $p = 1, 2, \infty$, the corresponding induced norms could be obtained as follows:

$$\|\mathbf{X}\|_1 = \max_j \sum_{i=1}^{m} |x_{ij}| \quad \text{(column sum)} \tag{2.19}$$

$$\|\mathbf{X}\|_2 = \max_i \sqrt{\lambda_i \left(\mathbf{X}^T \mathbf{X}\right)} \tag{2.20}$$

$$\|\mathbf{X}\|_\infty = \max_i \sum_{j=1}^{m} |x_{ij}| \quad \text{(row sum)} \tag{2.21}$$

The induced norms are also satisfying the following:

$$\|\mathbf{XY}\|_p \leq \|\mathbf{X}\|_p \|\mathbf{Y}\|_p, \quad \forall \mathbf{X} \in \mathbb{R}^{n \times m}, \mathbf{Y} \in \mathbb{R}^{m \times l} \tag{2.22}$$

2.3 Norms for Functions

Open and Closed Sets

Definition 2.5 A set $S \subset \mathbb{R}^n$ is called open, if for each $x \in S$ there exists and $\epsilon > 0$ such that the interval $(x - \epsilon, x + \epsilon)$ is contained in S. Such an interval is often called an ϵ-neighborhood of x, or simply a neighborhood of x.

Continuous Function

Definition 2.6 A function $f : \mathbb{R}^n \to \mathbb{R}^m$ is said to be continuous at a point $x \in \mathbb{R}^n \subseteq \mathbb{R}^m$, if for each $\epsilon > 0$, there exists a $\delta(\epsilon, x)$ such that for all $y \in \mathbb{R}^n$ satisfying $\|x - y\| < \delta \Rightarrow \|f(x) - f(y)\| < \epsilon$. A function f is continuous on $\mathbb{R}^n$ if it is continuous at every point in $\mathbb{R}^n$.

Definition 2.7 A function $f : \mathbb{R}^n \to \mathbb{R}^m$ is uniformly continuous on $\mathbb{R}^n \subseteq \mathbb{R}^m$ if for each ϵ, there exists a $\delta(\epsilon)$ such that for all $x, y \in \mathbb{R}^n$ satisfying $|x - y| < \delta(\epsilon)$, then $|f(x) - f(y)| < \epsilon$.

Definition 2.8 A function $f : [0, \infty) \to R$ is piecewise continuous on $[0, \infty)$ if f is continuous on any finite interval $[a, b] \subset [0, \infty)$ except at a finite number of points on each of these intervals.

Definition 2.9 A function $f : \mathbb{R}^n \rightarrow \mathbb{R}^m$ is said to be Lipschitz continuous if there exists a constant $L > 0$, which is sometimes called the Lipschitz constant, such that $|f(x) - f(y)| \leq L|x - y|$ for all $x, y \in \mathbb{R}^n$, where $\mathbb{R}^n \subseteq \mathbb{R}^m$.

A function f is continuous on a set of S if it is continuous at every point of S, and it is uniformly continuous on S if given $\epsilon > 0$, there is $\delta(\epsilon) > 0$ (dependent only on ϵ), such that the inequality holds for all $x, y \in S$.

Differentiable Function The derivative of function $f : \mathbb{R} \rightarrow \mathbb{R}$ at a point x is defined

$$\dot{f}(x) = \lim_{\delta x \rightarrow 0} \frac{f(x + \delta x) - f(x)}{\delta x} \tag{2.23}$$

A function $f : \mathbb{R}^n \rightarrow \mathbb{R}^m$ is continuously differentiable at point x (on a set S) if the partial derivatives $\partial f_i / \partial x_j$ exist and continuous at x (at every point of S) for $1 \leq i \leq m$, $1 \leq j \leq n$, and the Jacobian matrix $f : \mathbb{R}^n \rightarrow \mathbb{R}^m$ is defined as

$$J = \left[\frac{\partial f}{\partial x}\right] = \begin{bmatrix} \partial f_1 / \partial x_1 & \cdots & \partial f_1 / \partial x_n \\ \vdots & \ddots & \vdots \\ \partial f_m / \partial x_1 & \cdots & \partial f_m / \partial x_n \end{bmatrix} \in \mathbb{R}^{m \times n} \tag{2.24}$$

For a scalar function $f(x, y)$ that depends on x and y, the gradient with respect to x is defined as

$$\nabla_x f(x, y) = \left[\frac{\partial f}{\partial x}\right] = \left[\frac{\partial f}{\partial x_1}, \frac{\partial f}{\partial x_2}, \ldots, \frac{\partial f}{\partial x_n}\right] \tag{2.25}$$

If f is only a function of x, we denote $\nabla f(x)$.

Mean Value Theorem If a function $f : \mathbb{R}^n \rightarrow \mathbb{R}^m$ is differentiable at each point $x, y \in \mathbb{R}^n$, where $\mathbb{R}^n \subseteq \mathbb{R}^m$, such that the line segment $L(x, y) \in \mathbb{R}^n$, then there exists some $z \in L(x, y)$ such that

$$f(y) - f(x) = \left.\frac{\partial f(x)}{\partial x}\right|_{x=z} (y - x) \tag{2.26}$$

which is equivalent to $\frac{\partial f(x)}{\partial x} = \frac{f(y) - f(x)}{y - x}$.

Function Norms Given the time varying function $x(t) : \mathbb{R}_+ \rightarrow \mathbb{R}$, be continuous or piecewise continuous. The p-norm for $x(t)$ is defined as

$$\left\|x(t)\right\|_p = \left(\int_0^\infty \left|x(t)^p\right| dt\right)^{1/p}, \quad \text{for } p \in [1, \infty) \tag{2.27}$$

$$\left\|x(t)\right\|_\infty = \sup_{t \in [0, \infty)} \left|x(t)\right|, \quad \text{for } p = \infty \tag{2.28}$$

The function space over which the signal norm exists is define by letting $p = 1, 2, \infty$, the corresponding norm spaces are called L_1, L_2, L_∞, respectively, let $x(t) \in [0, \infty)$, which is define by

$$L_p = \left\{x(t) \in \mathbb{R}^n : \|x\|_p < \infty\right\} \tag{2.29}$$

for $p \in [1, \infty)$, that is, L_p is the set of all vector functions in $\mathbb{R}^n$ for which the p-norm is well defined (finite). From a signal point of view, the 1-norm, $\|x\|_1$, of the signal $x(t)$ is the integral of its absolute value, the square $\|x\|_2^2$ of the 2-norm is often called the energy of the signal $x(t)$, and the ∞-norm is its absolute maximum amplitude or peak value.

The following inequalities for signals will be useful:

- *Hölder's Inequality*: If scalar time varying function $x \in L_p$ and $y \in L_p$ for $p, q \in [1, +\infty)$ and $1/p + 1/q = 1$, then $xy \in L_1$ and $\|xy\|_1 \leq \|x\|_p \|y\|_q$.
- *Minkowski Inequality*: If scalar time varying function, x, $y \in L_p$ for $p \in [1, +\infty)$, then $x + y \in L_p$ and $\|x + y\|_p \leq \|x\|_p + \|y\|_p$.
- *Young's Inequality*: For scalar time varying function $x(t) \in \mathbb{R}$ and $y(t) \in \mathbb{R}$, it holds that $2xy \leq \frac{1}{\epsilon}x^2 + \epsilon y^2$, for any $\epsilon > 0$.
- *Completing the Square*: For scalar time functions $x(t) \in \mathbb{R}$ and $y(t) \in \mathbb{R}$, we have $-x^2 + 2xy = -x^2 + 2xy - y^2 + y^2 \leq y^2$.

Let $f(t) : \mathbb{R}_+ \to \mathbb{R}^n$, $f = [f_1(t), f_2(t), \ldots, f_n(t)]^T$ be a continuous or piecewise continuous vector function. Then, the corresponding p-norm spaces are defined as

$$L_p^n \triangleq \left\{f : \mathbb{R}_+ \to \mathbb{R}^n \;\middle|\; \|f\|_p = \int_0^\infty \|f\|^p \, dt < \infty, \text{ for } p \in [1, \infty)\right\} \tag{2.30}$$

$$L_\infty^n \triangleq \left\{f : \mathbb{R}_+ \to \mathbb{R}^n \;\middle|\; \|f\|_\infty = \sup_{t \in [0,\infty)} \|f\|_\infty < \infty\right\} \tag{2.31}$$

In some cases, during the period $[0, T]$, signals are bounded over finite time intervals, but may become infinity as time goes to infinity. Therefore, we only choose this period to extend L_p^n spaces by

$$L_b^n \triangleq \left\{f : \mathbb{R}_+ \to \mathbb{R}^n \;\middle|\; \|f\|_{pT} = \int_0^T \|f\|^p \, dt < \infty, \; \forall T \in \mathbb{R}_+\right\} \tag{2.32}$$

$$L_{b\infty}^n \triangleq \left\{f : \mathbb{R}_+ \to \mathbb{R}^n \;\middle|\; \|f\|_\infty = \sup_{t \leq T} \|f\| < \infty, \; \forall T \in \mathbb{R}_+\right\} \tag{2.33}$$

Definition and Properties of Sign Functions The signum function of a real number x is defined as follows:

$$\operatorname{sgn}(x) = \begin{cases} 1, & x > 0 \\ 0, & x = 0 \\ -1, & x < 0 \end{cases} \tag{2.34}$$

The properties of sign function are listed as follows.

(1) Any real number can be expressed as the product of its absolute value and its sign function:

$$x = \operatorname{sgn}(x) \cdot |x| \tag{2.35}$$

whenever x is not equal to 0, from Eq. (2.35), we have

$$\mathrm{sgn}(x) = \frac{x}{|x|} \tag{2.36}$$

(2) The signum function is the derivative of the absolute value function (up to the indeterminacy at zero): Note, the resultant power of x is 0, similar to the ordinary derivative of x. The numbers cancel and all we are left with is the sign of x,

$$\frac{d|x|}{dx} = \mathrm{sgn}(x) \tag{2.37}$$

(3) The signum function is differentiable with derivative 0 everywhere except at 0. It is not differentiable at 0 in the ordinary sense, but under the generalized notion of differentiation in distribution theory, the derivative of the signum function is two times δ function,

$$\frac{d\,\mathrm{sgn}(x)}{dx} = 2\delta \tag{2.38}$$

where

$$\delta(x) = \begin{cases} +\infty, & x = 0 \\ 0, & x \neq 0, \end{cases} \qquad \int_{-\infty}^{+\infty} \delta(x)\,dx = 1 \tag{2.39}$$

(4) For $k \geq 0$, a smooth approximation of the sign function is

$$\mathrm{sgn}(x) \approx \tanh(kx) \tag{2.40}$$

(5) For vectors, the sign function is defined as

$$\mathrm{sgn}(\mathbf{X}) \triangleq \left[\mathrm{sgn}(x_1)\ \mathrm{sgn}(x_2)\ \ldots\ \mathrm{sgn}(x_n)\right]^T \tag{2.41}$$

(6) In general, $\mathrm{sgn}(\mathbf{x}) \neq \mathrm{sgn}_1(x)$. However, $\mathrm{sgn}(\mathbf{x}) = \mathrm{sgn}(x)$ and $x^T \mathrm{sgn}(x) = x^T \mathrm{sgn}_1(x)$ if and only if one element of x is nonzero and the remaining elements of x are zero.
(7) The L_2-norm of $\mathrm{sgn}(x)$ is 1 and $\|\mathrm{sgn}(x)\| = \sqrt{n}$. From $x^T \mathrm{sgn}(x) = \sum_{j=1}^{n} |x_j|$ $= \|x\|_1$ and $x^T \mathrm{sgn}(x) = \frac{\|x\|^2}{\|x\|} = \|x\|$.
(8) While $\mathrm{sgn}(x)$ defines a unit vector in the direction of x, $\mathrm{sgn}(x)$ maps all the vectors that are in the same quadrant (excluding those vectors on the axes) into one vector, $\mathrm{sgn}(x)$ covers an n-dimensional unit ball. In other words, $\mathrm{sgn}(x)$ defines $3^n - 1$ vectors evenly distributed in the n-dimensional space.
(9) If $M \in \mathbb{R}^{n\times n}$ is positive definite (negative definite), i.e., $x^T Mx \geq 0$ (≤ 0), then $x^T M\,\mathrm{sgn}_1(x)$ is also positive (negative) and well defined because $x^T M\,\mathrm{sgn}_1(x)$ $= \frac{x^T Mx}{\|x\|}$ while no conclusion can be drawn for $x^T M\,\mathrm{sgn}(x)$.

2.4 Definitions

Consider the following nonautonomous system:

$$\dot{x} = f(t, x) \tag{2.42}$$

where $f : [0, \infty) \times D \to \mathbb{R}^n$ is piecewise continuous in t and locally Lipschitz in x on $[0, \infty) \times D$ and $D \subseteq \mathbb{R}^n$ is a domain that contains the origin $x = 0$.

Definition 2.10 The origin $x = 0$ is the equilibrium point of (2.42) if

$$f(t, 0) = 0, \quad \forall t \geq 0 \tag{2.43}$$

Definition 2.11 A continuous function $\alpha : [0, a) \to \mathbb{R}^+$ is said to belong to class K if it is strictly increasing and $\alpha(0) = 0$. It is said to belong to class K_∞ if $a = \infty$ and $\alpha(r) \to \infty$ as $r \to \infty$.

Definition 2.12 A continuous function $\beta : [0, a) \times \mathbb{R}^+ \to \mathbb{R}^+$ is said to belong to class KL if, for each fixed s, the mapping $\beta(r, s)$ belongs to class K with respect to r and, for each fixed r, the mapping $\beta(r, s)$ is decreasing with respect to s and $\beta(r, s) \to 0$ as $s \to \infty$. It is said to belong to class KL_∞ if, in addition, for each fixed s the mapping $\beta(r, s)$ belongs to class K_∞ with respect to r.

Definition 2.13 The equilibrium point $x = 0$ of (2.42) is

(1) stable if, for each $\varepsilon > 0$, there is $\delta = \delta(\varepsilon, t_0) > 0$ such that

$$\|x(t_0)\| < \delta \quad \Rightarrow \quad \|x(t)\| < \varepsilon, \quad \forall t \geq t_0 \geq 0 \tag{2.44}$$

(2) uniformly stable if, for each $\varepsilon > 0$, there is $\delta = \delta(\varepsilon) > 0$ independent of t_0 such that (2.44) is satisfied;
(3) unstable if it is not stable;
(4) asymptotically stable if it is stable and there is a positive constant $c = c(t_0)$ such that $x(t) \to 0$ as $t \to \infty$, for all $\|x(t_0)\| < c$;
(5) uniformly asymptotically stable if it is uniformly stable and there is a positive constant c, independent of t_0, such that for all $\|x(t_0)\| < c$, $x(t) \to 0$ as $t \to \infty$, uniformly in t_0; that is, for each $\eta > 0$, there is $T = T(\eta) > 0$ such that

$$\|x(t)\| < \eta, \quad \forall t \geq t_0 + T(\eta), \ \forall \|x(t_0)\| < c \tag{2.45}$$

(6) globally uniformly asymptotically stable (GUAS) if it is uniformly stable, $\delta(\varepsilon)$ can be chosen to satisfy $\lim_{\varepsilon \to \infty} \delta(\varepsilon) = \infty$, and, for each pair of positive numbers η and c, there is $T = T(\eta, c) > 0$ such that

$$\|x(t)\| < \eta, \quad \forall t \geq t_0 + T(\eta, c), \ \forall \|x(t_0)\| < c \tag{2.46}$$

Definition 2.14 The equilibrium point $x = 0$ of (2.42) is exponentially stable if there exist positive constants c, k, and λ such that

$$\|x(t)\| \leq k(\|x(t_0)\|)e^{-\lambda(t - t_0)}, \quad \forall t \geq t_0 \geq 0, \ \forall \|x(t_0)\| < c \tag{2.47}$$

and GES if (2.47) is satisfied for any initial state $x(t_0)$.

Definition 2.15 The equilibrium point $x = 0$ of (2.42) is K-exponentially stable if there exist positive constants c and λ and a class K function α such that

$$\|x(t)\| \leq \alpha(\|x(t_0)\|)e^{-\lambda(t-t_0)}, \quad \forall t \geq t_0 \geq 0, \ \forall \|x(t_0)\| < c \tag{2.48}$$

and globally K-exponentially stable if (2.48) is satisfied for any initial state $x(t_0)$.

Definition 2.16 The solutions of (2.42) are as follows:

(1) uniformly bounded if there exists a positive constant c, independent of $t_0 \geq 0$, and for every $a \in (0, c)$, there is $\beta = \beta(a) > 0$, independent of t_0, such that

$$\|x(t_0)\| \leq a \quad \Rightarrow \quad \|x(t)\| \leq \beta, \quad \forall t \geq t_0 \tag{2.49}$$

(2) globally uniformly bounded if (2.49) holds for an arbitrarily large a;
(3) uniformly ultimately bounded with ultimate bound b if there exist positive constants b and c, independent of $t_0 \geq 0$, and for every $a \in (0, c)$ there is $T = T(a, b) \geq 0$, independent of t_0, such that

$$\|x(t_0)\| \leq a \quad \Rightarrow \quad \|x(t)\| \leq b, \quad \forall t \geq t_0 + T \tag{2.50}$$

(4) globally uniformly ultimately bounded if (2.50) holds for an arbitrarily large a.

2.5 Lemmas and Theorems

Lemma 2.17 *The equilibrium point $x = 0$ of (2.42) is*

(1) *uniformly stable if and only if there exist a class K function α and a positive constant c, independent of t_0, such that*

$$\|x(t)\| \leq \alpha(\|x(t_0)\|), \quad \forall t \geq t_0 \geq 0, \ \forall x(t_0) \mid \|x(t_0)\| < c \tag{2.51}$$

(2) *uniformly asymptotically stable if and only if there exist a β function and a positive constant c, independent of t_0, such that*

$$\|x(t)\| \leq \beta(\|x(t_0)\|, t - t_0), \quad \forall t \geq t_0 \geq 0, \ \forall x(t_0) \mid \|x(t_0)\| < c \tag{2.52}$$

(3) *GUAS if and only if inequality (2.52) is satisfied with $\beta \in KL_\infty$ for any initial state $x(t_0)$.*

Lemma 2.18 *Assume that $d : \mathbb{R}^n \to \mathbb{R}^n$ satisfies*

$$P\left[\frac{\partial d}{\partial x}\right] + \left[\frac{\partial d}{\partial x}\right]^T P \geq 0, \quad \forall x \in \mathbb{R}^n \tag{2.53}$$

when $P = P^T > 0$. Then

$$(x - y)^T P(d(x) - d(y)) \geq 0, \quad \forall x, y \in \mathbb{R}^n \tag{2.54}$$

Theorem 2.19 *Let $D = \{x \in \mathbb{R}^n \mid \|x\| < r\}$ and $x = 0$ be an equilibrium point of (2.42). Let $V : D \times \mathbb{R}^n \to \mathbb{R}^+$ be a continuously differentiable function such that $\forall t \ge 0, \forall x \in D$,*

$$\begin{aligned} \gamma_1\big(\|x\|\big) &\le V(x,t) \le \gamma_2\big(\|x\|\big) \\ \frac{\partial V}{\partial t} + \frac{\partial V}{\partial x} f(t,x) &\le -\gamma_3\big(\|x\|\big) \end{aligned} \tag{2.55}$$

Then the equilibrium point $x = 0$ is

(1) *uniformly stable, if γ_1 and γ_2 are class K functions on $[0, r)$ and $\gamma_3 \ge 0$ on $[0, r)$;*
(2) *uniformly asymptotically stable, if γ_1, γ_2 and γ_3 are class K functions on $[0, r)$;*
(3) *exponentially stable if $\gamma_i(\rho) = k_i \rho^\alpha$ on $[0, r)$, $k_i > 0$, $\alpha > 0$, $i = 1, 2, 3$;*
(4) *globally uniformly stable if $D = \mathbb{R}^n$, γ_1 and γ_2 are class K_∞ functions, and $\gamma_3 \ge 0$ on $\mathbb{R}^+$;*
(5) *GUAS if $D = \mathbb{R}^n$, γ_1 and γ_2 are class K_∞ functions, and γ_3 is a class K function on $\mathbb{R}^+$;*
(6) *GES, if $D = \mathbb{R}^n$, $\gamma_i(\rho) = k_i \rho^\alpha$ on $\mathbb{R}^+$, $k_i > 0$, $\alpha > 0$, $i = 1, 2, 3$.*

Theorem 2.20 *Let $x = 0$ be an equilibrium point of (2.42) and suppose that f is locally Lipschitz in x and uniformly continuous in t. Let $V : \mathbb{R}^n \times \mathbb{R}^+ \to \mathbb{R}^+$ be a continuously differentiable function such that*

$$\begin{aligned} \gamma_1\big(\|x\|\big) &\le V(x,t) \le \gamma_2\big(\|x\|\big) \\ \dot{V} = \frac{\partial V}{\partial t} + \frac{\partial V}{\partial x} f(t,x) &\le -W(x) \le 0 \end{aligned} \tag{2.56}$$

for all $t \ge 0$ and $x \in \mathbb{R}^n$, where γ_1 and γ_2 are class K_∞ functions, and W is a continuous function. Then all solutions of (2.42) are globally uniformly bounded and satisfy

$$\lim_{t\to\infty} W\big(x(t)\big) = 0 \tag{2.57}$$

In addition, if $W(x)$ is positive definite, then the equilibrium point $x = 0$ is GUAS.

Theorem 2.21 *Let $V : [0, \infty) \times D \to \mathbb{R}$ be a continuously differentiable function and $D \in \mathbb{R}^n$ be a domain that contains the origin such that*

$$\begin{aligned} \alpha_1\big(\|x\|\big) &\le V(x,t) \le \alpha_2\big(\|x\|\big) \\ \frac{\partial V}{\partial t} + \frac{\partial V}{\partial x} f(t,x) &\le -W(x), \quad \forall \|x\| \ge \mu > 0 \end{aligned} \tag{2.58}$$

for all $t \ge 0$ and $x \in D$ where α_1 and α_2 are class K functions, and W is a continuously positive definite function. Take $r > 0$ such that $B_r \subset D$ and suppose that

$$\mu < \alpha_2^{-1}\big(\alpha_1(r)\big) \tag{2.59}$$

Then, there exists a class KL function β and for every initial state $x(t_0)$, satisfying $\|x(t_0)\| < \alpha_2^{-1}(\alpha_1(r))$, there is $T > 0$ (dependent on $x(t_0)$ and μ) such that the solutions of (2.42) *satisfies*

$$\begin{aligned} \|x(t)\| &\le \beta(\|x(t_o)\|, t - t_0), \quad \forall t_0 \le t \le t_0 + T \\ \|x(t)\| &< \alpha_1^{-1}(\alpha_2(\mu)), \quad \forall t \ge t_0 + T \end{aligned} \tag{2.60}$$

Moreover, if $D = \mathbb{R}^n$ and α_1 belongs to class K_∞, then (2.60) *holds for any initial state $x(t_0)$ with no restriction on how large μ is.*

2.6 Input-to-State Stability

Definition 2.22 The system

$$\dot{x} = f(t, x, u) \tag{2.61}$$

where f is piecewise continuous in t and locally Lipschitz in x and u, is said to be input-to-state stable (ISS) if there exist a class KL function β and a class K function γ, such that, for any $x(t_0)$ and for any input $u(\cdot)$ continuous and bounded on $[0, \infty)$, the solution exists for all $t \ge t_0 \ge 0$ and satisfies

$$\|x(t)\| \le \beta(\|x(t_0)\|, t - t_0) + \gamma\Big(\sup_{t_0 \le \tau \le t} \|u(\tau)\|\Big) \tag{2.62}$$

The following theorem establishes the equivalence between the existence of a Lyapunov-like function and the input-to-state stability.

Theorem 2.23 *Suppose that for the system* (2.61) *there exists a C^1 function $V : \mathbb{R}^+ \times \mathbb{R}^n \to \mathbb{R}^+$ such that for all $x \in \mathbb{R}^n$ and $u \in \mathbb{R}^m$,*

$$\begin{aligned} \gamma_1(\|x\|) \le V(t, x) &\le \gamma_2(\|x\|) \\ \|x\| \ge \rho(\|u\|) \quad \Rightarrow \quad \frac{\partial V}{\partial t} &+ \frac{\partial V}{\partial x} f(t, x, u) \le -\gamma_3(\|x\|) \end{aligned} \tag{2.63}$$

where γ_1, γ_2 and ρ are class K_∞ functions and γ_3 is a class-K function. Then the system (2.61) *is ISS with $\gamma = \gamma_1^{-1} \circ \gamma_2 \circ \rho$.*

Proof If $x(t_0)$ is in the set

$$R_{t_0} = \big\{x \in \mathbb{R}^n \mid \|x\| \le \rho\big(\sup\|u(\tau)\|_{\tau \ge \tau_0}\big)\big\} \tag{2.64}$$

then $x(t)$ remains within the set

$$S_{t_0} = \big\{x \in \mathbb{R}^n \mid \|x\| \le \gamma_1^{-1} \circ \gamma_2 \circ \rho\big(\sup\|u(\tau)\|_{\tau \ge \tau_0}\big)\big\} \tag{2.65}$$

for all $t \ge t_0$. Define $B = [t_0, T)$ as the time interval before $x(t)$ enters R_{t_0} for the first time. In view of the definition of R_{t_0}, we have

$$\dot{V} \le -\gamma_3 \circ \gamma_2^{-1}(V), \quad \forall t \in B \tag{2.66}$$

Then, there exists a class-KL function β_v such that $V(t) \le \beta_v(V(t_0), t - t_0)$, $\forall t \in B$, which implies

$$\|x(t)\| \le \gamma_1^{-1}\big(\beta_v\big(\gamma_2\big(\|x(t_0)\|\big), t - t_0\big)\big) := \beta\big(\|x(t_0)\|, t - t_0\big), \quad \forall t \in B \tag{2.67}$$

On the other hand, by (2.65), we conclude that

$$\|x(t)\| \le \gamma_1^{-1} \circ \gamma_2 \circ \rho\big(\sup\|u(\tau)\|_{\tau \ge \tau_0}\big) := \gamma\big(\sup\|u(\tau)\|_{\tau \ge \tau_0}\big) \tag{2.68}$$

for all $t \in [t_0, \infty]$. Then by (2.67) and (2.68),

$$\|x(t)\| \le \beta\big(\|x(t_0)\|, t - t_0\big) + \gamma\big(\sup\|u(\tau)\|_{\tau \ge \tau_0}\big), \quad \forall t \ge t_0 \ge 0 \tag{2.69}$$

By causality, we have

$$\|x(t)\| \le \beta\big(\|x(t_0)\|, t - t_0\big) + \gamma\big(\sup\|u(\tau)\|_{t_0 \le \tau \le t}\big), \quad \forall t \ge t_0 \ge 0 \tag{2.70}$$

A function V satisfying conditions (2.63) is called an ISS Lyapunov function. □

2.7 Lyapunov's Direct Method

This section presents an extension of the Lyapunov function concept, which is a useful tool to design an adaptive controller for nonlinear systems. Assuming that the problem is to design a feedback control law $\alpha(x)$ for the time-invariant system:

$$\dot{x} = f(x, u), \quad x \in \mathbb{R}^n, \; u \in \mathbb{R} \tag{2.71}$$

$$f(0, 0) = 0 \tag{2.72}$$

such that the equilibrium $x = 0$ of the closed loop system:

$$\dot{x} = f\big(x, \alpha(x)\big) \tag{2.73}$$

is globally asymptotically stable (GAS). We can take a function $V(x)$ as a Lyapunov candidate function, and require that its derivative along the solutions of (2.73) satisfy $\dot{V}(x) \le -W(x)$, where $W(x)$ is a positive definite function. We therefore need to find $\alpha(x)$ guarantee that for all $x \in \mathbb{R}^n$ such that

$$\frac{\partial V(x)}{\partial x} f\big(x, \alpha(x)\big) \le -W(x) \tag{2.74}$$

This is a difficult problem. A stabilizing control law for (2.72) may exist but we may fail to satisfy (2.74) because of a poor choice of $V(x)$ and $W(x)$. A system for which a good choice of $V(x)$ and $W(x)$ exists is said to possess a control Lyapunov function (CLF). For systems affine in the control:

$$\dot{x} = f(x) + g(x)u, \quad f(0) = 0 \tag{2.75}$$

the CLF inequality (2.74) becomes

$$\frac{\partial V}{\partial x} f(x) + \frac{\partial V}{\partial x} g(x)\alpha(x) \le -W(x) \tag{2.76}$$

If $V(x)$ is a CLF for (2.75), then a particular stabilizing control law $\alpha(x)$, smooth for all $x \neq 0$, is given by

$$u = \alpha(x) = \begin{cases} -\frac{\frac{\partial V}{\partial x} f(x) + \sqrt{(\frac{\partial V}{\partial x} f(x))^2 + (\frac{\partial V}{\partial x} g(x))^4}}{\frac{\partial V}{\partial x} g(x)}, & \frac{\partial V}{\partial x} g(x) \neq 0 \\ 0, & \frac{\partial V}{\partial x} g(x) = 0 \end{cases}$$

It should be noted that (2.76) can be satisfied only if

$$\frac{\partial V}{\partial x} g(x) = 0 \quad \Rightarrow \quad \frac{\partial V}{\partial x} f(x) < 0, \quad \forall x \neq 0 \tag{2.77}$$

and that in this case (2.77) gives

$$W(x) = \sqrt{\left(\frac{\partial V}{\partial x} f\right)^2 + \left(\frac{\partial V}{\partial x} g\right)^4} > 0, \quad \forall x \neq 0 \tag{2.78}$$

The main drawback of the CLF concept as a design tool is that for most nonlinear systems a CLF is not known. The task of finding an appropriate CLF may be as complex as that of designing a stabilizing feedback law.

2.8 Barbalat-Like Lemmas

This section presents lemmas that are useful in investigating the convergence of time-varying systems.

If a function $f \in L_1$ may not be bounded. On the converse, if a function f is bounded, it is not necessary that $f \in L_1$. However, if $f \in L_1 \cap L_\infty$, then $f \in L_p$ for all $p \in [1, \infty)$. Moreover, $f \in L_p$ could not lead to $f \to 0$ as $t \to \infty$. If f is bounded can also lead to $f \to 0$ as $t \to \infty$. However, we have the following results.

Lemma 2.24 (Barbalat's Lemma) *Consider the function* $\phi : \mathbb{R}^+ \to \mathbb{R}$. *If* ϕ *is uniformly continuous and* $\lim_{t\to\infty} \int_0^t \phi(\tau)\,d\tau$ *exists and is finite, then*

$$\lim_{t\to\infty} \phi(t) = 0 \tag{2.79}$$

Lemma 2.25 *Assume that a nonnegative scalar differentiable function* $f(t)$ *enjoys the following conditions*

$$\begin{aligned} &1. \left|\frac{d}{dt} f(t)\right| \le k_1 f(t) \\ &2. \int_0^\infty f(t)\,dt \le k_2 \end{aligned} \tag{2.80}$$

for all $t \ge 0$, *where* k_1 *and* k_2 *are positive constants, then* $\lim_{t\to\infty} f(t) = 0$.

Proof Integrating both sides of (2.80) gives

$$\begin{aligned} f(t) &\le f(0) + k_1 \int_0^t f(s)\,ds \le f(0) + k_1 k_2 \\ f(t) &\ge f(0) - k_1 \int_0^t f(s)\,ds \ge f(0) - k_1 k_2 \end{aligned} \tag{2.81}$$

These inequalities imply that $f(t)$ is a uniform bounded function. From (2.81) and the second condition in (2.80), we have that $f(t)$ is also bounded on the half axis $[0, \infty)$, i.e., $f(t) \le k_3$ with k_3 a positive constant. Hence, $|\frac{d}{dt} f(t)| \le k_1 k_3$. Now assume that $\lim_{t\to\infty} f(t) \ne 0$. Then there exists a sequence of points t_i and a positive constant ϵ such that $f(t_i) \ge \epsilon$, $t_i \to \infty$, $i \to \infty$, $|t_i - t_{i-1}| > 2\epsilon/(k_1 k_3)$ and moreover $f(s) \ge \epsilon/2$, $s \in L_i = [t_i - \epsilon/(2k_1k_3), t_i + \epsilon/(2k_1k_3)]$. Since the segments L_i and L_j do not intersect for any i and j with $i \ne j$, we have

$$\int_0^\infty f(t)\,dt \ge \int_0^T f(t)\,dt \ge \sum_{t_i \le T} \int_{L_i} f(t)\,dt \ge \frac{\epsilon}{2} \frac{\epsilon}{k_1 k_3} M(T) \tag{2.82}$$

where $M(T)$ is the number of points t_i not exceeding T. Since $\lim_{T\to\infty} M(T) = \infty$, the integral $\int_0^\infty f(t)\,dt$ is divergent. This contradicts Condition 2 in (2.80). This contradiction proves the lemma. □

Remark 2.26 Lemma 2.25 is different from Barbalat's Lemma 2.24. While Barbalat's Lemma assumes that $f(t)$ is uniformly continuous, Lemma 2.25 assumes that $|\frac{d}{dt} f(t)|$ is bounded by $k_1 f(t)$.

Corollary 2.27 *If $f(t)$ is uniformly continuous, such that $\int_0^\infty f(\tau)\,d\tau$ exists and is finite, then $f(t) \to 0$ as $t \to \infty$.*

Corollary 2.28 *If $f(t)$, $\dot{f}(t) \in L_\infty$, and $f(t) \in L_p$, for some $p \in [1, \infty)$, then $f(t) \to 0$ as $t \to \infty$.*

Corollary 2.29 *For the differentiable function $f(t)$, if $\lim_{t\to\infty} f(t) = k < \infty$ and $\ddot{f}(t)$ exists, then $\dot{f}(t) \to 0$ as $t \to \infty$.*

Lemma 2.30 *Consider a scalar system*

$$\dot{x} = -cx + p(t) \tag{2.83}$$

where $c > 0$ and $p(t)$ is a bounded and uniformly continuous function. If, for any initial time $t_0 > 0$ and any initial condition $x(t_0)$, the solution $x(t)$ is bounded and converges to 0 as $t \to \infty$, then

$$\lim_{t\to\infty} p(t) = 0 \tag{2.84}$$

Lemma 2.31 *Consider a first-order differential equation of the form*

$$\dot{x} = -\big(a(t) + f_1(\xi(t))\big)x + f_2(\xi(t)) \tag{2.85}$$

where f_1 and f_2 are continuous functions, and $\xi : [0, \infty) \to \mathbb{R}^m$ is a time-varying vector-valued signal that exponentially converges to zero and, for all $t \geq t_0 \geq 0$, satisfies

$$\left|f_i\left(\xi(t)\right)\right| \leq \gamma_i\left(\left\|\xi(t_0)\right\|\right)e^{-\sigma_i(t-t_0)} \tag{2.86}$$

where $\sigma_i, i = 1, 2$ and γ_i are class-K functions. If $a(t)$ enjoys the property that there is a constant σ_3 such that

$$\int_{t_1}^{t_2} a(\tau)d\tau \geq \sigma_3(t_2 - t_1), \quad \forall t_2 \geq t_1 \geq 0 \tag{2.87}$$

then there exist a class-K function γ and a constant $\sigma > 0$ such that

$$\left|x(t)\right| \leq \gamma\left(\left\|x\left(t_0, \xi(t_0)\right)\right\|\right)e^{-\sigma(t-t_0)} \tag{2.88}$$

2.9 Controllability and Observability of Nonlinear Systems

2.9.1 Controllability

This section deals with the controllability and observability properties of nonlinear systems described by linear time varying state–space representations. In particular, consider a nonlinear system defined by the state–space representation:

$$\dot{x}(t) = f(x) + \sum_{i=1}^{m} g_i(x)u_i, \quad x \in \Omega_x \subset \mathbb{R}^n \tag{2.89}$$

where $u = [u_1, u_2, \ldots, u_n]^T \in \Omega_u \subset \mathbb{R}^m$ is the input vector. The system (2.89) is defined to be controllable if there exists an admissible input vector $u(t)$ such that the state $x(t)$ can converge from an initial point $x(t_0) = x_0 \in \Omega_x$ to the final point $x(t_f) \in \Omega_x$ within a finite time interval $t_f - t_0$. The controllability means that the control system is with a set of input channels through which the input can excite the states effectively to converge to the destination x_f. Then, the controllability of (2.89) should mainly depend on the function forms of all $f(x)$ and $g_i(x)$. The controllability of the nonlinear system (2.89) is based on a useful mathematical concept called Lie algebra, which is defined as follows.

Definition 2.32 A Lie algebra over the real field $\mathbb{R}$ or the complex field $\mathbb{C}$ is a vector space $\mathbb{G}$ for which a bilinear map $(X, Y) \to [X, Y]$ is defined from $\mathbb{G} \times \mathbb{G} \to \mathbb{G}$ such that

$$[X, Y] = -[Y, X], \quad X, Y \in \mathbb{G} \tag{2.90}$$

$$\left[X, [Y, Z]\right] + \left[Y, [Z, X]\right] + \left[Z, [X, Y]\right] = 0, \quad X, Y, Z \in \mathbb{G} \tag{2.91}$$

From the above definition, a Lie algebra is a vector space where an operator $[\cdot]$ is installed, which is called a Lie bracket, can be defined arbitrarily as long as it satisfies two conditions (2.90) and (2.91) simultaneously. The condition (2.90) is often called a skew symmetric relation and obviously implies that $[X, Y] = 0$. The condition (2.91) is called the Jacobi identity, which reveals a closed loop cyclic relation among any three elements in a Lie algebra.

Define a special Lie algebra $\mathcal{E}$ that collects all n-dimensional differentiable vector fields in $\mathbb{R}^n$ along with a commutative derivative relation: For any two vector fields f and $g \in \mathbb{R}^n$, which are functions of $x \in \mathbb{R}^n$, we have

$$[f, g] = \frac{\partial g}{\partial x} f - \frac{\partial f}{\partial x} g \tag{2.92}$$

We can see the above equation satisfy the two conditions (2.90) and (2.91) of a Lie algebra.

It is easy to extend the above Lie bracket between two vector fields to higher order derivatives, a more compact notation may be defined based on *an adjoint operator*, that is, $[f, g] = \mathrm{ad}_f g$. This new notation treats the Lie bracket $[f, g]$ as vector field g operated on by an adjoint operator $\mathrm{ad}_f = [f, \cdot]$. Therefore, for an n-order Lie bracket ($n > 1$), one can simply write

$$\big[f, \ldots [f, g] \ldots\big] = \mathrm{ad}_f^n g \tag{2.93}$$

For a general control system given by (2.89), we define a control Lie algebra Δ, which is spanned by all up to order $(n - 1)$ Lie brackets among f and g_1 through g_m as

$$\Delta = \mathrm{span}\big\{g_1, \ldots, g_m, \mathrm{ad}_f g_1, \ldots, \mathrm{ad}_f g_m, \ldots, \mathrm{ad}_f^{n-1} g_1, \ldots, \mathrm{ad}_f^{n-1} g_m\big\} \tag{2.94}$$

With the control Lie algebra concept, we can show that the following theorem is true and is also a general effective testing criterion for system controllability.

Theorem 2.33 *The control system* (2.89) *is controllable if and only if* $\dim(\Delta) = \dim(\Omega_x) = n$.

Note that because each element in Δ is a function of x, the dimension of Δ may be different from one point to another. Thus, if the preceding condition of dimension is valid only in a neighborhood of a point in $\Omega_x \subset \mathbb{R}^n$, we say that the system (2.89) is locally controllable. On the other hand, if the condition of dimension can cover all of region Ω_x, then it is globally controllable.

2.9.2 Observability

Consider the observability for the following nonlinear system

$$\dot{x} = f(x), \qquad y = h(x) \tag{2.95}$$

where $y \in \Omega_y \subset \mathbb{R}^m$ is the output vector. This system is said to be observable if for each pair of distinct states x_1 and x_2, the corresponding outputs y_1 and y_2 are also distinguishable. Clearly, the observability can be interpreted as a testing criterion to check whether the entire system has sufficient output channels to measure (or observe) each internal state change. Intuitively, the observability should depend on the function forms of both $f(x)$ and $h(x)$.

We introduce a Lie derivative, which is virtually *a directional derivative* for a scalar field $\lambda(x)$, with $x \in \mathbb{R}^n$ along the direction of an n-dimensional vector field $f(x)$. The mathematical expression is given as

$$L_f\lambda(x) = \frac{\partial\lambda(x)}{\partial x} f(x) \tag{2.96}$$

Since $\frac{\partial\lambda(x)}{\partial x}$ is a $1 \times n$ gradient vector of the scalar $\lambda(x)$ and the norm of a gradient vector represents the maximum rate of function value changes, the product of the gradient and the vector field $f(x)$ in (2.95) becomes the directional derivative of $\lambda(x)$ along $f(x)$. Therefore, the Lie derivative of a scalar field defined by (2.96) is also a scalar field. If each component of a vector field $h(x) \in \mathbb{R}^m$ is considered to take a Lie derivative along $f(x) \in \mathbb{R}^n$, then all components can be acted on concurrently and the result is a vector field that has the same dimension as $h(x)$; its ith element is the Lie derivative of the ith component of $h(x)$. Namely, if $h(x) = [h_1(x), \dots, h_m(x)]^T$ and each component $h_i(x)$, $i = 1, \dots, m$ is a scalar field, then the Lie derivative of the vector field $h(x)$ is defined as

$$L_f h(x) = \begin{bmatrix} L_f h_1(x) \\ \vdots \\ L_f h_m(x) \end{bmatrix} \tag{2.97}$$

With the Lie derivative concept, we now define an observation space Ω_0 over $\mathbb{R}^n$ as

$$\Omega_0 = \text{span}\{h(x), L_f h(x), \dots, L_f^{n-1} h(x)\} \tag{2.98}$$

In other words, this space is spanned by all up to order $(n-1)$ Lie derivatives of the output function $h(x)$. Then, we further define an observability distribution, denoted by $d\Omega_0$, which collects "the gradient" vector of every component in Ω_0. Namely,

$$d\Omega_0 = \text{span}\left\{\frac{\partial\phi}{\partial x} \mid \phi \in \Omega_0\right\} \tag{2.99}$$

With these definitions, we can present the following theorem for testing the observability.

Theorem 2.34 *The system* (2.95) *is observable if and only if* $\dim(d\Omega_0) = n$.

Similarly to the controllability case, this testing criterion also has locally observable and globally observable cases, depending on whether the condition of dimension in the theorem is valid only in a neighborhood of a point or over the entire state region.

2.9.3 Brockett's Theorem on Feedback Stabilization

The following theorem, which is due to Brockett [21], gives a necessary condition for the existence of a stabilizing control law for the system

$$\dot{x} = f(x, u) \tag{2.100}$$

at an equilibrium point x_0 with x being the state and u being the control input.

Theorem 2.35 *Let the system* (2.100) *be given with* $f(x_0, 0) = 0$ *and* $f(x, u)$ *continuously differentiable in a neighborhood of* $(x_0, 0)$. *A necessary condition for the existence of a continuously differentiable control law that makes* $(x_0, 0)$ *asymptotically stable is that*

1. *the linearized system should have no uncontrollable modes associated with eigenvalues whose real part is positive*;
2. *there exists a neighborhood* Ω *of* $(x_0, 0)$ *such that for each* $\xi \in \Omega$ *there exists a control* u_ξ *defined on* $[0, \infty)$ *such that this control steers the solution of* $\dot{x} = f(x, u_\xi)$ *from* $x = \xi$ *at* $t = 0$ *to* $x = x_0$ *at* $t = \infty$;
3. *the mapping* $A \times \mathbb{R}^m \to \mathbb{R}^n$ *defined by* $(x, u) \to f(x, u)$ *should be onto an open set containing* 0.

Remark 2.36 If the system (2.100) is of the form

$$\dot{x} = f(x) + \sum_{i=1}^{m} u_i g_i(x) \tag{2.101}$$

then condition of Theorem 2.35 implies that the stabilization problem cannot have a solution if there is a smooth distribution D containing $f(\cdot)$ and $g_1(\cdot), \ldots, g_m(\cdot)$ with $\dim D < n$. One further special case: If the system (2.100) is of the form

$$\dot{x} = \sum_{i=1}^{m} u_i g_i(x), \quad x \in \Omega \subset \mathbb{R}^n \tag{2.102}$$

with the vectors $g_i(x)$ being linearly independent at x_0, then there exists a solution to the stabilization problem if and only if $m = n$. In this case, we must have as many control parameters as we have dimensions of x.

2.10 Lyapunov Theorems

The Lyapunov approach provides a rigorous method for addressing stability. The method is a generalization of the idea that if there is some "measure of energy" in a system, then we can study the rate of change of the energy of the system to ascertain stability. Here, we review several concepts that are used in Lyapunov stability theory.

Definition 2.37 A continuous function $V : \mathbb{R}^n \times \mathbb{R}_+ \to \mathbb{R}$ is a locally positive definite function if for some $\varepsilon > 0$ and some continuous, strictly increasing function $\alpha : \mathbb{R}_+ \to \mathbb{R}$,

$$V(0,t) = 0, \quad \text{and} \quad V(x,t) \geq \alpha\|x\|, \quad \forall t \geq 0 \tag{2.103}$$

Definition 2.38 A continuous function $V : \mathbb{R}^n \times \mathbb{R}_+ \to \mathbb{R}$ is a positive definite function if it satisfies the conditions of Definition 2.37 and, additionally, $\alpha(p) \to \infty$ as $p \to \infty$.

Definition 2.39 A continuous function $V : \mathbb{R}^n \times \mathbb{R}_+ \to \mathbb{R}$ is decrescent if for some $\varepsilon > 0$ and some continuous, strictly increasing function $\beta : \mathbb{R}_+ \to \mathbb{R}$,

$$V(x,t) \leq \beta\big(\|x\|\big), \quad \forall x \in \Omega, \ \forall t \geq 0 \tag{2.104}$$

Using these definitions, the following theorem allows us to determine stability for a system by studying an appropriate energy function. Roughly, this theorem states that when $V(x,t)$ is a locally positive definite function and $V(x,t) \leq 0$ then we can conclude stability of the equilibrium point. The time derivative of V is taken along the trajectories of the system:

$$\dot{V}(x,t) = \frac{dV}{dt}(x,t) = \frac{\partial V}{\partial t}(x,t) + \left[\frac{\partial V}{\partial x}(x,t)\right]^T f(t,x) \tag{2.105}$$

Theorem 2.40 (Lyapunov Theorem) *Any nonlinear dynamic system*

$$\dot{x} = f(x,t), \quad x(0) = x_0 \tag{2.106}$$

with an the equilibrium point at the origin, let Ω be a ball of size around the origin, i.e., $\Omega = \{x : \|x\| \leq \varepsilon, \ \varepsilon > 0\}$,

(1) *if for $x \in \Omega$, there exists a scalar function $V(x,t) > 0$ such that $-\dot{V}(x,t) \leq 0$, the origin of system is stable;*
(2) *if for $x \in \Omega$, there exists a scalar decrescent function $V(x,t) > 0$ such that $-\dot{V}(x,t) \leq 0$, the origin of system is uniformly stable;*
(3) *if for $x \in \Omega$, there exists a scalar function $V(x,t) > 0$ such that $-\dot{V}(x,t) \leq 0$, the origin of system is asymptotically stable;*
(4) *if for $x \in \Omega$ there exists a scalar decrescent function $V(x,t) > 0$ such that $-\dot{V}(x,t) \leq 0$, the origin of system is uniformly asymptotically stable;*
(5) *if and only if there exists an $\varepsilon > 0$ and a function $V(x,t)$ which satisfies $\gamma_1\|x\|^2 \leq V(x,t) \leq \gamma_2\|x\|^2$, $\dot{V} \leq -\gamma_3\|x\|^2$, $\|\frac{\partial V}{\partial x}(x,t)\| \leq \gamma_4\|x\|$ with some positive constants $\gamma_1, \gamma_2, \gamma_3, \gamma_4$ and $\|x\| \leq \varepsilon$, $x = 0$ is an exponentially stable equilibrium point of $x = f(x,t)$;*
(6) *if there exist a scalar decrescent function $V : \mathbb{R}^n \times \mathbb{R}_+ \mapsto \mathbb{R}$, and a time $t_0 \geq 0$, such that $\dot{V} \leq 0$, and for any sufficiently small positive real number r, there exists a non-origin point $x \in \mathbf{B}_r$ such that $V(x,t_0) \geq 0$, the origin of system is unstable.*

The function $V(x,t)$ is called the Lyapunov function.

The indirect method of Lyapunov uses the linearization of a system to determine the local stability of the original system.

Theorem 2.41 (Stability by Linearization) *Consider the system* $\dot{x} = f(x,t)$ *and define*

$$A(t) = \frac{\partial f(x,t)}{\partial x} \tag{2.107}$$

with $x = 0$ *to be the Jacobian matrix of* $f(x,t)$ *with respect to* x*, evaluated at the origin. It follows that for each fixed* t*,*

$$f_1(x,t) = f(x,t) - A(t)x \tag{2.108}$$

approaches zero as x *approaches zero. Assume*

$$\lim_{\|x\|\to 0} \sup_{t\geq 0} \frac{\|f_1(x,t)\|}{\|x\|} = 0 \tag{2.109}$$

Further, if 0 *is a uniformly asymptotically stable equilibrium point of*

$$\dot{z} = A(t)z \tag{2.110}$$

then it is a locally uniformly asymptotically stable equilibrium point of $\dot{x} = f(x,t)$.

Invariant Set Theorems Asymptotic stability of a control system is a very important property. However, the Lyapunov theorems are usually difficult to apply because frequently $\dot{V}$, the derivative of the Lyapunov function candidate, is only semi-definite. With the help of the invariant set theorems, asymptotic stability can still possibly be concluded for autonomous systems from LaSalle's invariance principle [113]. The concept of an invariant set is a generalization of the concept of equilibrium point.

Definition 2.42 (α Limit Set) The set $\Omega \in \mathbb{R}^n$ is the α limit set of a trajectory $\omega(t, x_0, t_0)$ if for every $y \in \Omega$, there exists a strictly increasing sequence of times T such that $\omega(T, x_0, t_0) \to y$ as $T \to \infty$.

Definition 2.43 A set $\Omega \in \mathbb{R}^n$ is said to be an invariant set of the dynamic system $\dot{x} = f(x)$ if for all $y \in \Omega$ and $t_0 > 0$, we have $\omega(t, y, t_0) \in \Omega$, $\forall t > t_0$.

Theorem 2.44 (LaSalle's Theorem) *Let* Ω *be a compact invariant set* $\Omega = x \in \mathbb{R}^n : \dot{V}(x) = 0$ *and* $V : \mathbb{R}^n \to \mathbb{R}$ *be a locally positive definite function such that on the compact set we have* $\dot{V}(x) \leq 0$*. As* $t \to \infty$*, the trajectory tends to the largest invariant set inside* Ω*; i.e., its* α *limit set is contained inside the largest invariant set in* Ω*. In particular, if* Ω *contains no invariant sets other than* $x = 0$*, then* 0 *is asymptotically stable.*

Corollary 2.45 *Given the autonomous nonlinear system*

$$\dot{x} = f(x), \quad x(0) = x_0 \tag{2.111}$$

and let the origin be an equilibrium point, $V(x) : \mathcal{N} \to \mathbb{R}$ be a continuously differentiable positive definite function on a neighborhood $\mathcal{N}$ of the origin, such that $\dot{V}(x) \leq 0$ in $\mathcal{N}$, then the origin is asymptotically stable if there is no solution that can stay forever in $S = x \in \mathcal{N} \mid \dot{V}(x) = 0$, other than the trial solution. The origin is globally asymptotically stable if $\mathcal{N} = \mathbb{R}^n$ and $V(x)$ is radially unbounded.

Other Stability Results

Definition 2.46 A system is said to be BIBO (bounded-input and bounded-output) stable iff for any bounded-input, the output is bounded, i.e., if for any

$$\|u\| < M < \infty \tag{2.112}$$

there exist finite $\alpha > 0$ and β such that

$$\|y\| \leq \alpha M + \beta \tag{2.113}$$

Theorem 2.47 *Let the closed-loop transfer function $H(s) \in \mathbb{R}^{n \times n}(s)$ be exponentially stable and strictly proper, and $h(t)$ be the corresponding impulse response. We have the following properties*:

(1) *For any minimal representation of $H(s)$ (i.e., any minimal state representation of the form $\dot{x} = Ax + Bu$, $y = Cx$ with $H(s) = C(sI - A)^{-1}B$), the equilibrium point $x = 0$ is globally, exponentially, uniformly (in t) stable.*
(2) *If $u \cup L_1^n$, then $y = h * u \in L_1^n \cap L_\infty^n$, $\dot{y} \in L_\infty^n$, y is absolutely continuous and $y(t) \to 0$ as $t \to \infty$.*
(3) *If $u \in L_2^n$, then $y = h * u \in L_2^n \cap L_\infty^n$, $\dot{y} \in L_2^n$, y is continuous and $y(t) \to 0$ as $t \to \infty$.*
(4) *If $u \in L_\infty^n$, then $y = h * u \in L_\infty^n \cap L_\infty^n$, $\dot{y} \in L_\infty^n$, y is continuous and $y(t) \to 0$ as $t \to \infty$.*
(5) *If $u \in L_\infty^n$, and if $u(t) \to u_\infty$ (a constant vector in $\mathbb{R}$) as $t \to \infty$, then $y \to H(0)u_\infty$, as $t \to \infty$ and the convergence is exponential.*
(6) *If $u \in L_p^n$ and $1 < p < \infty$, then, $y = h * u \in L_p^n$ and $\dot{y} \in L_p^n$.*

$h * u$ denotes the convolution product of h and u.

Lemma 2.48 *Let $e(t) = h * r$, where $h = L^{-1}(H(s))$ and $H(s)$ is an $n \times n$ strictly proper, exponentially stable transfer function. Then $r \in L_2^n \Rightarrow e \in L_2^n \cap L_\infty^n$, $\dot{e} \to L_2^n$, e is continuous and $e \to 0$ as $t \to \infty$. If, in addition, $r \to 0$ as $t \to \infty$, then $\dot{e} \to 0$ [32].*

Based on this lemma, we have the following corollaries, which are used in the proof of the stability properties of the closed-loop system [44].

Corollary 2.49 *If $H(s)$ is an exponentially stable and strictly proper diagonal matrix with*

$$H(s) \triangleq \operatorname{diag} \frac{s^m + n_i(s)}{s^{m+p} + d_i(s)} \tag{2.114}$$

where $n_i(s)$ and $d_i(s)$ are the remaining polynomial terms of s for the ith entry, then, $s^i H(s)$, $1 \le i < p$ are exponentially stable and strictly proper, too. Therefore, if $r \in L_2^n \Rightarrow e, \dot{e}, \ldots, e^{(p-1)} \in L_2^n \cap L_\infty^n$, and $e, \dot{e}, \ldots, e^{(p-1)}$ are continuous and $e, \dot{e}, \ldots, e^{(p-1)} \to 0$ as $t \to \infty$, then $e^{(p)} \to 0$.

Corollary 2.50 *If $H(s)$ is defined as above and $p = 2$, then, $sH(s)$ is exponentially stable and strictly proper, too. Therefore, $r \in L_2^n \Rightarrow e$ and $\dot{e} \in L_2^n \cap L_\infty^n$, $\ddot{e} \in L_2^n$, e and $\dot{e} \in L_2^n \cap L_\infty^n$ are continuous and e and $\dot{e} \to 0$ as $t \to \infty$. In addition, if $r \to 0$ as $t \to \infty$, then $\ddot{e} \to 0$.*

Theorem 2.51 *Given a linear time-invariant (LTI) system*

$$\dot{x}(t) = Ax(t) \tag{2.115}$$

the system is stable iff there exists a symmetric positive definite solution P_L to the Lyapunov equation

$$A^T P_L + P_L A = -Q_L \tag{2.116}$$

where Q_L is an arbitrary symmetric positive-definite matrix.

Theorem 2.52 *For a given stable LTI system*

$$\dot{x}(t) = Ax(t) \tag{2.117}$$

the system is exponentially uniformly stable.

Definition 2.53 A function $T : U_o \to \mathbb{R}^n$ is called a diffeomorphism if it is smooth, and if its inverse, T^{-1} exists and is smooth. If the region U_o is the whole space $\mathbb{R}^n$, then $T(x)$ is called a global diffeomorphism [113].

Definition 2.54 A nonlinear system

$$\dot{x} = f(x) + \sum_{i=1}^{n} g_i(x)u_i \tag{2.118}$$

$$= f(x) + G(x)u \tag{2.119}$$

is said to be feedback linearizable in a neighborhood U_o of the origin if there exist a diffeomorphism $T : U_o \to \mathbb{R}^n$ and a nonlinear feedback control law

$$u = M^{-1}(x)\big[v - F(x)\big] \tag{2.120}$$

such that the transformed state

$$z = T(x) \tag{2.121}$$

and the new input v satisfies the linear time invariant system

$$\dot{z} = Az + Bv \tag{2.122}$$

where (A, B) is a controllable linear system [113].

Consider the linear time-invariant system given by

$$\dot{x}(t) = Ax(t), \quad t \geq 0 \tag{2.123}$$

In this special case, Lyapunov theory is very complete, and we have the following theorem.

Theorem 2.55 *For linear time-invariant system* (2.123), *the following statements are equivalent*:

(i) *the system is asymptotically stable*;
(ii) *the system is exponential stable*; *matrix A is Hurwitz*;
(iv) *the Lyapunov equation*

$$A^T P + PA = -Q \tag{2.124}$$

has a unique solution $P > 0$ for any $Q > 0$; *and*
(v) *Eq.* (2.124) *has a unique solution $P > 0$ for some $Q > 0$*.

The last statement asserts the existence of a quadratic Lyapunov function

$$V(x) = x^T Px$$

where P is symmetric and positive definite.

For the discrete-time linear time-invariant system

$$x_{k+1} = Ax_k, \quad k \in \mathbf{N}_+ \tag{2.125}$$

a similar result may be stated as follows.

2.11 Notes and References

This chapter briefly provides the fundamental concepts and tools that will be used for control design and stability analysis of wheeled inverted pendulum in the coming chapters. The interested reader is referred to [67, 72, 74, 79, 81, 95] for a detailed and comprehensive coverage of the topics discussed in this chapter.

Chapter 3
Modeling of WIP Systems

3.1 Introduction

As shown in Fig. 3.1, the WIP vehicle consists of a pair of identical wheels, a chassis, two wheel actuators, an inverted pendulum, and a motion control unit. The chassis supports the pair of wheels and the inverted pendulum. The wheel actuators rotate the wheels with respect to the chassis. The motion control unit controls the wheel actuators so as to move the vehicle and stabilize the inverted pendulum. Under such a configuration, the vehicle is characterized by the ability to balance on the two wheels and spin on the spot.

Given the WIP robot, kinematics focus on the velocity relationships regarding the linear and angular velocities represented in the task space and in the joint space, while dynamics study the relationship between the forces applied on WIP and the resulted motion, i.e., how the forces acting on robots make these robots move by their kinematics. Mathematically, the kinematic equations establish a mapping between the Cartesian space positions/orientations and the joint space positions (translation or revolution). The velocity relationship is then determined by the Jacobian associated with this mapping, whereas each row of the Jacobian matrix is a vector of partial derivatives of a Cartesian position variable with respect to all the joint space position variables. It should be mentioned that the Jacobian is one of the most important quantities in the analysis and control of robot motion.

This chapter describes how to model the dynamic properties of chains of interconnected rigid bodies that form a WIP system, how to calculate the chains' time evolution under a given set of internal and external forces and/or desired motion specifications, and how to provide a means of designing prototype mobile inverted pendulum and testing control approaches without building the actual mobile inverted pendulum. The theoretical principle behind rigid body dynamics is that one extends Newton's laws for the dynamics of a point mass to rigid bodies. However, the interconnection of rigid bodies by means of revolute joints gives rise to new physical properties that do not exist for one single point mass or one single rigid body. More in particular, the topology of the kinematic chain determines to a large

Z. Li et al., *Advanced Control of Wheeled Inverted Pendulum Systems*,
DOI 10.1007/978-1-4471-2963-9_3, © Springer-Verlag London 2013

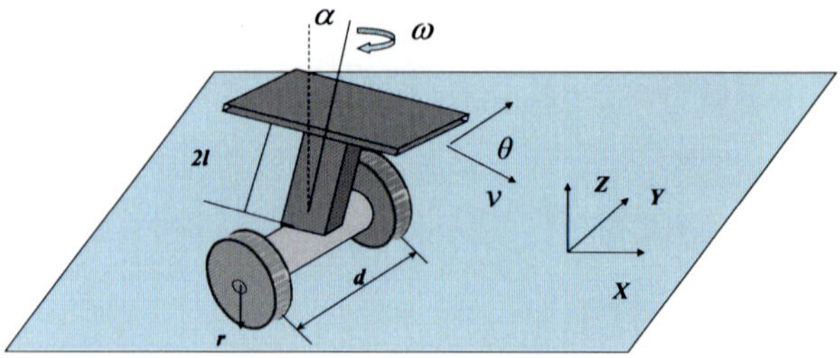

Fig. 3.1 Model of wheeled inverted pendulum

extent the minimal complexity of the computational algorithms that implement these physical properties.

Kinematic and dynamic modeling of WIP deals with the mathematical formulation of the kinematic and dynamic equations. Modeling helps us understanding the physical properties and is the basis to develop relevant control algorithms. In this chapter, to derive dynamics model we assume that all the physical parameters of the robot are known, e.g., dimensions of links, relative positions and orientations of connected parts, mass distributions of the links, joints and motors. While in practice, it is not straightforward to obtain accurate values for all these parameters. In addition, in this chapter we assume an ideal system with rigid bodies, joints without backlash and accurately modeled friction.

For the easy computation of kinematics model for mobile inverted pendulum, we decomposed mobile inverted pendulum into the mobile platform and the inverted pendulum.

3.2 Kinematics of the WIP Systems

For convenience of kinematics modeling, we decompose the WIP into a mobile platform and an inverted pendulum. Let the generalized coordinates of the mobile platform be $q = (x_o, y_o, \theta, \theta_r, \theta_l)$, where (x_o, y_o) is the center point of the driving wheels, θ is the heading angle of the mobile platform measured from x-axis, θ_r and θ_l are the angular positions of the two driving wheels, respectively. If the mobile platform of the WIP system satisfies the nonholonomic constraints without slipping, then the following constraint holds:

$$A(q)\dot{q} = 0 \tag{3.1}$$

where $A(q) \in \mathbb{R}^{3\times 5}$ is the matrix associated with the constraints.

As shown in Fig. 3.1, the motion and orientation of the WIP are achieved by two independent actuators which provide the torques to the wheels. The nonholonomic constraint states that the robot can only move in the direction normal to the axis of the driving wheels, i.e., the mobile base satisfies the conditions of pure rolling and nonslipping, therefore, the mobile platform is generally subject to three constraints. The first one is that the mobile robot can not move in the lateral direction, i.e.,

$$\dot{y}_o \cos\theta - \dot{x}_o \sin\theta = 0 \tag{3.2}$$

Equation (3.2) is a nonholonomic constraint that cannot be integrated analytically to result in a holonomic constraint among the configuration variables of the platform, namely x_o, y_o, and θ. As well known, the configuration space of the system is three dimensional (completely unrestricted), while the velocity space is two-dimensional. This constraint becomes

$$\dot{x}_o \cos\theta + \dot{y}_o \sin\theta + \frac{d}{2}\dot{\theta} = r\dot{\theta}_r \tag{3.3}$$

$$\dot{x}_o \cos\theta + \dot{y}_o \sin\theta - \frac{d}{2}\dot{\theta} = r\dot{\theta}_l \tag{3.4}$$

where d is the platform width and r is the radius of wheel.

Combining Eqs. (3.2), (3.3) and (3.4), we see that matrix $A(q)$ can be written as follows

$$A(q) = \begin{bmatrix} -\sin\theta & \cos\theta & 0 & 0 & 0 \\ \cos\theta & \sin\theta & \frac{d}{2} & -r & 0 \\ \cos\theta & \sin\theta & -\frac{d}{2} & 0 & -r \end{bmatrix} \tag{3.5}$$

Let m rank matrix $S(q) \in \mathbb{R}^{5\times 2}$ formed by a set of smooth and linearly independent vector fields spanning the null space of $A(q)$, i.e.,

$$S^T(q)A^T(q) = 0 \tag{3.6}$$

According to (3.1) and (3.6), it is possible to establish that

$$\dot{q} = S(q)v(t) \tag{3.7}$$

where $v(t) = [\dot{\theta}_r, \dot{\theta}_l]$ and matrix $S(q) \in \mathbb{R}^{5\times 2}$ is defined as follows

$$S(q) = \big[s_1(q)\ s_2(q)\big] = \begin{bmatrix} \frac{r}{2}\cos\theta & \frac{r}{2}\cos\theta \\ \frac{r}{2}\sin\theta & \frac{r}{2}\sin\theta \\ \frac{r}{d} & -\frac{r}{d} \\ 1 & 0 \\ 0 & 1 \end{bmatrix} \tag{3.8}$$

It is obvious that the matrix $S(q)$ are in the null space of matrix $A(q)$, that is, $S^T(q)A^T(q) = 0$. A distribution spanned by the columns of $S(q)$ can be described as

$$\Delta = \operatorname{span}\big\{s_1(q), s_2(q)\big\} \tag{3.9}$$

Remark 3.1 The number of holonomic or nonholonomic constraints can be determined by the involutivity of the distribution Δ. If the smallest involutive distribution containing Δ (denoted by Δ^*) spans the entire 5-dimensional space, all the constraints are nonholonomic. If $\dim(\Delta^*) = 5 - k$, then k constraints are holonomic and the others are nonholonomic.

To verify the involutivity of Δ, we compute the Lie bracket of $s_1(q)$ and $s_2(q)$

$$s_3(q) = [s_1(q)\ s_2(q)] = \frac{\partial s_2}{\partial q} s_1 - \frac{\partial s_1}{\partial q} s_2 = \begin{bmatrix} -\frac{r^2}{d} \sin\theta \\ \frac{r^2}{d} \cos\theta \\ 0 \\ 0 \\ 0 \end{bmatrix} \tag{3.10}$$

which is not in the distribution Δ spanned by $s_1(q)$ and $s_2(q)$. Therefore, at least one of the constraints is nonholonomic. We continue to compute the Lie bracket of $s_1(q)$ and $s_3(q)$

$$s_4(q) = [s_1(q)\ s_3(q)] = \frac{\partial s_3}{\partial q} s_1 - \frac{\partial s_1}{\partial q} s_3 = \begin{bmatrix} -\frac{r^3}{d^2} \cos\theta \\ -\frac{r^3}{d^2} \sin\theta \\ 0 \\ 0 \\ 0 \end{bmatrix} \tag{3.11}$$

which is linearly independent of $s_1(q)$, $s_2(q)$, and $s_3(q)$. However, the distribution spanned by $s_1(q)$, $s_2(q)$, $s_3(q)$ and s_4 is involutive. Therefore, we have

$$\Delta^* = \text{span}\{s_1(q), s_2(q), s_3(q), s_4(q)\} \tag{3.12}$$

It follows that two of the constraints are nonholonomic and the other one is holonomic.

To obtain the holonomic constraint, we subtract Eq. (3.3) from Eq. (3.4)

$$d\dot{\theta} = r(\dot{\theta}_r - \dot{\theta}_l) \tag{3.13}$$

Integrating the above equation and properly choosing the initial condition of $\theta(0) = \theta_r(0) = \theta_l(0)$, we have

$$\theta = \frac{r}{d}(\theta_r - \theta_l) \tag{3.14}$$

which is obviously a holonomic constraint equation. Thus, θ may be eliminated from the generalized coordinates.

The two nonholonomic constraints are

$$\dot{x}_o \sin\theta - \dot{y}_o \cos\theta = 0 \tag{3.15}$$

$$\dot{x}_o \cos\theta + \dot{y}_o \sin\theta = \frac{r}{2}(\dot{\theta}_r + \dot{\theta}_l) \tag{3.16}$$

The second nonholonomic constraint equation in the above is obtained by adding Eq. (3.3) from Eq. (3.4). It is understood that θ is now a shorthand notation for $r/d(\theta_r - \theta_l)$ rather than an independent variable. We may write these two constraint equations in the matrix form

$$A(q)\dot{q} = 0 \tag{3.17}$$

where the generalized coordinate vector q is now defined as

$$q = \begin{bmatrix} q_1 \\ q_2 \\ q_3 \\ q_4 \\ q_5 \end{bmatrix} = \begin{bmatrix} x_o \\ y_o \\ \theta \\ \theta_r \\ \theta_l \end{bmatrix} \tag{3.18}$$

The kinematics of this mechanism can be written as

$$\begin{bmatrix} \dot{x}_o \\ \dot{y}_o \\ \dot{\theta} \\ \dot{\theta}_r \\ \theta_l \end{bmatrix} = \begin{bmatrix} \frac{r\cos(\theta)}{2} & \frac{r\cos(\theta)}{2} \\ \frac{r\sin(\theta)}{2} & \frac{r\sin(\theta)}{2} \\ \frac{r}{d} & -\frac{r}{d} \\ 1 & 0 \\ 0 & 1 \end{bmatrix} \begin{bmatrix} \dot{\theta}_r \\ \dot{\theta}_l \end{bmatrix} \tag{3.19}$$

From the relationship between linear and angular velocities of the vehicle and its wheels' velocities, we have

$$\begin{bmatrix} \varphi_r \\ \varphi_l \end{bmatrix} = \begin{bmatrix} \frac{1}{r} & \frac{1}{d} \\ \frac{1}{r} & -\frac{1}{d} \end{bmatrix} \begin{bmatrix} v \\ \omega \end{bmatrix} \tag{3.20}$$

Substituting (3.20) into (3.19), we have

$$\begin{bmatrix} \dot{x}_o \\ \dot{y}_o \\ \dot{\theta} \\ \dot{\varphi}_r \\ \dot{\varphi}_l \end{bmatrix} = \begin{bmatrix} \cos\theta & 0 \\ \sin\theta & 0 \\ 0 & 1 \\ \frac{1}{r} & \frac{1}{d} \\ \frac{1}{r} & -\frac{1}{d} \end{bmatrix} \begin{bmatrix} v \\ \omega \end{bmatrix} \tag{3.21}$$

Skipping two equations from (3.21), we have

$$\begin{bmatrix} \dot{x}_c \\ \dot{y}_c \\ \dot{\theta} \end{bmatrix} = \begin{bmatrix} \cos\theta & 0 \\ \sin\theta & 0 \\ 0 & 1 \end{bmatrix} \begin{bmatrix} v \\ \omega \end{bmatrix} \tag{3.22}$$

3.3 Dynamics of WIP Systems

In general, the dynamics of wheeled inverted pendulum can be derived by two traditional formulations: the closed-form Lagrange–Euler formulation and forward–backward recursive Newton–Euler formulation. The Lagrange–Euler approach treats the WIP as a whole and performs the analysis using the Lagrangian function (the difference between the kinetic energy and the potential energy of the mobile robotic system), which compose of each link of the wheeled inverted pendulum. While Newton–Euler approach describe the combined translational and rotational dynamics of a rigid body with respect to the each link's center of mass. The dynamics of the whole wheeled inverted pendulum can be described by the forward–backward recursive dynamic equations. Therefore, two different kinds of formulations provide different insights to the physical meaning of dynamics. Dynamic analysis is to find the relationship between the generalized coordinates q and the generalized forces τ. A closed-form equation like the Lagrange–Euler formulation is preferred such that we can conduct the controllers to obtain the time evolution of the generalized coordinates.

Thus, in the following section, the Lagrange–Euler formulation will be discussed in detail from Sect. 3.3.1 to Sect. 3.3.4, which follows the description of the previous work [44]. Section 3.3.5 comes from lots of the previous works, such as [44, 113], etc.

3.3.1 Lagrange–Euler Equations

We briefly introduce the principle of virtual work since the Lagrange–Euler equations of motion are a set of differential equations that describe the time evolution of mechanical systems under holonomic constraints.

Consider a system consisting of l particles, with corresponding coordinates $r_1, r_2, \dots, r_l$ is subject to holonomic constraints as follows

$$f_i(r_1 \dots r_l) = 0, \quad i = 1, 2, \dots, m \tag{3.23}$$

The constraint implies a force (called constraint force) is produced, that hold this constraint forces. The system subject to constraints (3.23) has m fewer degree of freedom than the unconstrained system, then the coordinates of the l constraints are described in term of n generalized coordinates $q_1, q_2, \dots, q_n$ as

$$r_i = r_i(q), \quad i = 1, 2, \dots, l \tag{3.24}$$

where $q = [q_1, q_2, \dots, q_n]^T$ and $q_1, q_2, \dots, q_n$ are independent. To keep the discussion simple, l is assumed to be finite.

Differentiating the constraint function $f_i(\cdot)$ with respect to time, we obtain new constraint

$$\frac{d}{dt} f_i(r_1, r_2, \dots, r_l) = \frac{q\partial f_i}{\partial r_1}\frac{dr_1}{dt} + \dots + \frac{\partial f_i}{\partial r_l}\frac{dr_l}{dt} = 0 \tag{3.25}$$

The constraint of the form

$$\omega_1(r_1, \dots, r_l)dr_1 + \dots + \omega_k(r_1, \dots, r_k)dr_k = 0 \tag{3.26}$$

is called nonholonomic if it can not be integrated back to $f_i(\cdot)$.

Given (3.25), by definition a set of infinitesimal displacements $\Delta r_1, \dots, \Delta r_l$, that are consistent with the constraint

$$\frac{\partial f_i}{\partial r_1}\Delta r_1 + \dots + \frac{\partial f_i}{\partial r_l}\Delta r_l = 0 \tag{3.27}$$

are called virtual displacements, which can be precisely defined as follows with Eq. (3.24) holding

$$\delta r_i = \sum_{j=1}^{n} \frac{\partial r_i}{\partial q_j} \delta q_j, \quad i = 1, 2, \ldots, l \tag{3.28}$$

where $\delta q_1, \delta q_2, \ldots, \delta q_n$ of the generalized coordinates are unconstrained.

Consider a system of l-particle with the total force F_i, suppose that

- the system has a holonomic constraint, that is some of particles exposed to constraint force f_{ic};
- there are the external force f_{ie} to the particles; and
- the constrained system is in equilibrium;

then the work done by all forces applied to ith-particle along each set of virtual displacement is zero,

$$\sum_{i=1}^{l} F_i^T \delta r_i = 0 \tag{3.29}$$

If the total work done by the constraint forces corresponding to any set of virtual displacement is zero, that is

$$\sum_{i=1}^{l} f_{ic}^T \delta r_i = 0 \tag{3.30}$$

Substituting Eq. (3.31) into (3.30), we have

$$\sum_{i=1}^{l} f_{ie}^T \delta r_i = 0 \tag{3.31}$$

which expresses the principle of virtual work: if satisfying (3.30), the work done by external forces corresponding to any set of virtual displacements is zero. Suppose that each constraint will be in equilibrium and consider the fictitious additional force $\dot{p}_i$ for each constraint with the momentum of the ith constraint p_i. By substituting p_i with F_i in Eq. (3.29), and the constraint forces are eliminated as before by using the principle of virtual work, we can obtain

$$\sum_{i=1}^{l} f_{ie}^T \delta r_i - \sum_{i=1}^{l} \dot{p}_i \delta r_i = 0 \tag{3.32}$$

The virtual work by the force f_{ie} is expressed as

$$\sum_{i=1}^{l} f_{ie}^T \delta r_i = \sum_{j=1}^{n} \left(\sum_{i=1}^{l} f_{ie}^T \frac{\partial r_i}{\partial q_j} \right) \delta q_j = \sum_{j=1}^{n} \psi_j \delta q_j = \psi^T \delta q \tag{3.33}$$

where $\psi = [\psi_1, \psi_2 \ldots \psi_n]$, $\psi_j = \sum_{i=1}^{k} f_{ie}^T \frac{\partial r_i}{\partial q_j}$ is called the jth generalized force.

Considering $p_i = m_i \dot{r}_i$, the second summation in Eq. (3.32) becomes

$$\sum_{i=1}^{l} \dot{p}_i^T \delta r_i = \sum_{i=1}^{l}\sum_{j=1}^{n} m_i \ddot{r}_i^T \frac{\partial r_i}{\partial q_j}\delta q_j = \sum_{j=1}^{n} \eta_j \delta q_j = \eta^T \delta q \tag{3.34}$$

where $\eta = [\eta_1, \eta_2 \ldots \eta_n]^T$, and $\eta_j = \sum_{i=1}^{k} m_i \ddot{r}_i^T \frac{\partial r_i}{\partial q_j}$.

Using the chain-rule, we can obtain

$$\frac{\partial v_i}{\partial \dot{q}_j} = \frac{\partial r_i}{\partial q_j} \tag{3.35}$$

since

$$v_i = \dot{r}_i = \sum_{j=1}^{n} \frac{\partial r_i}{\partial q_j}\dot{q}_j \tag{3.36}$$

We can further obtain

$$\frac{d}{dt}\left[\frac{\partial r_i}{\partial q_j}\right] = \sum_{l=1}^{n} \frac{\partial^2 r_i}{\partial q_j \partial q_l}\dot{q}_l = \frac{\partial v_i}{\partial q_j} \tag{3.37}$$

Based on the product rule of differentiation, we have $m_i \ddot{r}_i^T \frac{\partial r_i}{\partial q_j} = \frac{d}{dt}[m_i \dot{r}_i^T] - m_i \dot{r}_i^T \frac{d}{dt}[\frac{\partial r_i}{\partial q_j}]$, and considering the above three relations, we can rewrite η_j as

$$\eta_j = \sum_{i=1}^{l} m_i \ddot{r}_i^T \frac{\partial r_i}{\partial q_j} = \sum_{i=1}^{l}\left(\frac{d}{dt}\left[m_i v_i^T \frac{\partial v_i}{\partial \dot{q}_j}\right] - m_i v_i^T \frac{\partial v_i}{\partial q_j}\right) \tag{3.38}$$

Let $K = \sum_{i=1}^{l} \frac{1}{2} m_i v_i^T v_i$ be the kinetic energy, considering (3.38), we can obtain $\eta_j = \frac{d}{dt}\frac{\partial K}{\partial \dot{q}_j} - \frac{\partial K}{\partial q_j}$, $j = 1, 2, \ldots, n$, rewriting the above equation in a vector form, we have

$$\eta = \frac{d}{dt}\frac{\partial K}{\partial \dot{q}} - \frac{\partial K}{\partial q} \tag{3.39}$$

Summary equations from (3.32), (3.33), (3.34), and (3.39), we have

$$\left[\frac{d}{dt}\frac{\partial K}{\partial \dot{q}} - \frac{\partial K}{\partial q} - \psi\right]^T \delta q = 0 \tag{3.40}$$

Define a scalar potential energy $P(q)$ such that $\psi = -\frac{\partial P}{\partial q}$, since the virtual displacement vector δq is unconstrained and its elements δq_j are independent, which leads to $\frac{d}{dt}\frac{\partial K}{\partial \dot{q}} - \frac{\partial K}{\partial q} - \psi = 0$, it can be written as

$$\frac{d}{dt}\frac{\partial L}{\partial \dot{q}} - \frac{\partial L}{\partial q} = 0 \tag{3.41}$$

where $L(q, \dot{q}) = K(q, \dot{q}) - P(q)$ is Lagrangian function.

Remark 3.2 Given generalized coordinates, the choice of Lagrangian is not unique for a particular set of equations of motion.

Remark 3.3 A necessary and sufficient condition that F be the gradient of some scalar function P, i.e., $F = -\frac{\partial P(r)}{\partial q}$, which in turn means that the generalized force is derivable from P by differentiating with respect to q.

If the generalized force ψ includes an external applied force and a potential field force, suppose there exists a vector τ and a scalar potential function $P(q)$ satisfying $\psi = \tau - \frac{\partial P}{\partial q}$, then, Eq. (3.41) can be written in the form

$$\frac{d}{dt}\frac{\partial L}{\partial \dot{q}} - \frac{\partial L}{\partial q} = \tau \tag{3.42}$$

Equations (3.41) and/or (3.42) are called the Lagrangian equations or Lagrange–Euler equations in the robotics literature.

3.3.2 Kinetic Energy

Consider the velocity of the point in base coordinates described by $v_i = \frac{dr_i}{dt} = \sum_{j=1}^{i}[\frac{\partial T_i^0}{\partial q_j}\dot{q}_j]r_i^i = \sum_{j=1}^{n}[\frac{\partial T_i^0}{\partial q_j}\dot{q}_j]r_i^i$ with $\frac{\partial T_i^0}{\partial q_j} = 0, \forall j > i$, we have

$$dK_i = \frac{1}{2}\,\text{trace}\left[\sum_{j=1}^{n}\sum_{k=1}^{n}\frac{\partial T_i^0}{\partial q_j}\left(r_i^i r_i^{iT}\,dm\right)\frac{\partial T_i^{0T}}{\partial q_k}\dot{q}_j\dot{q}_k\right] \tag{3.43}$$

Define the 4×4 pseudo-inertia matrix for the ith link as

$$J_i = \int_{\text{link } i} r_i^i r_i^{iT}\,dm \tag{3.44}$$

The total kinetic energy for the ith link can be expresses as

$$K_i = \int_{\text{link } i} dK_i = \frac{1}{2}\,\text{trace}\left[\sum_{j=1}^{n}\sum_{k=1}^{n}\frac{\partial T_i^0}{\partial q_j}J_i\frac{\partial T_i^{0T}}{\partial q_k}\dot{q}_j\dot{q}_k\right] \tag{3.45}$$

Consider the generalized coordinates as $r_i^i = [x\ \ y\ \ z\ \ 1]^T$, we can rewrite (3.44) as [44]

$$J_i = \begin{bmatrix} \frac{-I_{ixx}+I_{iyy}+I_{izz}}{2} & I_{ixy} & I_{ixz} & m_i\overline{x}_i \\ I_{ixy} & \frac{I_{ixx}-I_{iyy}+I_{izz}}{2} & I_{iyz} & m_i\overline{y}_i \\ I_{ixz} & I_{iyz} & \frac{I_{ixx}+I_{iyy}+I_{izz}}{2} & m_i\overline{z}_i \\ m_i\overline{x}_i & m_i\overline{y}_i & m_i\overline{z}_i & m_i \end{bmatrix} \tag{3.46}$$

where $I_{ixx} = \int(y^2+z^2)\,dm$, $I_{iyy} = \int(x^2+z^2)\,dm$, $I_{izz} = \int(x^2+y^2)\,dm$, $I_{ixy} = \int xy\,dm$, $I_{ixz} = \int xz\,dm$, $I_{iyz} = \int yz\,dm$, and $m_i\overline{x}_i = \int x\,dm$, $m_i\overline{y}_i = \int y\,dm$,

$m_i\bar{z}_i = \int z dm$ with m_i as the total mass of the ith link, and $\bar{r}_i^i = [\bar{x}\ \bar{y}\ \bar{z}\ 1]^T$ as the center of mass vector of the ith link from the ith link coordinate frame and expressed in the ith link coordinate frame.

Therefore, the total kinetic energy can be written as

$$\begin{aligned} K(q,\dot{q}) &= \frac{1}{2}\sum_{i=1}^{n} \text{trace}\left[\sum_{j=1}^{n}\sum_{k=1}^{n} \frac{\partial T_i^0}{\partial q_j} J_i \frac{\partial T_i^{0T}}{\partial q_k}\dot{q}_j\dot{q}_k\right] \\ &= \frac{1}{2}\sum_{j=1}^{n}\sum_{k=1}^{n} m_{jk}\dot{q}_j\dot{q}_k = \frac{1}{2}\dot{q}^T M(q)\dot{q} \end{aligned} \tag{3.47}$$

where the jkth element m_{jk} of the $n \times n$ inertia matrix $M(q)$ is defined as

$$m_{jk}(q) = \sum_{i=1}^{n} \text{trace}\left[\frac{\partial T_i^0}{\partial q_j} J_i \frac{\partial T_i^{0T}}{\partial q_k}\right] \tag{3.48}$$

3.3.3 Potential Energy

The total potential energy of the robot is therefore expressed as

$$P(q) = -\sum_{i=1}^{n} P_i \tag{3.49}$$

where P_i is the potential energy of the ith link with mass m_i and center of gravity $\bar{r}_i^i$ expressed in the coordinates of its own frame, the potential energy of the link is given $P_i = -m_i g^T T_i^0 \bar{r}_i^i$, the gravity vector is expressed in the base coordinates as $g = [g_x\ g_y\ g_z\ 0]^T$.

3.3.4 Lagrangian Equations

Consider the kinetic energy $K(q,\dot{q})$ and the potential energy $P(q)$ can be expressed as $K(q,\dot{q}) = \frac{1}{2}\dot{q}^T M(q)\dot{q}$, $P(q) = -\sum_{i=1}^{n} m_i g T_i^0 \bar{r}_i^i$, so the Lagrangian function $L(q,\dot{q}) = K(q,\dot{q}) - P(q)$ is thus given by

$$L(q,\dot{q}) = \frac{1}{2}\dot{q}^T M(q)\dot{q} - P(q) \tag{3.50}$$

We can obtain

$$\frac{\partial L}{\partial \dot{q}_k} = \sum_{j=1}^{n} m_{kj}\dot{q}_j \tag{3.51}$$

$$\frac{d}{dt}\frac{\partial L}{\partial \dot{q}_k} = \sum_{j=1}^{n} m_{kj}\ddot{q}_j + \sum_{j=1}^{n} \frac{d}{dt} m_{kj}\dot{q}_j \tag{3.52}$$

$$= \sum_{j=1}^{n} m_{kj}\ddot{q}_j + \sum_{j=1}^{n}\sum_{i=1}^{n} \frac{\partial m_{kj}}{\partial q_i}\dot{q}_i\dot{q}_j \tag{3.53}$$

$$\frac{\partial L}{\partial q_k} = \frac{1}{2}\sum_{i=1}^{n}\sum_{j=1}^{n} \frac{\partial m_{ij}}{\partial q_k}\dot{q}_i\dot{q}_j - \frac{\partial P}{\partial q_k} \tag{3.54}$$

for $k = 1, 2, \dots, n$.

Considering the symmetry of the inertia matrix, we have

$$\sum_{i=1}^{n}\sum_{j=1}^{n} \frac{\partial m_{kj}}{\partial q_i}\dot{q}_i\dot{q}_j = \frac{1}{2}\sum_{i=1}^{n}\sum_{j=1}^{n}\left[\frac{\partial m_{kj}}{\partial q_i}\dot{q}_i\dot{q}_j + \frac{\partial m_{ki}}{\partial q_i}\dot{q}_j\dot{q}_i\right]$$

$$= \frac{1}{2}\sum_{i=1}^{n}\sum_{j=1}^{n}\left[\frac{\partial m_{kj}}{\partial q_i} + \frac{\partial m_{ki}}{\partial q_j}\right]\dot{q}_i\dot{q}_j \tag{3.55}$$

The Lagrange–Euler equations can then be written as

$$\sum_{j=1}^{n} m_{kj}\ddot{q}_j + \sum_{i=1}^{n}\sum_{j=1}^{n}\left(\frac{\partial m_{kj}}{\partial q_i} - \frac{1}{2}\frac{\partial m_{ij}}{\partial q_k}\right)\dot{q}_i\dot{q}_j + \frac{\partial P}{\partial q_k}$$

$$= \sum_{j=1}^{n} m_{kj}\ddot{q}_j + \frac{1}{2}\sum_{i=1}^{n}\sum_{j=1}^{n}\left[\frac{\partial m_{ki}}{\partial q_j} + \frac{\partial m_{kj}}{\partial q_i} - \frac{\partial m_{ij}}{\partial q_k}\right]\dot{q}_i\dot{q}_j$$

$$= \sum_{j=1}^{n} m_{kj}\ddot{q}_j + \sum_{i=1}^{n}\sum_{j=1}^{n} c_{ijk}\dot{q}_i\dot{q}_j = \tau_k \tag{3.56}$$

where c_{ijk} is the Christoffel symbols (of the first kind) defined as

$$c_{ijk}(q) \triangleq \frac{1}{2}\left[\frac{\partial m_{kj}(q)}{\partial q_i} + \frac{\partial m_{ki}(q)}{\partial q_j} - \frac{\partial m_{ij}(q)}{\partial q_k}\right] \tag{3.57}$$

Define $g_k(q) = \frac{\partial P(q)}{\partial q_k}$, then the Lagrange–Euler equations can be written as

$$\sum_{j=1}^{n} m_{kj}(q)\ddot{q}_j + \sum_{i=1}^{n}\sum_{j=1}^{n} c_{ijk}(q)\dot{q}_i\dot{q}_j + g_k(q) = \tau_k, \quad k = 1, 2, \dots, n \tag{3.58}$$

It is common to write the above equations in matrix form

$$M(q)\ddot{q} + C(q, \dot{q})\dot{q} + G(q) = \tau \tag{3.59}$$

where kjth element of $C(q, \dot{q})$ defined as

$$c_{kj} = \sum_{i=1}^{n} c_{ijk}\dot{q}_i = \sum_{i=1}^{n}\frac{1}{2}\left[\frac{\partial m_{kj}}{\partial q_i} + \frac{\partial m_{ki}}{\partial q_j} - \frac{m_{ij}}{\partial q_k}\right]\dot{q}_i \tag{3.60}$$

To facilitate the understanding of the control problems, and to help design controllers for the above systems, it is essential to have a thorough study of the mathematical properties of the system.

3.3.5 Properties of Mechanical Dynamics

There are some properties summarized for the dynamics of wheeled inverted pendulum, which is convenient for controller design.

Property 3.4 The inertia matrix $M(q)$ is symmetric, i.e., $M(q) = M^T(q)$.

Property 3.5 The inertia Matrix $M(q)$ is uniformly positive definite, and bounded below and above, i.e. $\exists 0 < \alpha \le \beta < \infty$, such that $\alpha I_n \le M(q) \le \beta I_n$, $\forall q \in \mathbb{R}^n$, where I_n is the $n \times n$ identity matrix.

Property 3.6 The inverse of inertia matrix $M^{-1}(q)$ exists, and is also positive definite and bounded.

Property 3.7 Centrifugal and Coriolis forces $C(q, \dot{q})\dot{q}$ is quadratic in $\dot{q}$.

Property 3.8 It may be written in $C(q, \dot{q})\dot{q} = C_1(q)C_2[\dot{q}\dot{q}] = C_3(q)[\dot{q}\dot{q}] + C_4(q)[\dot{q}^2]$, where $[\dot{q}\dot{q}] = [\dot{q}_1\dot{q}_2, \dot{q}_1\dot{q}_3, \dots, \dot{q}_{n-1}\dot{q}_n]^T$ and $[\dot{q}^2] = [\dot{q}_1^2, \dot{q}_2^2, \dots, \dot{q}_n^2]^T$.

Property 3.9 Given two n-dimensional vectors x and y, the matrix $C(q, \dot{q})$ defined by Eq. (3.59) implies that $C(q, x)y = C(q, y)x$.

Property 3.10 The 2-norm of $C(q, \dot{q})$ satisfies the inequality $\|C(q, \dot{q})\| \le k_c(q)\|\dot{q}\|$, where $k_c(q) = \frac{1}{2} \max_{q \in \mathbb{R}^n} \sum_{k=1}^{n} \|C_k(q)\|$. For revolute robots, k_c is a finite constant since the dependence of $C_k(q)$, $k = 1, 2, \dots, n$, on q appears only in terms of sine and cosine functions of their entries.

Property 3.11 Gravitational force $G(q)$ can be derived from the gravitational potential energy function $P(q)$, i.e. $G(q) = \partial P(q)/\partial q$, and is also bounded, i.e., $\|G(q)\| \le k_{G(q)}$, where $k_{G(q)}$ is a scalar function which may be determined for any given WIP. For revolute joints, the bound is a constant independent of q whereas for prismatic joints, the bound may depend on q.

Property 3.12 The dependence of $M(q)$, $C(q, \dot{q})$ and $G(q)$ on q will appear only in terms of sine and cosine functions in their entries, so that $M(q)$, $C(q, \dot{q})$ and $G(q)$ have bounds that are independent of q.

Property 3.13 By defining each coefficient as a separate parameter, the dynamics can be written in the linear in the parameters (LIPs) form

$$M(q)\ddot{q} + C(q,\dot{q})\dot{q} + G(q) = Y(q,\dot{q},\ddot{q})P \tag{3.61}$$

where $Y(q,\dot{q},\ddot{q})$ is an $n \times r$ matrix of known functions known as the regressor matrix, and P is an r dimensional vector of parameters.

Remark 3.14 The above equation can also be written as

$$M(q)\ddot{q}_r + C(q,\dot{q})\dot{q}_r + G(q) = \Phi(q,\dot{q},\dot{q}_r,\ddot{q}_r)P \tag{3.62}$$

where $\dot{q}_r$ and $\ddot{q}_r$ are the corresponding n-dimensional vectors.

Property 3.15 The matrix $N(q,\dot{q})$ defined by $N(q,\dot{q}) = \dot{M}(q) - 2C(q,\dot{q})$ is skew-symmetric, i.e., $n_{kj}(q,\dot{q}) = -n_{jk}(q,\dot{q})$, if $C(q,\dot{q})$ is defined using the Christoffel symbols.

Property 3.16 Since $M(q)$ and therefore $\dot{M}(q)$ are symmetric matrices, the skew-symmetry of the matrix $\dot{M}(q) - 2C(q,\dot{q})$ can also be seen from the fact that $\dot{M}(q) = C(q,\dot{q}) + C^T(q,\dot{q})$.

Property 3.17 The system is passive from τ to $\dot{q}$.

Property 3.18 Even though the skew-symmetry property of $N(q,\dot{q})$ is guaranteed if $C(q,\dot{q})$ is defined by the Christoffel symbols, it is always true that $\dot{q}^T[\dot{M}(q) - 2C(q,\dot{q})]\dot{q} = 0$.

Property 3.19 The system is feedback linearizable, i.e., there exists a nonlinear transformation such that the transformed system is a linear controllable system.

3.3.6 Dynamics of Wheeled Inverted Pendulum

Consider two-wheeled inverted pendulum shown in Fig. 3.1. The total kinetic energy can be written as

$$\begin{aligned} K = {} & \frac{1}{2}Mv^2 + M_w v^2 + \frac{1}{2}m(v + l\cos\alpha\dot{\alpha})^2 + \frac{1}{2}m(-l\sin\alpha\dot{\alpha})^2 \\ & + I_w\left(\frac{v}{r}\right)^2 + \frac{1}{2}I_M\dot{\alpha}^2 + \frac{1}{2}I_p\omega^2 + 2\left(M_w + \frac{I_w}{r^2}\right)d^2\omega^2 \end{aligned} \tag{3.63}$$

and the total potential energy is

$$U = mgl(1 - \cos\alpha) \tag{3.64}$$

which is subjected to the following constraints $\dot{x}\cos\theta - \dot{y}\sin\theta = 0$. Using the Lagrangian approach like the above section, we can choose $q = [x,\theta,\alpha]$, and $\dot{q} = [v, w, \dot{\alpha}]^T$ as generalized coordinations, then we have

$$\frac{d}{dt}\left(\frac{\partial L}{\partial \dot{x}}\right) - \frac{\partial L}{\partial x} = \frac{\tau_l}{r} + \frac{\tau_r}{r} + d_l + d_r \tag{3.65}$$

$$\frac{d}{dt}\left(\frac{\partial L}{\partial \dot{\theta}}\right) - \frac{\partial L}{\partial \theta} = 2d\left(\frac{\tau_l}{r} - \frac{\tau_r}{r} + d_l - d_r\right) \tag{3.66}$$

$$\frac{d}{dt}\left(\frac{\partial L}{\dot{\alpha}}\right) - \frac{\partial L}{\partial \alpha} = 0 \tag{3.67}$$

Let

$$\begin{aligned} L = K - U = {} & \frac{1}{2}Mv^2 + M_w v^2 + \frac{1}{2}m(v + l\cos\alpha\dot{\alpha})^2 + \frac{1}{2}m(-l\sin\alpha\dot{\alpha})^2 \\ & + I_w\left(\frac{v}{r}\right)^2 + \frac{1}{2}I_M\dot{\alpha}^2 + \frac{1}{2}\left(I_p + 2\left(M_w + \frac{I_w}{r^2}\right)d^2\right)\omega^2 \\ & - mgl(1 - \cos\alpha) \end{aligned} \tag{3.68}$$

Then, we have

$$\frac{\partial L}{\partial v} = Mv + 2M_w v + m(v + l\dot{\alpha}\cos\alpha) + 2\frac{I_w}{r^2}v \tag{3.69}$$

$$\frac{d}{dt}\left(\frac{\partial L}{\partial v}\right) = \left(M + 2M_w + m + 2\frac{I_w}{r^2}\right)\dot{v} + ml\left(-\dot{\alpha}^2\sin\alpha + \ddot{\alpha}\cos\alpha\right) \tag{3.70}$$

$$\frac{\partial L}{\partial x} = 0 \tag{3.71}$$

Then, we have

$$\frac{\partial L}{\partial \omega} = \left(I_p + 2\left(M_w + \frac{I_w}{r^2}\right)d^2\right)\omega \tag{3.72}$$

$$\frac{d}{dt}\left(\frac{\partial L}{\partial \omega}\right) = I_p\dot{\omega} + 2\left(M_w + \frac{I_w}{r^2}\right)d^2\dot{\omega} \tag{3.73}$$

$$\frac{\partial L}{\partial \theta} = 0 \tag{3.74}$$

Moreover, we have

$$\frac{\partial L}{\partial \dot{\alpha}} = mlv\cos\alpha + ml^2\dot{\alpha} + I_M\dot{\alpha} \tag{3.75}$$

$$\frac{d}{dt}\left(\frac{\partial L}{\partial \dot{\alpha}}\right) = ml\dot{v}\cos\alpha + mlv(-\sin\alpha)\dot{\alpha} + ml^2\ddot{\alpha} + I_M\ddot{\alpha} \tag{3.76}$$

$$\begin{aligned} \frac{\partial L}{\partial \alpha} = {} & m(v + l\dot{\alpha}\cos\alpha)(-l\dot{\alpha}\sin\alpha) + m(l\dot{\alpha}\sin\alpha)(l\dot{\alpha}\cos\alpha) \\ & + mgl\sin\alpha \end{aligned} \tag{3.77}$$

Then we can obtain the dynamics equation as

$$\begin{aligned} & \left(M + 2M_w + m + 2\frac{I_w}{r^2}\right)\dot{v} + ml\ddot{\alpha}\cos\alpha - ml\dot{\alpha}^2\sin\alpha \\ & = \frac{\tau_l}{r} + \frac{\tau_r}{r} + d_l + d_r \end{aligned} \tag{3.78}$$

$$\left(I_p + 2\left(M_w + \frac{I_w}{r^2}\right)d^2\right)\dot{\omega} = 2d\left(\frac{\tau_l}{r} - \frac{\tau_r}{r} + d_l - d_r\right) \tag{3.79}$$

$$ml\dot{v}\cos\alpha + \left(ml^2 + I_M\right)\ddot{\alpha} - mgl\sin\alpha = 0 \tag{3.80}$$

3.4 Newton–Euler Approach

Consider the mobile inverted pendulum shown in Fig. 3.1, we give Newton–Euler to describe the time evolution of mechanical systems under nonholonomic constraints [35]. The table of forces and moments acting on the vehicle is shown in Table 3.1.

With an assumption that there is no slip between the wheels and the ground, balancing forces and moments acting on the left wheel results in the following equations of motion of the left wheel:

$$I_w\ddot{\theta}_l = F_l - H_l r \tag{3.81}$$

$$M_w\ddot{x}_l = d_l - F_l + H_l \tag{3.82}$$

Similarly, for the right wheel we have

$$I_w\ddot{\theta}_r = F_r - H_r r \tag{3.83}$$

$$M_w\ddot{x}_r = d_r - F_r + H_r \tag{3.84}$$

Balancing forces acting on the pendulum on the x-axis direction and moments about the center point of gravity O results in

$$-ml\cos(\alpha)\ddot{\alpha} + ml\dot{\alpha}^2\sin(\alpha) - m\ddot{x} = F_p \tag{3.85}$$

$$ml^2\ddot{\alpha} + ml\cos(\alpha)\ddot{x} - mgl\sin(\alpha) = M_p \tag{3.86}$$

where ml^2 is the moment of inertia of the pendulum with respect to the y-axis.

Balancing forces acting on the platform along the x-axis direction, and moments about the z-axis results in

$$M\ddot{x} = F_l + F_r + F_p \tag{3.87}$$

$$I_M\ddot{\alpha} = -M_p \tag{3.88}$$

Balancing the moments acting on the platform and pendulum about the z-axis gives

$$I_M\ddot{\theta} = d(F_l - F_r) \tag{3.89}$$

The relationship between the displacement of the wheel along the x-axis and the rotational angle of the wheel about the y-axis is

$$\begin{cases} \theta_l = \frac{x_l}{r} \\ \theta_r = \frac{x_r}{r} \end{cases} \Rightarrow \quad \theta_l - \theta_r = \frac{x_l - x_r}{r} \tag{3.90}$$

On the other hand, the relationship between the rotational angle (heading angle) θ

Table 3.1 The parameters and variables of wheeled inverted pendulum

F_l, F_r	Interacting forces between the left and right wheels and the platform
H_l, H_r	Friction forces acting on the left and right wheels
τ_l, τ_r	Torques provided by wheel actuators acting on the left and right wheels
d_l, d_r	External forces acting on the left and right wheels
θ_l, θ_r	Rotational angles of the left and right wheels
x_l, x_r	Displacements of the left and right wheels along the x-axis
α	Tilt angle of the pendulum
θ	Heading angle of the vehicle
M_w	Mass of the wheel
I_w	Moment of inertia of the wheel with respect to the y-axis
r	Radius of the wheel
m	Mass of the pendulum
g	Gravity acceleration
l	Distance from the point O to the center of gravity, CG, of the pendulum
d	Distance between the left and right wheels along the y-axis
M	Mass of the platform
I_M	Moment of inertia of the platform about the y-axis
I_p	Moment of inertia of the platform and pendulum about the z-axis
F_p	Interacting force between the pendulum and the platform on the x-axis
M_p	Interacting moment between the pendulum and the platform about the y-axis
v	the forward velocity of the mobile platform
ω	the rotation velocity of the mobile platform, and $\omega = \dot{\theta}$

of the vehicle about the z-axis and the displacement of the wheels along the x-axis is

$$\theta = \frac{x_l - x_r}{d} \tag{3.91}$$

By subtracting Eqs. (3.81) and (3.82) from Eqs. (3.83) and (3.84), respectively, we have

$$I_w(\ddot{\theta}_l - \ddot{\theta}_r) = \tau_l - \tau_r - (H_l - H_r)r \tag{3.92}$$

$$M_w(\ddot{x}_l - \ddot{x}_r) = d_l - d_r - (F_l - F_r) + H_l - H_r \tag{3.93}$$

By substituting Eqs. (3.90) and (3.97) into Eqs. (3.92) and (3.93), we have

$$I_w \frac{d}{r} \ddot{\theta} = \tau_l - \tau_r - (H_l - H_r)r \tag{3.94}$$

$$dM_w \ddot{\theta} = d_l - d_r - (F_l - F_r) + H_l - H_r \tag{3.95}$$

Dividing both sides of Eq. (3.94) by r then adding with Eq. (3.95) result in

$$d\left(\frac{I_w}{r^2} + M_w\right)\ddot{\theta} = \frac{\tau_l - \tau_r}{r} - (F_l - F_r) + d_l - d_r \tag{3.96}$$

Multiplying both sides of Eq. (3.96) by d then adding with Eq. (3.89), then we obtain

$$\ddot{\theta} = \frac{d}{rI_\theta}(\tau_l - \tau_r) + \frac{d}{I_\theta}(d_l - d_r) \tag{3.97}$$

where $I_\theta = I_p + d^2(M_w + \frac{I_w}{r^2})$.

The relationship between displacement of the wheel along the x-axis and the rotational angle of the wheel about the y-axis (see Eq. (3.97)) gives

$$\theta_l + \theta_r = \frac{x_l + x_r}{r} \tag{3.98}$$

On the other hand, the relationship between displacement of the vehicle and wheels along the x-axis is

$$x = \frac{x_l + x_r}{2} \tag{3.99}$$

Considering (3.81)–(3.84), we have

$$I_w(\ddot{\theta}_l + \ddot{\theta}_r) = \tau_l + \tau_r - (H_l + H_r)r \tag{3.100}$$

$$M_w(\ddot{x}_l + \ddot{x}_r) = d_l + d_r - (F_l + F_r) + H_l + H_r \tag{3.101}$$

By substituting Eqs. (3.98) and (3.99) into Eqs. (3.100) and (3.101), we obtain

$$\frac{2I_w}{r}\ddot{x} = \tau_l + \tau_r - (H_l + H_r)r \tag{3.102}$$

$$2M_w\ddot{x} = d_l + d_r - (F_l + F_r) + H_l + H_r \tag{3.103}$$

Integrating the above equations, we have

$$2\left(\frac{I_w}{r^2} + M_w\right)\ddot{x} = \frac{\tau_l + \tau_r}{r} - (F_l + F_r) + d_l + d_r \tag{3.104}$$

By subtracting Eq. (3.85) from Eq. (3.87), we obtain

$$(M + m)\ddot{x} + ml\cos(\alpha)\ddot{\alpha} - ml\dot{\alpha}^2\sin(\alpha) = F_l + F_r \tag{3.105}$$

Adding Eq. (3.104) with Eq. (3.105) results in

$$ml\cos(\alpha)\ddot{\alpha} + \left(M + m + 2\left(\frac{I_w}{r^2} + M_w\right)\right)\ddot{x} = ml\dot{\alpha}^2\sin(\alpha) + \frac{\tau_l + \tau_r}{r} + d_l + d_r \tag{3.106}$$

Adding Eq. (3.86) with Eq. (3.88) gives

$$\left(ml^2 + I_M\right)\ddot{\alpha} + ml\cos(\alpha)\ddot{x} = mgl\sin(\alpha) \tag{3.107}$$

From the above derivation, we can obtain the dynamics equation of motion of the two-wheeled mobile vehicle with an inverted pendulum as Eqs. (3.97), (3.106), (3.107), which are identical with Eqs. (3.78), (3.79), and (3.80).

3.5 Conclusion

In this chapter, we first describe the kinematics and dynamics model for wheeled inverted pendulum. The Lagrange–Euler equations of motion and Newton–Euler approach have been discussed respectively. Based on the Lagrange–Euler formulation, the dynamics for a general wheeled inverted pendulum have been presented, which incorporates the dynamic interactions between the mobile platform and the inverted pendulum. The structural properties of robots, which are useful for controller design, have also been briefly summarized. Finally, the dynamic equations for wheeled inverted pendulum have been derived in a step to step manner.

Chapter 4
Linear Control

4.1 Introduction

Although WIP systems are intrinsically nonlinear and their dynamics are described by nonlinear differential equations, it is often possible to obtain a linearized model of the system. Furthermore, if the system operates around an operating point, and the signals involved are small, a linear model that approximates the nonlinear system in the region of operation can be obtained. Several techniques for the design of controllers and analysis of nonlinear systems through linearization were applied before.

In [59], motion control was proposed using linear state–space model. In [56], dynamics were derived using a Newtonian approach and the control was design by the equations linearized around an operating point. In [106], the dynamic equations were studied, with the pendulum pitch and the rotation angles of the two wheels as the variables of interest, and a linear controller was designed for stabilization under the consider of its robustness in [107]. In [19], a linear stabilizing controller was derived by a planar model without considering vehicle yaw. However, the above control laws are designed on the linearized dynamics which only exhibits desirable behavior around the operating point, and do not have global applicability. In [75], the exact dynamics of two-wheeled inverted pendulum were investigated, and linear feedback control was developed on the dynamic model. In [103], a two-level velocity controller via partial feedback linearized and a stabilizing position controller were derived, however, the controller design is not robust with respect to parameter uncertainties.

Wheeled inverted pendulums are intrinsically nonlinear and their dynamics will be described by nonlinear differential equations. The presence of nonlinearities in systems complicates analysis. Despite this, for the case of controller design, it is often possible to obtain a linearized model of the system. If the system operates around an operating point, and the signals involved are small signals, a linear model that approximates the nonlinear system in the region of operation can be obtained. In this chapter, we first derive the linearized dynamic model. Then several techniques for the design of controllers and analysis techniques for linear systems can then

Z. Li et al., *Advanced Control of Wheeled Inverted Pendulum Systems*,
DOI 10.1007/978-1-4471-2963-9_4, © Springer-Verlag London 2013

be applied. The control methods applied here are the PID control and the Linear Quadratic Regulator optimization method used for obtaining an optimal feedback gain. Since the approach of H_∞, optimal control has been widely discussed for robustness and its capability of disturbance attenuation in nonlinear systems. Therefore, by linearizing the dynamics of wheeled inverted pendulum and obtaining a precise knowledge of the structure of the entire plant model, robust H_∞ controls, including Riccati-like H_∞, and LMI-based H_∞, are applied to the linearized model with a small perturbation. The corresponding simulation examples are presented for each approach.

4.2 Linearization of the WIP Dynamics

Before the design of either controller, linear mathematical models must first be obtained. The linearization may be done by expanding the nonlinear function into a Taylor series about the operating point and neglecting the higher order terms of the expansion [100].

The nonlinear state space equations of the WIP can be written as

$$\dot{x} = f(x, u) \tag{4.1}$$

$$u = Kx \tag{4.2}$$

$$y = Cx \tag{4.3}$$

where $x \in \mathbb{R}^5$, $u \in \mathbb{R}^2$, and $y \in \mathbb{R}^5$.

Considering (4.1), the equilibrium point of the system may be found by solving

$$f(x, u) = 0 \tag{4.4}$$

The equilibrium points are found to be at $(x_0, u_0) = (0, 0)$. This is followed by doing a Taylor series expansion of (4.1) about the equilibrium point, which results in

$$\begin{aligned}
\dot{x} &= f(x, u) \\
&= f(x_0, u_0) + \left[\frac{\partial f}{\partial x_1}(x_1 - x_0) + \cdots + \frac{\partial f}{\partial x_5}(x_5 - x_{05}) + \frac{\partial f}{\partial u_1}(u_1 - u_0) \right. \\
&\quad \left. + \frac{\partial f}{\partial u_2}(u_1 - u_{02}) \right] + \frac{1}{2!}\left[\frac{\partial^2 f}{\partial x_1^2}(x_1 - x_0)^2 + \cdots + \frac{\partial^2 f}{\partial x_5^2}(x_5 - x_{05})^2 \right. \\
&\quad \left. + \frac{\partial^2 f}{\partial u_1^2}(u_1 - u_0)^2 + \frac{\partial^2 f}{\partial u_2^2}(u_1 - u_{02})^2 \right] + \cdots
\end{aligned} \tag{4.5}$$

The partial derivatives are evaluated at the operating point, and neglecting the higher order derivatives yields

$$\begin{aligned}
\dot{x} - f(x_0, u_0) = &\left[\frac{\partial f}{\partial x_1}(x_1 - x_0) + \cdots + \frac{\partial f}{\partial x_5}(x_5 - x_{05}) \right. \\
&\left. + \frac{\partial f}{\partial u_1}(u_1 - u_0) + \frac{\partial f}{\partial u_2}(u_1 - u_{02}) \right]
\end{aligned} \tag{4.6}$$

It is noted that the matrix C in Eq. (4.3) is already constant and it therefore does not need to be linearized. The resulting linearized state space representation of the system is given by

$$\dot{x} = Ax + Bu \tag{4.7}$$

$$y = Cx \tag{4.8}$$

where A and B are constant matrices defined as

$$A = \begin{bmatrix} \frac{\partial f_1}{\partial x_1} & \cdots & \frac{\partial f_1}{\partial x_5} \\ \vdots & \ddots & \vdots \\ \frac{\partial f_5}{\partial x_5} & \cdots & \frac{\partial f_5}{\partial x_5} \end{bmatrix} \tag{4.9}$$

$$B = \begin{bmatrix} \frac{\partial f_1}{\partial u_1} & \frac{\partial f_1}{\partial u_2} \\ \vdots & \vdots \\ \frac{\partial f_5}{\partial u_1} & \frac{\partial f_5}{\partial u_2} \end{bmatrix} \tag{4.10}$$

Following the linearization of the system, the controllability and observability matrices of the resulting system given by $M = (B \; AB \; A^2B \; \ldots \; A^4B)$ and $O = (C \; CA \; CA^2 \; \ldots \; CA^4)^T$ are calculated using MATLAB. Both the matrices are found to have full rank of 4. This implies that the linearized system is both controllable and observable.

Using the dynamics model presented in Chap. 3, the dynamic model of the vehicle is linearized around the point $\alpha = \dot{\alpha} = 0$, $\theta = \dot{\theta} = 0$. Ignoring the nonholonomic constraints (x_0, y_0), we define the following 5-dimensional vector of state variables

$$x = \begin{bmatrix} x_1 \\ x_2 \\ x_3 \\ x_4 \\ x_5 \end{bmatrix} = \begin{bmatrix} \alpha \\ \dot{\alpha} \\ v \\ \theta \\ \dot{\theta} \end{bmatrix} \tag{4.11}$$

Using the dynamic model, the linearized vehicle model can be expressed in the following state space form

$$\begin{bmatrix} \dot{x}_1 \\ \dot{x}_2 \\ x_3 \\ x_4 \\ x_5 \end{bmatrix} = \begin{bmatrix} 0 & 1 & 0 & 0 & 0 \\ t_1 & 0 & 0 & 0 & 0 \\ t_2 & 0 & 0 & 0 & 0 \\ 0 & 0 & 0 & 0 & 1 \\ 0 & 0 & 0 & 0 & 0 \end{bmatrix} \begin{bmatrix} x_1 \\ x_2 \\ x_3 \\ x_4 \\ x_5 \end{bmatrix} + \begin{bmatrix} 0 & 0 \\ t_3 & 0 \\ t_4 & 0 \\ 0 & 0 \\ 0 & t_5 \end{bmatrix} \begin{bmatrix} u_1 \\ u_2 \end{bmatrix} \tag{4.12}$$

where

$$t_1 = \frac{(m + 2m_w + M)mlg}{(m + 2m_w + M)I + (m + M + 2m_w)l^2}$$

$$t_2 = -\frac{m^2gl^2}{(M + 2m_w)I + (M + 2m_w)ml^2}$$

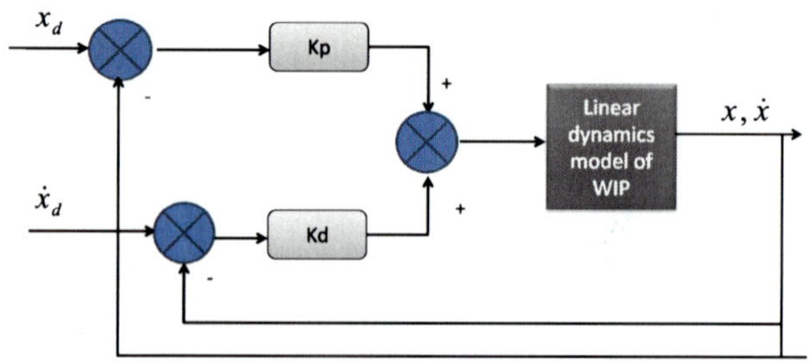

Fig. 4.1 Block diagram of PD controller with WIP

$$t_3 = -\frac{ml}{(M + 2m_w + m)I + (M + 2m_w)ml^2}$$
$$t_4 = \frac{1}{r}\frac{I + ml^2}{(M + 2m_w + m)I + (M + 2m_w)ml^2}$$
$$t_5 = \frac{2d}{r(2I_w + I_p)}$$
$$u_1 = \tau_1 + \tau_2$$
$$u_2 = \tau_1 - \tau_2$$

The coefficients in the matrices in (4.12) are constant but known. The state–space model (4.12) of the vehicle consists of two decoupled subsystems—the first of these which comprises the first three equations is a “pendulum”-type system, and the second is rotation system. We assume that the state variables are measured and available for feedback control.

4.3 PD Control Design

Consider the linearized state system (4.7), let the desired state x_d, and the actual state x, and define $e = x_d - x$, consider the control as

The control law is given by

$$u = K(x_d - x) \tag{4.13}$$

where K is designed coefficient. A PD controller is used for each displacement and displacement rate pair of the WIP system, and the block diagram of the closed loop system is shown in Fig. 4.1.

Introducing the control (4.13), the closed loop state space equation can be rewritten as

$$\dot{x} = Ax + BK(x_d - x) = (A - BK)x + BKx_d \tag{4.14}$$

and the closed loop characteristic equation is

$$\det\bigl(sI - (A - BK)\bigr) = 0 \tag{4.15}$$

Therefore, for the closed loop system with poles at the desired pole locations defined by Eq. (4.15), the elements of K must be chosen such that the desired poles location is within the negative plane. Two conditions must be satisfied for the successful placement of all the closed loop poles. The first requires the ability to measure all the state variables in the system, or the capability of estimating certain states with the use of a state observer. The other requirement is for the system to be completely state controllable. This condition would allow the arbitrary placement of the closed loop poles. For the linear wheeled inverted pendulum system, it had been shown earlier that the system is both controllable and observable. This satisfies both of the conditions stated above.

4.4 LQR based Optimal Control Design

In the design of Linear Quadratic Regulator (LQR) controllers [100], it is necessary to be able to select appropriate values for the gain matrix K in the linear control law, such that a cost function is minimized. The control law and cost function for a system with real vectors are

$$u = -Ke \tag{4.16}$$

The cost function can be described as

$$J = \int_0^\infty \left(e^T Qe + u^T Ru\right) dt \tag{4.17}$$

The cost function is a function of the matrix of states, $e \in \mathbb{R}^{n\times 1}$, and the matrix of inputs, $u \in \mathbb{R}^{m\times 1}$, in the MIMO case. $Q \in \mathbb{R}^{n\times n}$ and $R \in \mathbb{R}^{m\times m}$ are symmetric positive definite matrices. Matrices Q and R are chosen depending on the amount of cost the designer wants to attribute to each element of state or input. Allocating a larger cost for certain parameters (e and u) would imply that their performance would be placed at a higher priority over the others. For example, if the input is allocated a high relative cost factor, the controller would be optimized in a way such that the control task is attained with as little input as possible.

To solve for the matrix K, the following reduced matrix Riccati Equation is solved for the matrix $P \in \mathbb{R}^{n\times m}$

$$A^T P + PA - PBR^{-1}B^T P + Q = 0 \tag{4.18}$$

where P is a positive definite matrix that always exists when the closed loop system is stable (i.e., the matrix $(A - BK)$ is stable), and R is positive matrix. With the solution for P, which, the matrix K may be obtained from

$$K = R^{-1}B^T P \tag{4.19}$$

Then, the we can design the control from the above equation.

4.5 H_∞ Control

H_∞ is a robust control, which can realize disturbance attenuation. For the system (4.12), we can rewrite as

$$\dot{x} = Ax + B_1 u + B_2 \omega \tag{4.20}$$

$$z = Cx + D_1 \omega + D_2 \tag{4.21}$$

$$y = x \tag{4.22}$$

where ω is external disturbance vector, z is the performance signal, and $D_1 = 0$.

The objective of the control is list as follows:

(1) If $x = 0$ is the close loop operation point, and for any initial state $x(0) \in \mathbb{R}^5$, $x(t) \to 0$; (2) for any disturbance $\omega \in L^2_\infty$, the close loop can realize disturbance attenuation, that is,

$$\int_0^\infty \left[q_1 \alpha^2(t) + q_2 \dot{\alpha}^2(t) + q_3 \theta^2(t) + q_4 \dot{\theta}^2(t) + q_5 \left(\int_0^t v \, dt \right)^2 + q_6 v^2 + p u^2 \right] dt < \int_0^\infty \omega^2(t) \, dt \tag{4.23}$$

where $q_i \geq 0$, $i = 1, \ldots, 6$, and $p > 0$ is coefficient, let

$$C_1 = \begin{bmatrix} \sqrt{q_1} & 0 & 0 & 0 & 0 & \\ 0 & \sqrt{q_2} & 0 & 0 & 0 & 0 \\ 0 & 0 & \sqrt{q_3} & 0 & 0 & 0 \\ 0 & 0 & 0 & \sqrt{q_4} & 0 & 0 \\ 0 & 0 & 0 & 0 & \sqrt{q_5} & 0 \\ 0 & 0 & 0 & 0 & 0 & \sqrt{q_6} \\ 0 & 0 & 0 & 0 & 0 & 0 \end{bmatrix}, \quad D_2 = \begin{bmatrix} 0 \\ 0 \\ 0 \\ 0 \\ 0 \\ 0 \\ \sqrt{p} \end{bmatrix} \tag{4.24}$$

which is equivalent to

$$\|z\|_2 = \|\omega\|_2 \tag{4.25}$$

If we define the transfer function as $T_\omega(s)$ for ω and z as

$$\|T_\omega\|_\infty = \sup_{\omega \neq 0} \frac{\|z\|_2}{\|\omega\|_2} \tag{4.26}$$

then the disturbance attenuation performance is equivalent to $\|T_\infty\|_\infty < 1$.

4.5.1 Riccati-Based H_∞ Control

Theorem 4.1 *Consider the system* (4.12), *the control law is designed as follows*:

$$u = -\frac{1}{2}R^{-1}B^T P x \tag{4.27}$$

where the symmetric positive definite matrix P satisfies the following Riccati-like equation

$$A^T P + PA + Q + PB\left(\frac{1}{\gamma^2}I - R^{-1}\right)B^T P = 0 \tag{4.28}$$

where γ is the attenuation levels, $Q = Q^T > 0$ is a prescribed weighting matrix and R is a gains matrix. Then the proposed control law can guarantee the global stability and robustness of the closed-loop system and achieve the following H_∞ tracking performance:

$$\frac{1}{2}\int_0^\infty \|x\|^2\,dt \le \frac{2}{\lambda_{\min(Q)}}V(0) + \frac{\gamma^2}{\lambda_{\min}(Q)}\int_0^t \bar{d}^2\,dt \tag{4.29}$$

where $\lambda_{\min}(Q)$ is the minimal eigenvalue of Q.

Proof Choose a Lyapunov function as

$$V = \frac{1}{2}x^T P x \tag{4.30}$$

The time derivative of V along the error trajectory is

$$\dot{V} = \frac{1}{2}x^T\left(A^T P + PA\right)x + \omega^T B_2^T P x \tag{4.31}$$

In summary, by completing the squares and using the control input in (4.27) and the Riccati-like equation in (4.28) the derivative $\dot{V}$ can be bounded as

$$\begin{aligned}\dot{V} &\le \frac{1}{2}x^T\left(A^T P + PA + PB\left(\frac{1}{\gamma^2}I - R^{-1}\right)B^T P\right)x + \frac{1}{2}\gamma^2\omega^T\omega \\ &\le -\frac{1}{2}x^T Q x + \frac{1}{2}\gamma^2\omega^T\omega\end{aligned} \tag{4.32}$$

Integrating the above inequality (4.32) yields

$$V(t) - V(0) \le -\frac{1}{2}\int_0^t x^T Q x + \frac{\gamma^2}{2}\int_0^t \|\omega\|^2\,dt \tag{4.33}$$

Since $V(t) \ge 0$, we obtain the following L_2 criterion:

$$\frac{1}{2}\int_0^t x^T Q x\,dt \le V(0) + \frac{\gamma^2}{2}\int_0^t \|\omega\|^2\,dt \tag{4.34}$$

The knowledge that Q is a positive definite matrix and the fact that $\lambda_{\min}(Q)x^T x \le x^T Q x$ imply

$$\frac{1}{2}\int_0^t \|x\|^2\,dt \le \frac{2}{\lambda_{\min}(Q)}V(0) + \frac{\gamma^2}{\lambda_{\min}(Q)}\int_0^t \|\omega\|^2\,dt \tag{4.35}$$

Using Barbalat's lemma, one can see that the tracking error converges to zero in an infinite time. □

Remark 4.2 The inequality (4.33) reveals that the integrated squared error of x is less than or equal to the sum of $V(0)$ and the integrated squared error of x. Since the $V(0)$ is finite, if x is squared integrable then we can conclude that x will approach to zero.

Remark 4.3 From (4.34), if the systems starts with the initial condition $V(0)=0$, i.e., $x(0)=0$, the L_2 gain must satisfy

$$\sup\frac{\|x\|}{\|\omega\|} \le \frac{\gamma}{\sqrt{\lambda_{\min}(Q)}} \tag{4.36}$$

where $\|x\|^2 = \int_0^t x^T x\,dt$ and $\|\omega\|^2 = \int_0^t \|\omega\|^2\,dt$. Inequality (4.35) indicates that when smaller attenuation levels γ is specified, a better tracking performance can be achieved.

4.5.2 LMI-Based H_∞ Control

The H_∞ state-feedback control objective is a synthesis of a state-feedback that renders the underlying system L_2-gain $< \gamma$. For this, we design the state-feedback controller of the form

$$u = Kx \tag{4.37}$$

where the gain matrix K is yet to be determined. Using (4.37) in (4.12), the closed-loop system admits the following

$$\dot{x} = (A + B_1K)x + B_2\omega z = (C_1 + D_2K)x \tag{4.38}$$

A solution to the tracking problem within the H_∞ control context amounts now to the determination of a gain matrix K which is stabilizing (internally) the closed-loop system, and in addition, achieves L_2-gain $< \gamma$.

The attenuation of external disturbances is guaranteed considering the H_1 performance related to the state $x(t)$ as follows

$$\int_0^t x^T Qx\,dt \le \gamma^2\int_0^t \left(u^T u + \omega^T\omega\right)dt \tag{4.39}$$

The control law is based on the classical structure as

$$u = Kx \tag{4.40}$$

where K are gain matrices with appropriate dimension.

Theorem 4.4 *Assume there is a gain-matrix K such that the closed-loop system* (4.38) *is dissipative with respect to the supply rate $\gamma^2\|\omega\|^2 - \|z\|^2$, with the energy function given by* (4.31), *that is,*

$$\dot{V} + \frac{1}{2}\left(z^T z - \gamma^2\omega^2\omega\right) < 0, \quad \forall x, \omega \tag{4.41}$$

Then the closed-loop system (4.38) *is globally asymptotically* (*internally, i.e.,* $\omega = 0$) *stable, and it has L_2-gain $< \gamma$.*

Proof Consider the following candidate Lyapunov function:

$$V = \frac{1}{2}X^T P X \tag{4.42}$$

with $P = P^T > 0$, The stability of the closed-loop model (4.38) is satisfied under the H_∞ performance (4.39) with the attenuation level γ if

$$\dot{V} + x^T Q x - \gamma^2\omega^T\omega \le 0 \tag{4.43}$$

The condition (4.43) leads to

$$x^T(t)\left[A^T P + PA + Q\right]x + \omega^T P x + x^T P\omega - \gamma^2\omega^T\omega \le 0 \tag{4.44}$$

or equivalently

$$\begin{bmatrix} x \\ \omega \end{bmatrix}^T \begin{bmatrix} A^T P + PA + Q & P \\ P & -\gamma^2 I \end{bmatrix} \begin{bmatrix} x \\ \omega \end{bmatrix} \le 0 \tag{4.45}$$

□

4.6 Backstepping

Considering the linearized model (4.12), we can decouple the state–space model (4.12) into the rotation subsystem, which composes of the first two equations, and the pendulum-type subsystem, which consists of the left three equations. First, a control for the pendulum part of the system is designed. Consider the pendulum-subsystem of (4.12) for the tilt angle α as

$$\dot{x}_1 = x_2 \tag{4.46}$$

$$\dot{x}_2 = f(x_1) + bu_1 \tag{4.47}$$

where $f(x_1)$ and b are known coefficients from (4.12). The control objective is to regulate the tilt angle α to zero. The procedure of backstepping control is design as:

Step 1 Define the state error as

$$z_1 = x_1 - z_d \tag{4.48}$$

where z_d is reference signal. Then

$$\dot{z}_1 = \dot{x}_1 - \dot{z}_d = x_2 - \dot{z}_d \tag{4.49}$$

Define a virtual signal as

$$\alpha_1 = -c_1 z_1 + \dot{z}_d \tag{4.50}$$

where $c_1 > 0$. Define $z_2 = x_2 - \alpha_1$. Choose the Lyapunov function candidate as

$$V_1 = \frac{1}{2} z_1^2 \tag{4.51}$$

Then, we have

$$\dot{V}_1 = z_1 \dot{z}_1 = z_1 (x_2 - \dot{z}_d) = z_1 (z_2 + \alpha_1 - \dot{z}_d) \tag{4.52}$$

Substitute (4.50) into (4.52), we have

$$\dot{V}_1 = -c_1 z_1^2 + z_1 z_2 \tag{4.53}$$

If $z_2 = 0$, then $\dot{V}_1 \le 0$.

Step 2 Define Lyapunov function as

$$V_2 = V_1 + \frac{1}{2} z_2^2 \tag{4.54}$$

Since

$$\dot{z}_2 = \dot{x}_2 - \dot{\alpha}_1 = f(x_1) + bu + c_1 \dot{z}_1 - \ddot{z}_d \tag{4.55}$$

then

$$\dot{V}_2 = \dot{V}_1 + z_2 \dot{z}_2 = -c_1 \dot{z}_1^2 + z_1 z_2 + z_2 \big(f(x_1) + bu + c_1 \dot{z}_1 + \ddot{z}_d \big) \tag{4.56}$$

In order to $\dot{V}_2 \le 0$, we choose

$$u = \frac{1}{b} \big(-f(x_1) - c_2 z_2 - z_1 - c_1 \dot{z}_1 + \ddot{z}_d \big) \tag{4.57}$$

where $c_2 > 0$, then we have

$$\dot{V}_2 = -c_1 \dot{z}_1^2 - c_2 \dot{z}_2^2 \le 0 \tag{4.58}$$

The stability properties of (4.58) follows from (4.54). Since V_2 is positive definite and $\dot{V}_2$ is negative semidefinite, we have that z_1, z_2 uniformly bounded, and also that $z_1 \to 0$, $z_2 \to 0$ as $t \to \infty$. Using (4.57), we conclude that the control (4.57) is also bounded. From (4.46) and (4.47), it follows that x_1 and x_2 are also bounded and converge to zero.

Remark 4.5 Once the tilt angle α has been stabilized, in order to drive the vehicle with a desired velocity v_d, the vehicle speed can be stabilized, by completing the control (4.57) with an additional term, as follows

$$u_1 = u_1 - K_1 \operatorname{sat}(v - v_d) \tag{4.59}$$

where K_1 is positive gains.

Remark 4.6 For the rotation subsystem, in order to drive the vehicle with a desired direction angle θ_d, the vehicle rotation speed can be stabilized, the control input u_2 can be designed as

$$u_2 = -K_2(\theta - \theta_d) \tag{4.60}$$

where K_2 is positive gains.

4.7 Simulation Studies

4.7.1 PD Control

From the model described previously, there are a number of dynamic parameters in the system, which will result in variations of the state matrices. To include the influences of parameter uncertainty into simulation. The linearized WIP with the initial conditions $\alpha(0) = -10$ degree, $\dot{\alpha}(0) = 0$, $v(0) = 0$, $\theta(0) = \dot{\theta}(0) = 0.0$, and $M = 10$ kg, $m_w = 0.5$ kg, $m = 5.0$ kg, $2l = 0.5$ m, $d = 0.5$ m, $r = 0.5$ m, $I_p = 2.0$ kg m^2, it is easy to have the linear state equation as

$$\begin{aligned} A &= \begin{bmatrix} 0 & 1.0000 & 0 & 0 & 0 & 0 \\ 38.4000 & 0 & 0 & 0 & 0 & 0 \\ 0 & 0 & 0 & 1.0000 & 0 & 0 \\ -3.0000 & 0 & 0 & 0 & 0 & 0 \\ 0 & 0 & 0 & 0 & 0 & 1 \\ 0 & 0 & 0 & 0 & 0 & 0 \end{bmatrix} \\ B &= \begin{bmatrix} 0 & 0 \\ -0.2449 & 0 \\ 0 & 0 \\ 0.0816 & 0 \\ 0 & 0 \\ 0 & 0.5000 \end{bmatrix} \end{aligned} \tag{4.61}$$

In the simulation, A and BK are 6×6 matrices, there should be 6 poles, if we choose the pole point as $[10 - 10i, -10 + 10i, -10, -20, -10, -20]$, it is easy to obtain $K = [-1.0592 \times 1.0e^4, -0.1723 \times 1.0e^4, -1.8027 \times 1.0e^4, -0.4507 \times 1.0e^4, 400, 60]^T$, the initial condition of $x(0) = [-10/57.3, 0, 0.20, 0, 0.5, 0]^T$, then, we can obtain the response of the system output shown in Figs. 4.2–4.6.

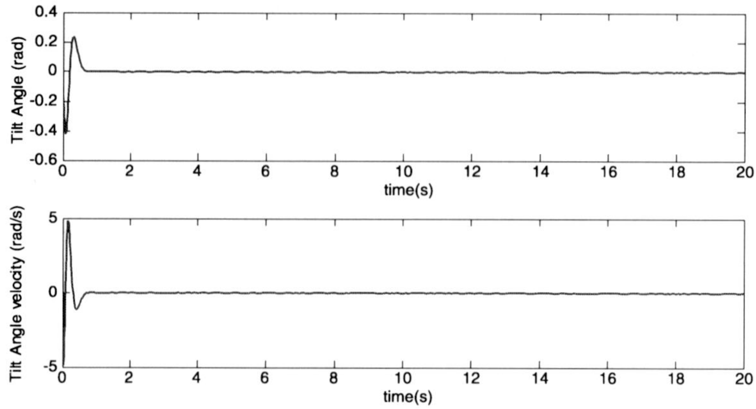

Fig. 4.2 The tile angle and its velocity

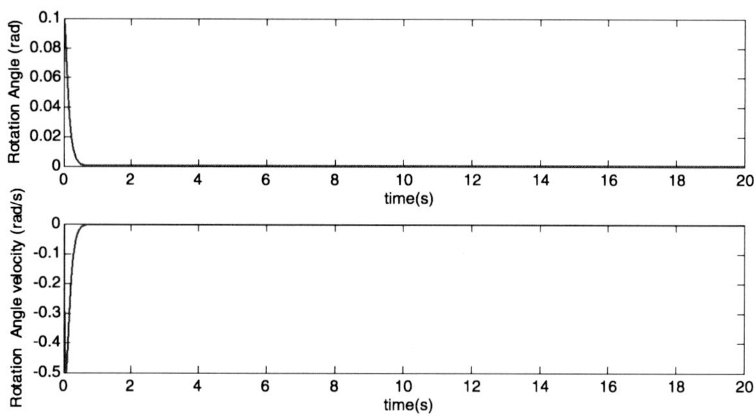

Fig. 4.3 The rotation angle and its velocity

From the simulation plots, it was observed when matrices A and B are constants over time, the system is stable and able to follow the reference inputs despite the longer settling time. However, it is obvious that the system will be stable only when the uncertainties vary over a small range of values, and the stability of the controller to perform according to the desired specifications will then depend on whether the uncertainties stay within certain bounds.

4.7.2 LQR Control

A controller based on LQR design is used for the linearized WIP, with the initial conditions $\alpha(0) = -10$ degree, $\dot{\alpha}(0) = 0$, $v(0) = 0$, $\theta(0) = \dot{\theta}(0) = 0.0$, and $M = 10$ kg,

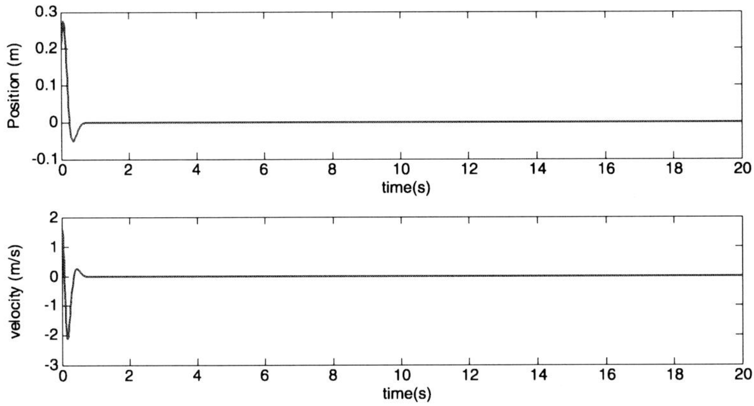

Fig. 4.4 The position and its velocity

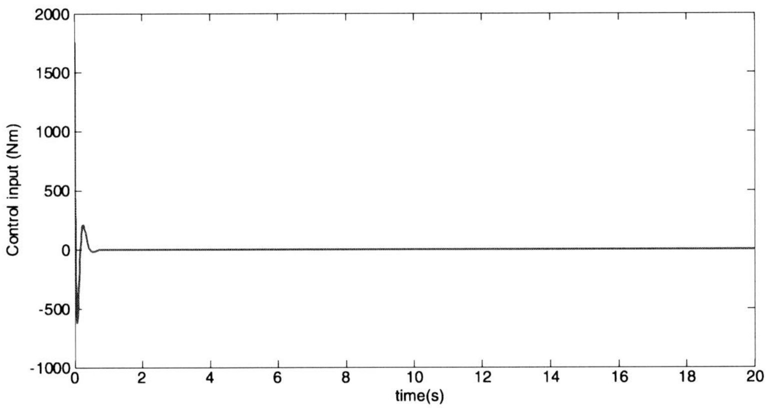

Fig. 4.5 The input u_1

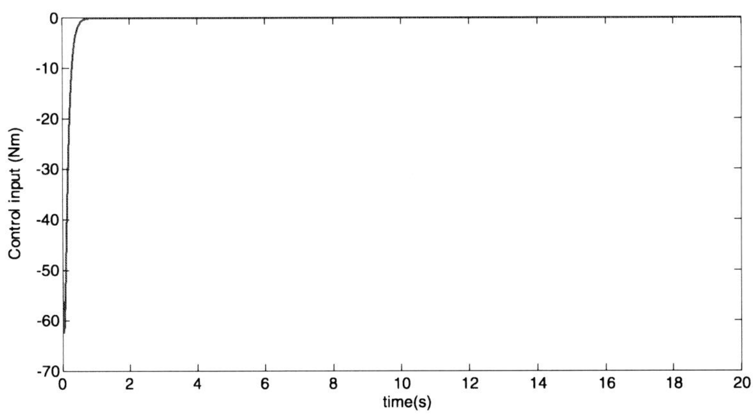

Fig. 4.6 The input u_2

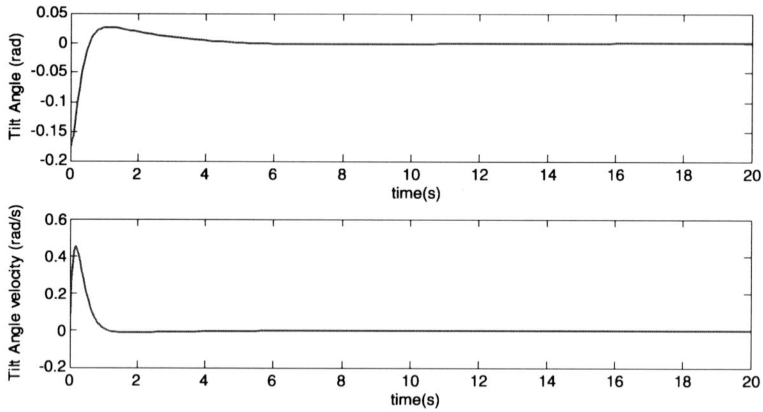

Fig. 4.7 The tile angle and its velocity

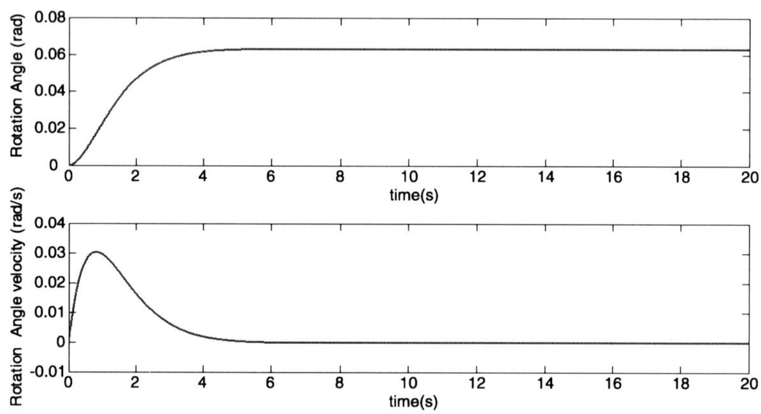

Fig. 4.8 The rotation angle and its velocity

$m_w = 0.5$ kg, $m = 5.0$ kg, $2l = 0.5$ m, $d = 0.5$ m, $r = 0.5$ m, $I_p = 2.0$ kg m^2, the friction coefficient $\mu = 0.005$. In order to examine the effects of having different performance indices on the system response, the simulation is performed for Q and R, using LQR design, we select $Q = \text{diag}[100, 10, 1, 1, 1, 1]$, $R = 0.1$, using (4.19), we have $K = [-352.4578, -58.3972, -3.1623, -11.6735, 3.1623, 4.7591]^T$. The response of the system to each set of Q and R's are shown in Figs. 4.7–4.11.

4.7.3 H_∞-Like Riccati Control

For the linearized WIP, with the initial conditions $\alpha(0) = 30$ degree, $\dot{\alpha}(0) = 0.2$ degree/s, $v(0) = 0$, $\theta(0) = \dot{\theta}(0) = 0.0$, and $M = 4$ kg, $m_w = 0.5$ kg, $m = 2.0$ kg,

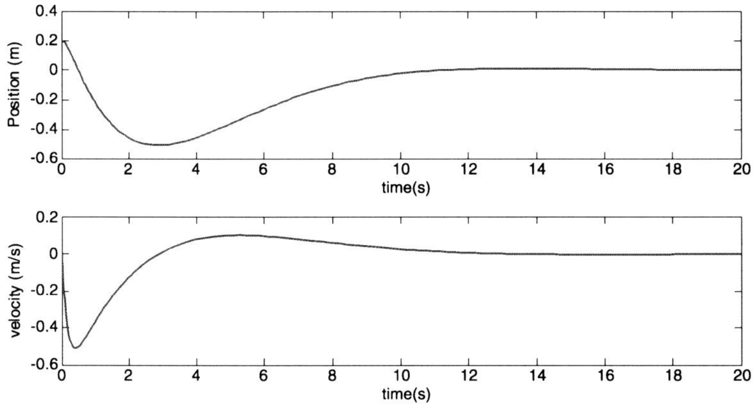

Fig. 4.9 The position and its velocity

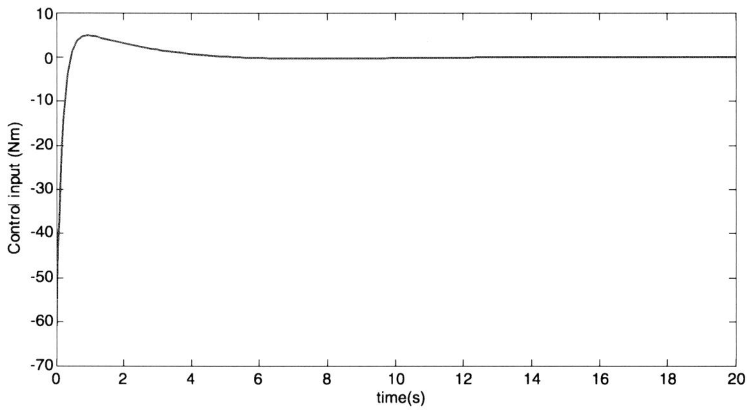

Fig. 4.10 The input u_1

$2l = 0.5$ m, $d = 0.5$ m, $r = 0.5$ m, $I_p = 2.0\ \mathrm{kg\,m^2}$. From (4.23), we choose $q_1 = q_2 = \cdots = q_6 = 1.0$, and $p = 1$. For solving Riccati equation (4.28), we choose $B = [B_1, B_2]$, and

$$B_1 = \begin{bmatrix} 0 \\ 0 \\ 0.1 \\ 0.1 \\ 0 \\ 0.1 \end{bmatrix}, \qquad B_2 = \begin{bmatrix} 0 & 0 \\ -0.2449 & 0 \\ 0 & 0 \\ 0.0816 & 0 \\ 0 & 0 \\ 0 & 0.5000 \end{bmatrix}$$

and $R = \mathrm{diag}[-1, 1, -1, 1]$, we can obtain the gain as $K = [185.1260, 2.0189, 31.0858, 11.7312, -0.7217, -1.8783]^T$, then the corresponding responses are listed in Figs. 4.12–4.16.

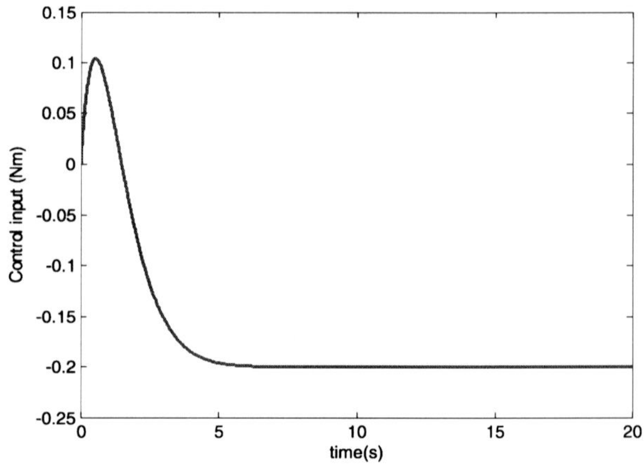

Fig. 4.11 The input u_2

4.7.4 LMI-Based H_∞ Control

For the linearized WIP, with the initial conditions $\alpha(0) = 30$ degree, $\dot{\alpha}(0) = 0.2$ degree/s, $v(0) = 0$, $\theta(0) = \dot{\theta}(0) = 0.0$, and $M = 10$ kg, $m_w = 0.5$ kg, $m = 5.0$ kg, $2l = 0.5$ m, $d = 0.5$ m, $r = 0.5$ m, $I_p = 2.0\ \text{kg}\,\text{m}^2$. From the system parameters chosen above, we can solve (4.45) using the LMI toolbox in the MATLAB and obtain $K = [1175.8, 10.8, 192.4, 55.5, -1.0258, -1.9419]$, therefore the system output are listed in Figs. 4.17–4.21.

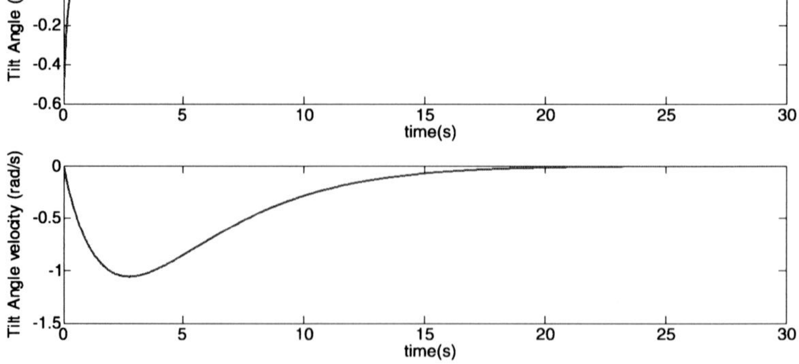

Fig. 4.12 The tile angle and its velocity

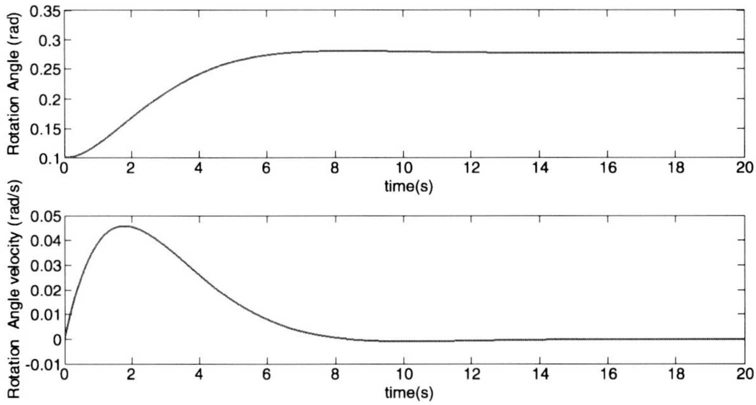

Fig. 4.13 The rotation angle and its velocity

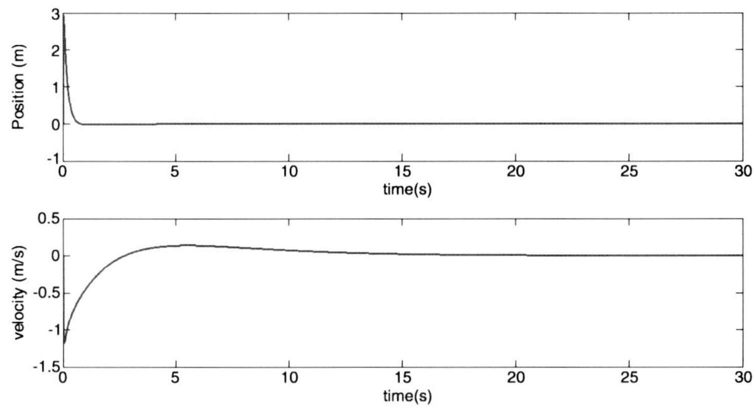

Fig. 4.14 The position and its velocity

4.7.5 Backstepping Control

Consider the linearized WIP (4.12), choose the desired trajectory of α as $z_d = 0.1\sin(\pi t)$, the control parameters $c_1 = c_2 = 35$, using the control (4.57), we can achieve the following output in Figs. 4.22–4.25.

4.8 Conclusion

In this chapter, we first present the linear model of wheel inverted pendulum, then PD, LQR, H_∞ and backstepping controllers have been implemented for WIP. There

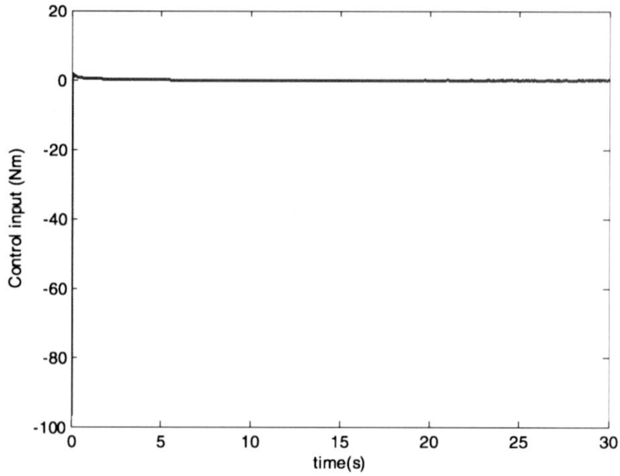

Fig. 4.15 The input u_1

are several ways to design a PD controller. In this chapter, the pole placement method was used. This technique allows the designer to specify the response of the closed loop system, but lacks the insight into robustness performance of the system. The response of the single WIP system is also examined for the case when its dynamics parameters vary with time. Comparisons are made between the performance of an LQR controller and PD controller. Finally, H_∞ and backstepping controllers are employed to verify the effectiveness of the linear control approaches.

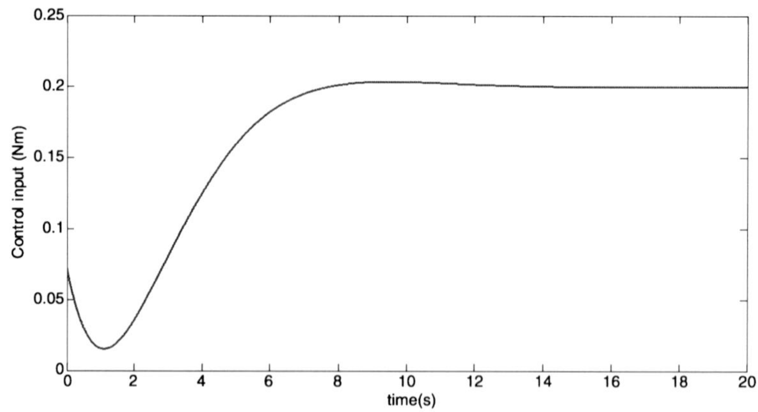

Fig. 4.16 The input u_2

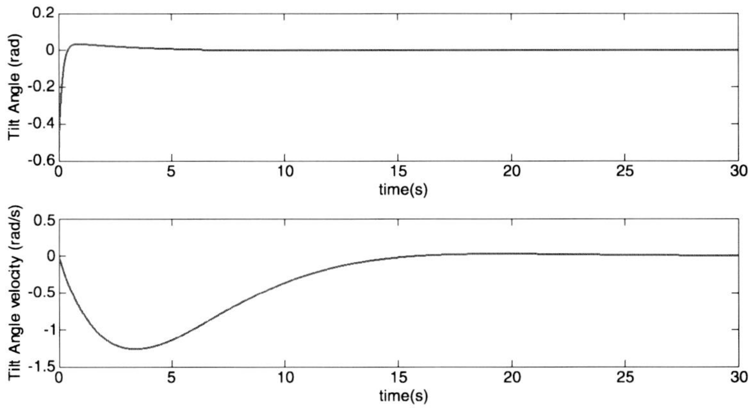

Fig. 4.17 The tile angle and its velocity

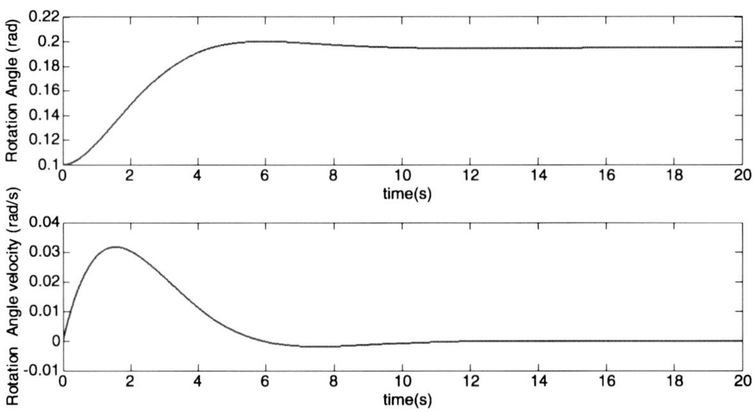

Fig. 4.18 The rotation angle and its velocity

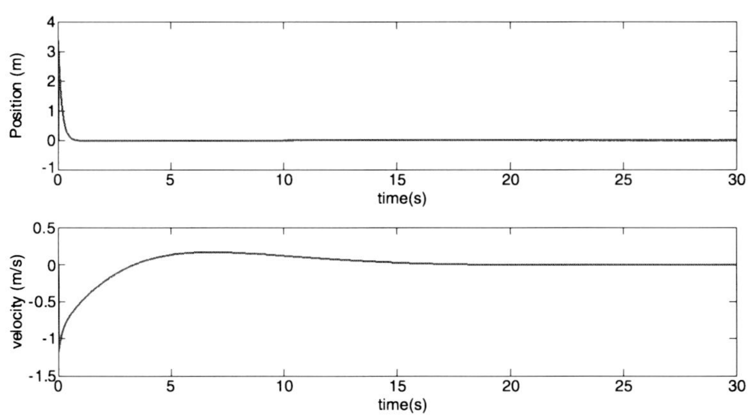

Fig. 4.19 The position and its velocity

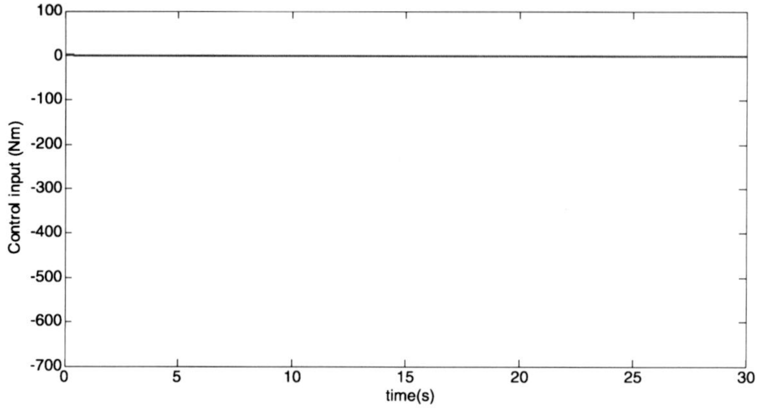

Fig. 4.20 The input u_1

Fig. 4.21 The input u_2

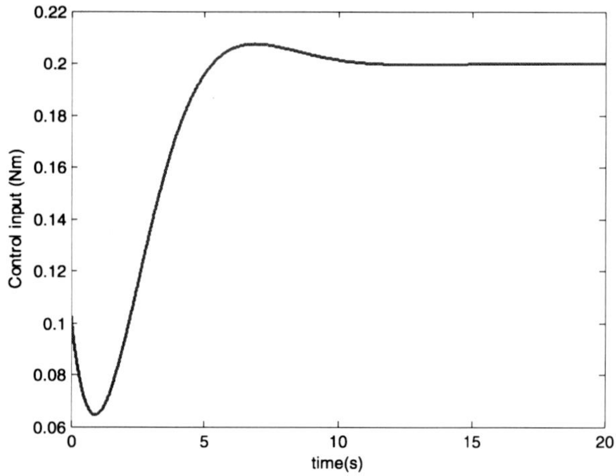

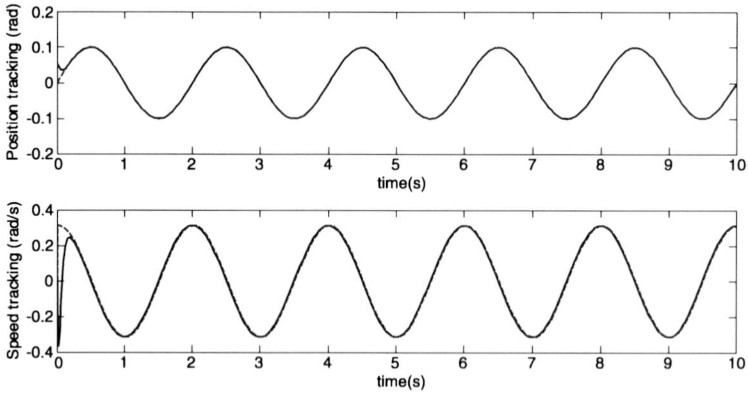

Fig. 4.22 The tile angle and its velocity

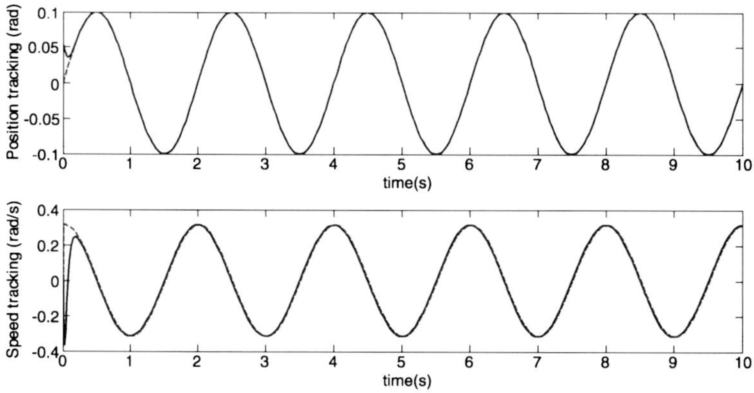

Fig. 4.23 The rotation angle and its velocity

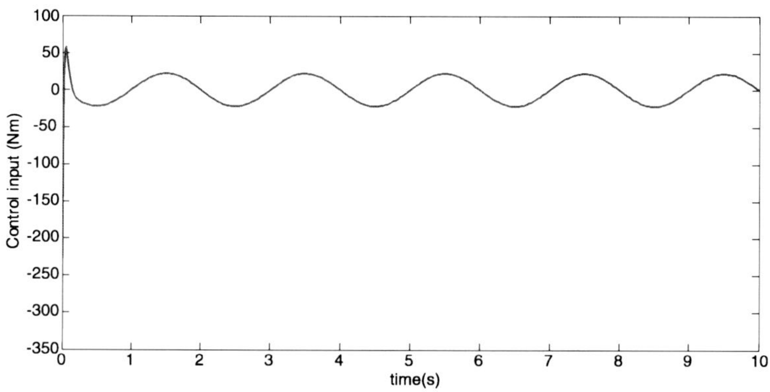

Fig. 4.24 The input u_1

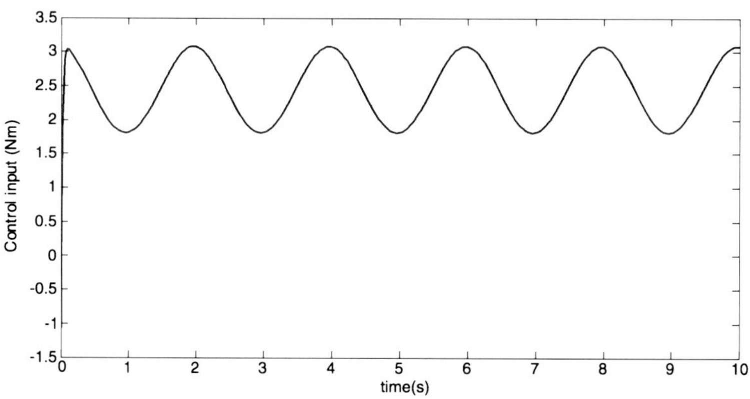

Fig. 4.25 The input u_2

Chapter 5
Nonlinear Control

5.1 Introduction

Motion of wheeled inverted pendulums is governed by under-actuated configuration i.e., the number of control inputs is less than the number of degrees of freedom to be stabilized [68], which makes it difficult to apply the conventional robotics approach for controlling Euler–Lagrange systems. Due to these reasons, increasing effort has been made towards control design that guarantee stability and robustness for mobile wheeled inverted pendulums. Compared with many other robot control schemes, the model-based computed torque control method is effective and has very good performance [134]. The underlying idea of this control scheme is to use the existing knowledge of the robot's model to linearize and decouple the robot dynamics so that the motion of each joint can be individually controlled using other well-developed linear control strategies. However, this control scheme has to overcome two difficulties. The first is that quick computation of the dynamic model is necessary. The second serious problem is that an accurate WIP-car's dynamic model is required as the computed-torque method is not robust in the presence of modeling uncertainties.

Feedback linearization is an approach to nonlinear control design which has attracted a great deal of research interest in recent years. The central idea of the approach is to algebraically transform a nonlinear system dynamics into a (fully or partly) linear one, so that linear control techniques can be applied. This differs entirely from conventional linearization (i.e., Jacobian linearization) in that feedback linearization is achieved by exact state transformations and feedback, rather than by linear approximations of the dynamics. Base on the idea of feedback linearization, model-based control is presented in this chapter. If accurate knowledge of the dynamic model is available, the model-based control can provide an effective solution to the problem. However, the performance of model-based approaches (non-adaptive and adaptive) is sensitive to the accuracy of the model representation. This is the limitation of model-based control.

Z. Li et al., *Advanced Control of Wheeled Inverted Pendulum Systems*,
DOI 10.1007/978-1-4471-2963-9_5,

5.2 Preliminaries

Assumption 5.1 The mobile wheeled inverted pendulum is subject to known nonholonomic constraints.

Remark 5.1 In actual implementation, we can adopt the methods of producing enough friction between the wheels of the mobile platform and the ground such that the assumption of nonholonomic constraints holds.

5.3 System Dynamics

Consider the following wheeled inverted pendulum dynamics described by Lagrangian formulation:

$$M(q)\ddot{q} + V(q,\dot{q})\dot{q} + F(\dot{q}) + G(q) + D = B(q)\tau + f \tag{5.1}$$

where $q = [q_1, q_2, q_3, q_4]^T = [x, y, \theta, \alpha]^T \in \mathbb{R}^4$ is the vector of generalized coordinates with x, y as the position coordinates, θ as the heading angle, and α as the tilt angle as shown in Fig. 3.1. $M(q) \in \mathbb{R}^{4\times 4}$ is the inertia matrix, $V(q,\dot{q})\dot{q} \in \mathbb{R}^4$ is the vector of Coriolis and centrifugal forces, $F(\dot{q}) \in \mathbb{R}^4$ is the vector of friction forces, $G(q) \in \mathbb{R}^4$ is the vector of gravitational forces, $D \in \mathbb{R}^4$ is the vector of the disturbances in the environment, $B(q) \in \mathbb{R}^{4\times 2}$ is the matrix of control coefficients, $\tau \in \mathbb{R}^2$ is the vector of control inputs, $f = J^T\lambda \in \mathbb{R}^3$ denotes the vector of constraint forces, $J^T \in \mathbb{R}^3$ is Jacobian matrix, and λ are Lagrangian multipliers corresponding to the nonholonomic constraints, respectively. If q is partitioned into $q_\upsilon = [x, y, \theta]^T$ and α, we obtain

$$M(q) = \begin{bmatrix} M_\upsilon & M_{\upsilon\alpha} \\ M_{\alpha\upsilon} & M_\alpha \end{bmatrix}, \qquad V(q,\dot{q}) = \begin{bmatrix} V_\upsilon & V_{\upsilon\alpha} \\ V_{\alpha\upsilon} & V_\alpha \end{bmatrix}, \qquad G(q) = \begin{bmatrix} G_\upsilon \\ G_\alpha \end{bmatrix}$$

$$D(t) = \begin{bmatrix} d_\upsilon \\ d_\alpha \end{bmatrix}, \qquad B(q) = \begin{bmatrix} B_\upsilon & 0 \\ 0 & B_\alpha \end{bmatrix}, \qquad \tau = \begin{bmatrix} \tau_\upsilon \\ 0 \end{bmatrix}$$

where M_υ and M_α describe the inertia matrices for the mobile platform and the inverted pendulum, respectively, $M_{\upsilon\alpha}$ and $M_{\alpha\upsilon}$ are the coupling inertia matrices of the mobile platform and the inverted pendulum, and V_υ and V_α denote the Centripetal and Coriolis torques for the mobile platform and the inverted pendulum, respectively. $V_{\upsilon\alpha}$ and $V_{\alpha\upsilon}$ are the coupling Centripetal and Coriolis torques of the mobile platform and the inverted pendulum. G_υ and G_α are the gravitational torque vectors for the mobile platform and the inverted pendulum, respectively. τ_υ is the control input vector for the mobile platform, and $d_\upsilon(t)$ and $d_\alpha(t)$ denote the external disturbances on the mobile platform and the inverted pendulum, respectively.

The control objective is to design a controller that ensures the tracking errors of $q_\upsilon(t)$ and $\alpha(t)$ from their respective desired trajectories $q_{\upsilon d}(t)$ and $\alpha_d(t)$ to be within a small neighborhood of zero, i.e.,

$$|q_v(t) - q_{vd}(t)| \le \epsilon_1 \tag{5.2}$$

$$|\alpha(t) - \alpha_d(t)| \le \epsilon_2 \tag{5.3}$$

where $\epsilon_i > 0$, $i = 1, 2$. Ideally, ϵ_i should be the threshold of measured noise. At the same time, all the closed loop signals are to be kept bounded.

Assumption 5.2 The desired trajectories $q_{vd}(t)$ and $\alpha_d(t)$ and their time derivatives up to the 3rd order are continuously differentiable and bounded for all $t \ge 0$.

The vehicle subjected to nonholonomic constraints can be expressed as

$$J\dot{q}_v = 0 \tag{5.4}$$

where $J = [\cos(\theta), \sin(\theta), 0]$ is the kinematic constraint matrix. The effect of the constraints can be viewed as a restriction of the dynamics on the manifold Ω_n as

$$\Omega_n = \{(q_v, \dot{q}_v) \mid J\dot{q}_v = 0\} \tag{5.5}$$

The generalized constraint forces for the nonholonomic constraints can be given by

$$f = J^T\lambda \tag{5.6}$$

Assume that the annihilator of the co-distribution spanned by the covector fields J is an 1-dimensional smooth nonsingular distribution Δ on $\mathbb{R}^2$. This distribution Δ is spanned by a set of smooth and linearly independent vector fields $H_1(q)$ and $H_2(q)$, i.e., $\Delta = \text{span}\{H_1(q), H_2(q)\}$, which in the local coordinates satisfy the following relation

$$H^T J^T = 0 \tag{5.7}$$

where $H = [H_1(q), H_2(q)] \in \mathbb{R}^{3\times 2}$. Note that $H^T H$ is of full rank. Constraint equation (5.4) implies the existence of vector $\dot{\eta} = [\omega, v]^T \in \mathbb{R}^2$ such that

$$\dot{q}_v = H(q)\dot{\eta} \tag{5.8}$$

Considering (5.57) and its derivative, letting $\zeta = [\eta, \alpha]^T$, and multiplying H^T by both sides of (5.1) to eliminate J^T, the dynamics of wheeled inverted pendulum can be expressed as

$$\mathbf{M}\ddot{\zeta} + \mathbf{V}\dot{\zeta} + \mathbf{F}(\dot{\zeta}) + \mathbf{G} + \mathbf{D} = \mathbf{B}\tau \tag{5.9}$$

$$\mathbf{M} = \begin{bmatrix} H^T M_v H & H^T M_{v\alpha} \\ M_{\alpha v} H & M_\alpha \end{bmatrix}, \qquad \mathbf{V} = \begin{bmatrix} H^T M_v \dot{H} + H^T V_v H & H^T V_{v\alpha} \\ M_{\alpha v}\dot{H} + V_{\alpha v} H & V_\alpha \end{bmatrix}$$

$$\mathbf{G} = \begin{bmatrix} H^T G_v \\ G_\alpha \end{bmatrix}, \qquad \mathbf{F} = \begin{bmatrix} H^T F_v \\ F_\alpha \end{bmatrix}, \qquad \mathbf{D} = \begin{bmatrix} H^T d_v \\ d_\alpha \end{bmatrix}, \qquad \mathbf{B}\tau = \begin{bmatrix} H^T B_v \tau_v \\ 0 \end{bmatrix}$$

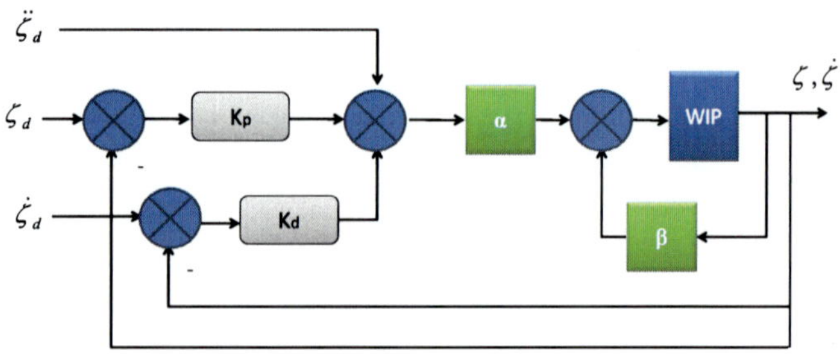

Fig. 5.1 The feedback linearization loop

5.4 Nonlinear Feedback Linearization

The feedback linearization method involves the use of feedback augmentation of the plant in order to cancel away the nonlinearities and render the augmented system into a simple linear double integrator system. Conventional linear control laws, such as a PD controller, may then be implemented as an outer loop to the feedback linearization loop. Figure 5.1 shows the block diagram of such a system with a PD controller.

Consider the reduced dynamics for mobile inverted pendulum described in (5.9), let $\Phi = \mathbf{V}\dot{\zeta} + \mathbf{F}(\dot{\zeta}) + \mathbf{G} + \mathbf{D}$. The implementation of the feedback linearization here assumes knowledge of the system matrices $\mathbf{M}$, Φ, and $\mathbf{B}$. Another assumption is that the system is observable and the states can either be measured or obtained using an estimator. The control law is designed as

$$\tau = \alpha\mu + \beta \tag{5.10}$$

with $\alpha = (\mathbf{M}^{-1}\mathbf{B})^T((\mathbf{M}^{-1}\mathbf{B})(\mathbf{M}^{-1}\mathbf{B})^T)^{-1}$, $\beta = \mathbf{B}^T(\mathbf{B}\mathbf{B}^T)^{-1}\Phi$, $\mu = K_p e + K_d\dot{e} + \ddot{\zeta}_d$ with $e = \zeta_d - \zeta$. With this control law, the augmented system becomes a double integrator system given by

$$\ddot{\zeta} = \mu \tag{5.11}$$

Therefore,

$$K_p e + K_d\dot{e} + \ddot{e} = 0 \tag{5.12}$$

where K_p and K_d are control gains with appropriate dimension. The values of K_p and K_d are chosen such that the error dynamics of the entire closed loop system have a given damping and a given response rise time.

5.5 Model Based Control Design

By exploiting the physical properties of wheeled inverted pendulum embedded in the dynamics of $\mathbf{M}(\zeta)$, $\mathbf{V}(\zeta, \dot{\zeta})$, $\mathbf{F}(\dot{\zeta})$, $\mathbf{D}$, and $\mathbf{G}(\zeta)$, one can design a controller delivering better performance than without which is explained below. According to the structure of the dynamics of the wheeled inverted pendulum, we know that:

$$\mathbf{M}(\zeta) = \begin{bmatrix} m_{11} & 0 & 0 \\ 0 & m_{22} & m_{23}(\zeta_3) \\ 0 & m_{23}(\zeta_3) & m_{33} \end{bmatrix}, \quad \mathbf{V}(\zeta,\dot{\zeta}) = \begin{bmatrix} v_{11} & 0 & v_{13} \\ 0 & 0 & v_{23}(\zeta,\dot{\zeta}) \\ v_{31} & 0 & 0 \end{bmatrix}$$

$$\mathbf{F}(\dot{\zeta}) = \begin{bmatrix} 0 \\ f_2(\dot{\zeta}) \\ f_3(\dot{\zeta}) \end{bmatrix}, \quad \mathbf{D} = \begin{bmatrix} d_1 \\ d_2 \\ d_3 \end{bmatrix}, \quad \mathbf{G}(\zeta) = \begin{bmatrix} g_1(\zeta) \\ g_2(\zeta) \\ g_3(\zeta) \end{bmatrix}, \quad \mathbf{B}\tau = \begin{bmatrix} \tau_1 \\ \tau_2 \\ 0 \end{bmatrix} \tag{5.13}$$

where m_{11}, m_{23}, m_{33} are unknown constants, $m_{22}(\zeta_3)$, $v_{11}(\zeta,\dot{\zeta})$, $v_{13}(\zeta,\dot{\zeta})$, $v_{23}(\zeta,\dot{\zeta})$, $v_{31}(\zeta,\dot{\zeta})$, $f_2(\dot{\zeta})$, $f_3(\dot{\zeta})$, $g_1(\zeta)$, $g_2(\zeta)$, $g_3(\zeta)$, and d_1, d_2, d_3 are unknown functions.

Property 5.2 Knowledge of the structure of the wheeled inverted pendulum dynamics in (5.13) can be obtained through physical understanding, such as the fact that vertical inertia is decoupled from the rotational inertia about the vertical axis, leading to the zero entries for m_{12}, m_{21}, and m_{13}, m_{31} in $M_1(\zeta)$.

Property 5.3 The inertia matrix $\mathbf{M}$ is positive definite, from which we know that the terms m_{11} and $\frac{m_{22}m_{33}-m_{23}^2(\zeta_3)}{m_{33}}$ are positive.

Property 5.4 The matrix $\dot{\mathbf{M}} - 2\mathbf{V}$ is skew-symmetric.

From (5.1) and (5.13), we can observe that $\ddot{\zeta}_2$ and $\ddot{\zeta}_3$ are coupled. Thus, control design directly based on (5.1) is difficult. Making use of the physical properties just mentioned plus a simple manipulation of (5.13), we can obtain three subsystems: ζ_1-subsystem, ζ_2-subsystem, and ζ_3-subsystem, respectively, as follows

$$m_{11}\ddot{\zeta}_1 + v_{11}\dot{\zeta}_1 + v_{13}\dot{\zeta}_3 + g_1 + d_1 = \tau_1 \tag{5.14}$$

$$\frac{m_{22}m_{33} - m_{23}^2(\zeta_3)}{m_{33}}\ddot{\zeta}_2 + v_{23}(\zeta,\dot{\zeta})\dot{\zeta}_3 + \frac{m_{23}(\zeta_3)}{m_{33}}\big(-v_{31}\dot{\zeta}_1 - f_3(\dot{\zeta}) - g_3 - d_3\big) + d_2 + f_2 + g_2 = \tau_2 \tag{5.15}$$

$$\frac{m_{22}m_{33} - m_{23}^2(\zeta_3)}{m_{22}}\ddot{\zeta}_3 + v_{31}\dot{\zeta}_1 + f_3(\dot{\zeta}) + g_3 + \frac{m_{23}(\zeta_3)}{m_{22}}\big(\tau_2 - v_{23}(\zeta,\dot{\zeta})\dot{\zeta}_3 - d_2 - f_2 - g_2\big) + d_3 = 0 \tag{5.16}$$

From the above equations, we can further obtain:

$$m_{11}\ddot{\zeta}_1 + v_{11}\dot{\zeta}_1 + v_{13}\dot{\zeta}_3 + g_1 + d_1 = \tau_1 \tag{5.17}$$

$$\frac{m_{22}m_{33} - m_{23}^2(\zeta_3)}{m_{33}}\ddot{\zeta}_2 - \frac{m_{23}(\zeta_3)}{m_{33}}(v_{31}\dot{\zeta}_1 + f_3 + g_3 + d_3)$$
$$- \big(\tau_2 - v_{23}(\zeta,\dot{\zeta})\dot{\zeta}_3 - d_2 - f_2 - g_2\big) = 0 \tag{5.18}$$

$$\frac{m_{22}m_{33} - m_{23}^2(\zeta_3)}{m_{23}(\zeta_3)}\ddot{\zeta}_3 - v_{23}(\zeta,\dot{\zeta})\dot{\zeta}_3$$
$$+ \frac{m_{22}}{m_{23}(\zeta_3)}(v_{31}\dot{\zeta}_1 + f_3 + g_3 + d_3) - (f_2 + g_2 + d_2) = -\tau_2 \tag{5.19}$$

Using the above three equations, we can analyze and design controllers for each subsystem. For clarity, define the tracking errors and the filtered tracking errors as

$$e_j = \zeta_j - \zeta_{jd} \tag{5.20}$$
$$r_j = \dot{e}_j + \Lambda_j e_j \tag{5.21}$$

where Λ_j is a positive number, $j = 1, 3$. Then, based on Lemma 2.48, we only need to study the properties of r_j to study the stability of e_j and $\dot{e}_j$.

In addition, the following computable signals are defined:

$$\dot{\zeta}_{jr} = \dot{\zeta}_{jd} - \Lambda_j e_j \tag{5.22}$$
$$\ddot{\zeta}_{jr} = \ddot{\zeta}_{jd} - \Lambda_j \dot{e}_j \tag{5.23}$$

With the system decomposed into three sub-systems, controller design can be carried out in three parts as well. In this section, we describe how the controller is designed for each sub-system using the physical model of the system.

ζ_1-Subsystem Since $\dot{\zeta}_1 = \dot{\zeta}_{1r} + r_1$ and $\ddot{\zeta}_1 = \ddot{\zeta}_{1r} + \dot{r}_1$, Eq. (5.17) becomes

$$m_{11}\dot{r}_1 = \tau_1 - (m_{11}\ddot{\zeta}_{1r} + v_{11}\dot{\zeta}_1 + v_{13}\dot{\zeta}_3 + g_1 + d_1) \tag{5.24}$$

Under the assumption that the dynamics of ζ_1-subsystem are known and no external disturbances present, that is, $d_1 = 0$, we propose the following model-based control

$$\tau_1 = -k_{p1}r_1 - k_{i1}\int_0^t r_1\,ds + (m_{11}\ddot{\zeta}_{1r} + v_{11}\dot{\zeta}_1 + v_{13}\dot{\zeta}_3 + g_1) \tag{5.25}$$

The above control law compensates the dynamic model of the system on the right-hand side of (5.24), and implements a simple linear PI controller. To analyze the closed loop stability for the ζ_1-subsystem, consider the following Lyapunov function:

$$V_1 = \frac{m_{11}}{2}r_1^2 + \frac{1}{2}\left(\int_0^t r_1\,ds\right)k_{i1}\int_0^t r_1\,ds \tag{5.26}$$

Its time derivative is given by

$$\dot{V}_1 = r_1\left(m_{11}\dot{r}_1 + k_{i1}\int_0^t r_1\,ds\right) \tag{5.27}$$

Substituting (5.24) into (5.27), we have

$$\dot{V}_1 = r_1\left[\tau_1 - (m_{11}\ddot{\zeta}_{1r} + \upsilon_{11}\dot{\zeta}_1 + \upsilon_{13}\dot{\zeta}_3 + g_1) + k_i\int_0^t r_1\,ds\right] \tag{5.28}$$

Then, we have

$$\dot{V}_1 = -k_{p1}r_1^2 \le 0$$

Integrating both sides of the above equation gives

$$V_1(t) - V_1(0) \le -\int_0^t k_{p1}r_1^2\,ds \le 0 \tag{5.29}$$

From above all, r_1 converges to the origin as $t \to \infty$. Thus, V_1 is bounded which implies that $r_1 \in L_\infty$. From (5.29), we have

$$\int_0^t k_{p1}r_1^2\,ds \le V_1(0) - V_1(t) \tag{5.30}$$

which leads to $r_1 \in L_2$. From $r_1 = \dot{e}_1 + \Lambda_1 e_1$, it can be obtained that $e_1, \dot{e}_1 \in L_\infty$. As we have established $e_1, \dot{e}_1 \in L_\infty$, from Assumption 5.2, we conclude that $\zeta_1, \dot{\zeta}_1, \dot{\zeta}_{1r}, \ddot{\zeta}_{1r} \in L_\infty$.

Therefore, all the signals on the right-hand side of (5.24) are bounded, and we can conclude that $\dot{r}_1$ and therefore $\ddot{\zeta}_1$ are bounded. Thus, $r_1 \to 0$ as $t \to \infty$. Consequently, we have $e_1 \to 0$, $\dot{e}_1 \to 0$ as $t \to \infty$.

ζ_3-Subsystem Since $\dot{\zeta}_3 = \dot{\zeta}_{3r} + r_3$, $\ddot{\zeta}_3 = \ddot{\zeta}_{3r} + \dot{r}_3$, Eq. (5.19) becomes

$$\begin{aligned}
&\frac{m_{22}m_{33} - m_{23}^2(\zeta_3)}{m_{23}(\zeta_3)}\dot{r}_3 + \tilde{\upsilon}_{23}(\zeta,\dot{\zeta})r_3 \\
&\quad = -\tau_2 - \left(\frac{m_{22}m_{33} - m_{23}^2(\zeta_3)}{m_{23}(\zeta_3)}\ddot{\zeta}_{3r} - \hat{\upsilon}_{23}r_3 - \upsilon_{23}(\zeta,\dot{\zeta})\dot{\zeta}_{3r}\right) \\
&\qquad - \frac{m_{23}(\zeta_3)}{m_{33}}(\upsilon_{31}\dot{\zeta}_1 + f_3 + d_3 + g_3) + (f_2 + g_2 + d_2)
\end{aligned} \tag{5.31}$$

where we decompose $\upsilon_{23} = \hat{\upsilon}_{23} - \tilde{\upsilon}_{23}$, such that $\frac{d}{dt}(\frac{m_{22}m_{33}-m_{23}^2(\zeta)}{m_{23}(\zeta_3)}) - 2\tilde{\upsilon}_{23} = 0$. Let

$$\begin{aligned}
\mu_3 = {} & \frac{m_{22}m_{33} - m_{23}^2(\zeta_3)}{m_{23}(\zeta_3)}\ddot{\zeta}_{3r} - \hat{\upsilon}_{23}r_3 - \upsilon_{23}(\zeta,\dot{\zeta})\dot{\zeta}_{3r} \\
& + \frac{m_{23}}{m_{33}}(-\upsilon_{31}\dot{\zeta}_1 - f_3 - g_3 - d_3) + (f_2 + g_2 + d_2)
\end{aligned}$$

Under the assumption that the dynamics of ζ_3-subsystem are known and no external disturbances present, that is, $d_2 = d_3 = 0$, consider the following model-based control laws

$$\tau_2 = -k_{p3}r_3 - k_{i1}\int_0^t r_3\,ds - \mu_3 \tag{5.32}$$

Again, the above control law compensates the dynamic model of the system on the right-hand side of (5.31), and implements a simple linear PI controller. To analyze the closed loop stability for the ζ_3-subsystem, consider the following Lyapunov function:

$$V_3 = \frac{1}{2}\frac{m_{22}m_{33} - m_{23}^2(\zeta_3)}{m_{23}(\zeta_3)} r_3^2 + \frac{1}{2}\left(\int_0^t r_3\, ds\right) k_i \int_0^t r_3\, ds \tag{5.33}$$

Its time derivative is given by

$$\dot{V}_3 = r_3\left(\frac{m_{22}m_{33} - m_{23}^2(\zeta_3)}{m_{23}(\zeta_3)}\dot{r}_3 + k_i \int_0^t r_3\, ds\right) \tag{5.34}$$

where Property 5.4 has been used. Substituting (5.31) into (5.34), we have

$$\begin{aligned} \dot{V}_3 = r_3\Bigg[&-\tau_2 - \left(\frac{m_{22}(q)m_{33} - m_{23}^2(\zeta_3)}{m_{23}(\zeta)}\ddot{\zeta}_{3r} - \hat{v}_{23}r_3 - v_{23}(\zeta, \dot{\zeta})\dot{\zeta}_{3r}\right) \\ &+ \frac{m_{23}(\zeta_3)}{m_{33}}(-v_{31}\dot{\zeta}_1 - f_3 - g_3) + (f_2 + g_2)\Bigg] \end{aligned} \tag{5.35}$$

Therefore, we have

$$\dot{V}_3 = -k_{p3}r_3^2 \le 0$$

Similar to (5.29) and (5.30), we can have

$$V_3(t) - V_3(0) \le -\int_0^t k_{p3}r_3^2\, ds \le 0 \tag{5.36}$$

Therefore, all the signals on the right hand side of (5.32) are bounded, and we conclude that $\dot{r}_3$ and therefore $\ddot{\zeta}_3$ are bounded. Thus, $r_3 \to 0$ as $t \to \infty$ can be obtained. Consequently, we have $e_3 \to 0$, $\dot{e}_3 \to 0$ as $t \to \infty$.

ζ_2-Subsystem For system (5.17)–(5.19) under the control laws (5.25) and (5.32), the ζ_2-subsystem (5.18) can be rewritten as

$$\dot{\varphi} = f(\xi, \varphi, u) \tag{5.37}$$

where $\varphi = [\zeta_2, \dot{\zeta}_2]^T$, $\xi = [\zeta_1, \zeta_3, \dot{\zeta}_1, \dot{\zeta}_3]^T$, $u = [\tau_1, \tau_2]^T$. Then the zero dynamics of the system can be addressed as [42]

$$\dot{\varphi} = f\big(0, \varphi, u^*(0, \varphi)\big) \tag{5.38}$$

where $u^* = [\tau_1^*, \tau_2^*]^T$.

Assumption 5.3 System (5.17), (5.18), (5.19) is hyperbolically minimum-phase, i.e., zero dynamics (5.38) is exponentially stable. In addition, assume that the control input u is designed as a function of the states (ξ, φ) and the reference signal satisfying Assumption 5.2, and the function $f(\xi, \varphi, u)$ is Lipschitz in ξ, i.e., there exists Lipschitz constants L_ξ and L_f for $f(\xi, \varphi, u)$ such that

$$\big\| f(\xi, \varphi, u) - f(0, \varphi, u_\varphi)\big\| \le L_\xi \|\xi\| + L_f \tag{5.39}$$

where $u_\varphi = u^*(0, \varphi)$.

By the converse theorem of Lyapunov [60], there exists a Lyapunov function $V_0(\varphi)$ which satisfies

$$\sigma_2\|\varphi\|^2 \leq V_0(\varphi) \leq \sigma_1\|\varphi\|^2 \tag{5.40}$$

$$\frac{\partial V_0}{\partial \varphi} f(0, \varphi, u_\varphi) \leq -\lambda_a\|\varphi\|^2 \tag{5.41}$$

$$\left\| \frac{\partial V_0}{\partial \varphi} \right\| \leq \lambda_b\|\varphi\| \tag{5.42}$$

where σ_1, σ_2, λ_a and λ_b are positive constants.

From the previous stability analysis about the ζ_1-subsystem (5.17) and the ζ_3-subsystem (5.19), we know that $\zeta_1, \zeta_3, \dot{\zeta}_1, \dot{\zeta}_3$ are bounded. Accordingly, ξ are bounded. We assume that

$$\|\xi\| \leq \|\xi\|_{\max} \tag{5.43}$$

where $\|\xi\|_{\max}$ is a positive constant.

Lemma 5.5 *For the internal dynamics $\dot{\varphi} = f(\xi, \varphi, u)$ of the system, if Assumptions 5.2 is satisfied, then there exist positive constants L_η and T_0 such that*

$$\left\|\varphi(t)\right\| \leq L_\varphi, \quad \forall t > T_0. \tag{5.44}$$

Proof There exists a Lyapunov function $V_0(\varphi)$. Differentiating $V_0(\varphi)$ along (5.17), (5.18), (5.19) yields

$$\begin{aligned} \dot{V}_0(\varphi) &= \frac{\partial V_0}{\partial \varphi} f(\xi, \varphi, u) \\ &= \frac{\partial V_0}{\partial \varphi} f(0, \varphi, u_\varphi) + \frac{\partial V_0}{\partial \varphi}\left[f(\xi, \varphi, u) - f(0, \varphi, u_\varphi)\right] \end{aligned} \tag{5.45}$$

Noting that (5.39)–(5.42) and (5.45) can be written as

$$\dot{V}_0(\varphi) \leq -\lambda_a\|\varphi\|^2 + \lambda_b\|\varphi\|\left(L_\xi\|\xi\| + L_f\right) \tag{5.46}$$

Noting (5.43), we have

$$\dot{V}_0(\varphi) \leq -\lambda_a\|\varphi\|^2 + \lambda_b\|\varphi\|\left(L_\xi\|\xi\|_{\max} + L_f\right) \tag{5.47}$$

Therefore, $\dot{V}_0(\eta) \leq 0$, whenever

$$\|\varphi\| \geq \frac{\lambda_b}{\lambda_a}\|\varphi\|\left(L_\xi\|\xi\|_{\max} + L_f\right) \tag{5.48}$$

By letting $L_\varphi = \frac{\lambda_b}{\lambda_a}\|\varphi\|(L_\xi\|\xi\|_{\max} + L_f)$, we conclude that there exists a positive constant T_0, such that (5.44) holds. □

Theorem 5.6 *Consider the system* (5.17–5.19) *with Assumptions* 5.2, *under the action of control laws* (5.25) *and* (5.32). *For each compact set Ω_{10}, where $(\zeta_1(0), \dot{\zeta}_1(0)) \in \Omega_{10}$, each compact set Ω_{30}, where $(\zeta_3(0), \dot{\zeta}_3(0)) \in \Omega_{30}$, each compact set Ω_{30}, where $(\zeta_3(0), \dot{\zeta}_3(0)) \in \Omega_{30}$, the tracking errors r_1 and r_3 converge to the origin. and all the signals in the closed loop system are bounded.*

Proof From the results (5.29) and (5.36), it is clear that the tracking errors r_1 and r_3 converge to the origin. From Lemma 2.48, we know that e_1, $\dot{e}_1$, e_3, and $\dot{e}_3$ are also bounded. From the boundedness of ζ_{1d} and ζ_{3d} in Assumption 5.2, we know that ζ_1 and ζ_3 are bounded. Since $\dot{\zeta}_{1d}$ and $\dot{\zeta}_{3d}$ are also bounded, it follows that $\dot{\zeta}_1$ and $\dot{\zeta}_3$ are bounded. We know that the ζ_2-subsystem (5.18) is stable, and ζ_2 and $\dot{\zeta}_2$ are bounded. This completes the proof. □

5.6 Model-Based Disturbance Rejection Control

At present, the external shock vibrations or disturbances always appear when the wheeled inverted pendulums move outside. So as to avoid the malfunction of severe shock and disturbance, in practice, accelerometers have to be frequently selected to measure acceleration signal to inject to the feedforward controller for disturbance rejection. Therefore, the disturbance rejection control is presented based on the identification of transfer function from the accelerometer.

In [25], nonlinear disturbance observer for robotic manipulators was derived and overcomes the disadvantages of existing observer designed by linear system techniques. In [26], a disturbance estimation based on the VSS equivalent control was proposed for the minimum-phase dynamical systems (with respect to the relationship between the disturbance and output) with arbitrary relative degrees. In [91], a nonlinear disturbance observer (NDO) coupled with a sliding-mode fuzzy neural network (SFNN) with a feedback-error-learning (FEL) strategy, was proposed for a class of time-varying nonlinear systems with unknown disturbances. In [133], parameter adaptation and disturbance observer (DOB) had been merged into one control design with theoretically guaranteed performance.

In these works, the disturbances are not explicitly assumed to be fast time-varying, the performances of the existing control methods are unclear if the fast time-varying disturbances can be handled. Motivated by the work in [51], where a novel disturbance observer based on integral filters was proposed for disturbance rejection in servo mechanisms, therefore, in this section, we investigate control design for wheeled inverted pendulum with arbitrarily fast time-varying disturbances. The main novelties in this section are listed as follows:

(i) control design for wheeled inverted pendulum through disturbance rejection;
(ii) exponential observer design for arbitrarily fast time-varying disturbances in wheeled inverted pendulum for mobile applications.

Assume the dynamical parameters of WIP are known ahead, while the friction forces and fast-varying disturbances in the system are unknown, we need to estimate those uncertainties to design an effective controller. In this section, we make use of model-based control combing disturbance observer.

Give the desired trajectory ζ_d, and determine a control law to ensure the tracking error of ζ from their respective desired trajectories ζ_d to be within a small neighborhood of zero, i.e., $\|\zeta - \zeta_d\| \le \epsilon$, where $\epsilon > 0$. Ideally, ϵ should be the threshold of measurable noise. For the impact phase, we should guarantee the system stability during the transition phase.

Consider (5.17) and (5.19), we can rewrite them as

$$\mathcal{M}\ddot{\xi} + \mathcal{V}(\zeta, \dot{\zeta})\dot{\xi} + \mathcal{G} + \mathcal{D} = \mathcal{B}\tau \tag{5.49}$$

where $\mathcal{M} = \begin{bmatrix} m_{11} & 0 \\ 0 & \frac{m_{33}-m_{23}^2(\zeta_3)}{m_{23}(\zeta_3)} \end{bmatrix}$, $\mathcal{V}(\zeta, \dot{\zeta}) = \begin{bmatrix} v_{11} & v_{13} \\ (m_{22}/m_{23}(\zeta_3))v_{31}(\zeta,\dot{\zeta}) & -v_{23}(\zeta,\dot{\zeta}) \end{bmatrix}$, $\xi = [\zeta_1, \zeta_3]^T$, $\mathcal{B} = \begin{bmatrix} 1 & 0 \\ -1 & 0 \end{bmatrix}$, $\mathcal{G} + \mathcal{D} = \begin{bmatrix} g_1+d_1 \\ \frac{m_{22}}{m_{23}(\zeta_3)}(f_3+g_3+d_3)-(f_2+g_2+d_2) \end{bmatrix}$.

Let $e_1 = \zeta_1 - \zeta_{1d}$, $e_3 = \zeta_3 - \zeta_{3d}$, and $e = [e_1, e_3]^T$, we can rewrite (5.49) as

$$\begin{aligned} &\mathcal{M}\ddot{e} + \mathcal{V}(\zeta, \dot{\zeta})\dot{e} = \mathcal{B}\tau - \omega + \chi \\ &\omega = \mathcal{M}\ddot{\xi}_d + \mathcal{V}(\zeta_d, \dot{\zeta}_d)\dot{\xi}_d + \mathcal{G} + \mathcal{D} \\ &\chi = \big[\mathcal{V}(\zeta_d, \dot{\zeta}_d) - \mathcal{V}(\zeta, \dot{\zeta})\big]\dot{\xi}_d \end{aligned} \tag{5.50}$$

From Sect. 5.5, we know the dynamics of mobile inverted pendulum can be decoupled into ζ_1, ζ_3 subsystem and zero-dynamics ζ_2 subsystem. According to the definition of ω, $\chi \in \mathbb{R}^2$, we denote ω_k and χ_k, $k = 1, 2$, as the kth elements of Ξ and χ, respectively, which corresponds to the kth equation in the dynamics.

We define the kth component of (5.50) as

$$m_k\ddot{e}_k + v_k(\zeta, \dot{\zeta})\dot{e}_k = u_k + \omega_k + \chi_k \tag{5.51}$$

with

$$\begin{aligned} &\omega_k = -m_k\ddot{\xi}_{kd} - v_k(\zeta_d, \dot{\zeta}_d)\dot{\xi}_{kd} - g_k - d_k \\ &v_1 = v_{11}e_1 + v_{12}e_2 \\ &v_2 = v_{21}e_1 + v_{22}e_2 \\ &\chi_k = \big[\big(v_k(\zeta_d, \dot{\zeta}_d) - v_k(\zeta, \dot{\zeta})\big)\big]\dot{\xi}_{kd} \end{aligned}$$

From Property 5.3, we know that ω_k is upper-bounded.

We give the following lemma [51] for the disturbance observer adopted in the control.

Lemma 5.7 [51] *Considering the following integral filters,*

$$\dot{\eta}(t) = -\alpha\eta(t) + (\alpha - \beta)e^{-\beta t}\eta(0) + \int_0^t e^{-\alpha(t-s)}\omega(s)\,ds \tag{5.52}$$

$$\dot{\hat{\eta}}(t) = -\alpha\hat{\eta}(t) + (\alpha - \beta)e^{-\beta t}\hat{\eta}(0) + \int_0^t e^{-\alpha(t-s)}\hat{\omega}(s)\,ds \tag{5.53}$$

where α and β are positive constants, $\eta(0)$ and $\hat{\eta}(0)$ are initial values, we have the following conclusions:

(i) *The signal $\hat{w}(t)$ can converge to it true value $w(t)$ exponentially, i.e., $|\tilde{\omega}(t)| \le \delta e^{-\sigma t}$, where $\tilde{\omega}(t) = \hat{\omega}(t) - \omega(t)$, where δ is a positive constant, and σ is a positive constant.*

(ii) *The signal $\hat{w}(t)$ can be obtained from the following integral equation*

$$\int_0^t e^{\alpha s}\hat{\omega}(s)\,ds = \alpha e^{\alpha s}\int_0^t e^{-\alpha s}\psi(s)\,ds + \psi(t) \tag{5.54}$$

where

$$\begin{aligned}\psi(t) &= (\beta-\gamma)e^{(\alpha+\gamma-\beta)t}\big(\hat{\eta}(0)-\eta(0)\big)+\beta e^{(\alpha-\beta)t}\eta(0)\\ &\quad+\big(1-e^{\gamma t}\big)\int_0^t e^{\alpha s}mv(s)\,ds\\ &\quad-\big[\alpha+(\gamma-\alpha)e^{\gamma t}\big]\int_0^t\int_0^s e^{\alpha r}mv(r)\,dr\,ds-\big(1-e^{\gamma t}\big)\theta(t)\end{aligned} \tag{5.55}$$

are computable signals, with $v_k = -k_{pk}e_k - k_{dk}\dot{e}_k$, $\theta(t) = m_{kj}e_k^{\alpha t}\dot{e}_k(t) - m_{kj}\dot{e}_k(0) - m_{kj}\alpha\int_0^t e_k^{\alpha s}\dot{e}_K(s)\,ds$, and constant $\gamma > 0$ is a design parameter.

Proof The lemma can be proved easily by following the materials in [9]. It is given here as a lemma for clarity and completeness.

(i) Consider the following differential equation

$$\dot{\hat{\eta}}(t) - \dot{\eta}(t) + \gamma\big(\hat{\eta}(t)-\eta(t)\big) + e^{-\gamma t}\dot{\eta}(t) = 0 \tag{5.56}$$

Its solution is

$$\hat{\eta}(t) = \big(1-e^{-\gamma t}\big)\eta(t) + e^{-\gamma t}\eta(0)$$

It means that $\hat{\eta}(t)$ converges to $\eta(t)$ exponentially. From (5.52) and (5.53), we can obtain that

$$\eta(t) = e^{-\alpha t}\int_0^t e^{\alpha s}\left[(\alpha-\beta)e^{-\beta s}\eta(0) + \int_0^s e^{-\alpha(t-r)}\omega(r)\,dr\right]ds \tag{5.57}$$

$$\dot{\eta}(t) = (\alpha-\beta)e^{-\beta t}\eta(0) + \int_0^t e^{-\alpha(t-s)}\omega(s)\,ds \tag{5.58}$$

$$\hat{\eta}(t) = e^{-\alpha t}\int_0^t e^{\alpha s}\left[(\alpha-\beta)e^{-\beta s}\hat{\eta}(0) + \int_0^s e^{-\alpha(t-r)}\hat{\omega}(r)\,dr\right]ds \tag{5.59}$$

$$\dot{\hat{\eta}}(t) = (\alpha-\beta)e^{-\beta t}\hat{\eta}(0) + \int_0^t e^{-\alpha(t-s)}\hat{\omega}(s)\,ds \tag{5.60}$$

Substituting (5.57)–(5.60) into (5.56) results in

$$\int_0^t e^{\alpha s}\tilde{\omega}(s)\,ds + (\gamma-\alpha)\int_0^t\int_0^s e^{\alpha r}\tilde{\omega}(s)\,dr\,ds = \phi(t) \tag{5.61}$$

where

$$\begin{aligned}\phi(t) &= (\beta-\gamma)e^{(\alpha-\beta)t}\big[\hat{\eta}(0)-\eta(0)\big] + \beta e^{(\alpha-\beta-\gamma)t}\eta(0) - e^{-\gamma t}\int_0^t e^{\alpha s}\omega(s)\,ds\\ &\quad+\alpha e^{-\gamma t}\int_0^t\int_0^s e^{\alpha r}\omega(r)\,dr\,ds\end{aligned} \tag{5.62}$$

Define the following variable

$$\chi(t) = \int_0^t \int_0^s e^{\alpha r} \tilde{\omega}(r)\, dr\, ds \tag{5.63}$$

Differentiating both sides of (5.63) and combining (5.61), we obtain

$$\dot{\chi}(t) = -(\gamma - \alpha)\chi(t) + \phi(t)$$

Its solution is

$$\chi(t) = e^{-(\gamma-\alpha)t} \int_0^t e^{(\gamma-\alpha)s} \phi(s)\, ds \tag{5.64}$$

Differentiating both sides of (5.64) twice, we have

$$\ddot{\chi}(t) = (\gamma - \alpha)^2 e^{-(\gamma-\alpha)t} \int_0^t e^{(\gamma-\alpha)s} \phi(s)\, ds - (\gamma - \alpha)\phi(t) + \dot{\phi}(t) \tag{5.65}$$

where

$$\begin{aligned}\dot{\phi}(t) &= (\alpha - \beta)(\beta - \gamma)e^{(\alpha-\beta)t}\big[\hat{\eta}(0) - \eta(0)\big] \\ &\quad + \beta(\alpha - \beta - \gamma)e^{(\alpha-\beta-\gamma)t}\eta(0) - e^{(\alpha-\gamma)t}\omega(t) \\ &\quad + (\alpha + \gamma)e^{-\gamma t} \int_0^t e^{\alpha s}\omega(s)\, ds - \gamma\alpha e^{-\gamma t} \int_0^t \int_0^s e^{\alpha r}\omega(r)\, dr\, ds\end{aligned} \tag{5.66}$$

Differentiating both sides of (5.63) twice, we have

$$\ddot{\chi}(t) = e^{\alpha t}\tilde{\omega}(t) \tag{5.67}$$

Combining (5.65) and (5.67), we have

$$\tilde{\omega}(t) = (\gamma - \alpha)^2 e^{-\gamma t} \int_0^t e^{(\gamma-\alpha)s}\phi(s)\, ds - e^{-\alpha t}(\gamma - \alpha)\phi(t) + e^{-\alpha t}\dot{\phi}(t) \tag{5.68}$$

Since $\omega(t)$ is upper-bounded, and from (5.62), (5.66) and (5.68), we can see that there exists a nonnegative function $\delta(t)$ such that $|\tilde{\omega}(t)| \le \delta(t)e^{-\sigma t}$ with $\sigma = \min(\beta, \gamma)$.

(ii) From (5.77), we have

$$\omega(t) = \mathcal{M}\ddot{e} + \mathcal{V}(\zeta, \dot{\zeta})\dot{e} + \nu + \hat{\omega}(t) \tag{5.69}$$

with $\nu = [\nu_1, \nu_2]^T$. Noting that

$$\int_0^t e^{\alpha s}\mathcal{M}\ddot{e}(s)\, ds = \theta(t)$$

Substituting (5.69) into (5.61), we have

$$\int_0^t e^{\alpha s}\hat{\omega}(s)\, ds - \alpha \int_0^t \int_0^s e^{\alpha s}\hat{\omega}(r)\, dr\, ds = \psi(t) \tag{5.70}$$

To simplify (5.70), we define the following signal

$$\zeta = \int_0^t \int_0^s e^{\alpha s} \hat{\omega}(r)\, dr\, ds \tag{5.71}$$

Its first derivative is

$$\dot{\zeta} = \int_0^t e^{\alpha s} \hat{\omega}(s)\, ds \tag{5.72}$$

Substituting (5.71) and (5.72) into (5.70), we have

$$\dot{\zeta} - \alpha \zeta = \psi \tag{5.73}$$

Its solution is

$$\zeta = e^{\alpha s} \int_0^t e^{-\alpha s} \psi\, ds \tag{5.74}$$

Substituting (5.74) to (5.73), we have

$$\int_0^t e^{\alpha s} \hat{\omega}(s)\, ds = \alpha e^{\alpha s} \int_0^t e^{-\alpha s} \psi(s)\, ds + \psi(t) \tag{5.75}$$

As such, we can numerically solve $\hat{w}$ from (5.75) using existing integral equation solving methods in [12, 13]. The proof is completed. □

Considering the disturbance observer in Lemma 5.7, we propose the following control for the mobile inverted pendulum as

$$u_k = -k_{pk} e_k - k_{dk} \dot{e}_k - \hat{\omega}_k - \chi_k \tag{5.76}$$

where $k_{pk} > 0$, $k_{dk} > 0$. Substituting Eq. (5.76) into Eq. (5.51) leads to the following closed-loop dynamics

$$m_k \ddot{e}_k + \upsilon_k(\zeta, \dot{\zeta}) \dot{e}_k = -k_{pk} e_k - k_{dk} \dot{e}_k - \hat{\omega}_k + \omega_k \tag{5.77}$$

Let $\nu_k = -k_{pk} e_k - k_{dk} \dot{e}_k$.

Theorem 5.8 *Consider the mechanical system described by* (5.9) *and its decentralized dynamics model* (5.51), *using the control law* (5.76), *the following hold for any* $(\zeta(0), \dot{\zeta}(0))$: e_k *and* $\dot{e}_k$ *converge to* 0 *as* $t \to \infty$; *and* τ *are bounded for all* $t \geq 0$.

Proof To facilitate the control design, consider the following Lyapunov function as

$$V = \frac{1}{2} e^T K_p e + \frac{1}{2} \dot{e}^T \mathbf{M} \dot{e} \tag{5.78}$$

The derivative of V along (5.51) is given by

$$
\begin{aligned}
\dot{V}_1 &= e^T K_p \dot{e} + \frac{1}{2}\left[\dot{e}^T \dot{\mathbf{M}} \dot{e} + \ddot{e}^T \mathbf{M} \dot{e} + \dot{e}^T \mathbf{M} \ddot{e}\right] \\
&= e^T K_p \dot{e} + \frac{1}{2} \dot{e} \dot{M} \dot{e} + \dot{e} M \ddot{e} \\
&= e^T K_p \dot{e} + \sum_{k=1}^{2} \dot{e}_k \left[\frac{1}{2} \dot{m}_k \dot{e}_k - \upsilon_k \dot{e}_k + u_k - \omega_k + \chi_k\right]
\end{aligned} \tag{5.79}
$$

Consider Property 5.3, and integrating (5.76) into (5.79), we have

$$
\begin{aligned}
\dot{V} &\leq \sum_{k=1}^{2} k_{pk} e_k \dot{e}_k + \sum_{k=1}^{2} \dot{e}_k [u_k + \omega_k + \chi_k] \\
&= \sum_{k=1}^{2} k_{pk} e_k \dot{e}_k + \sum_{k=1}^{2} \dot{e}_k [-k_{pk} e_k - k_{dk} \dot{e}_k - \hat{\omega}_k + \omega_k] \\
&= -\sum_{k=1}^{2} k_{dk} \dot{e}_k^2 - \sum_{k=1}^{2} \dot{e}_k \hat{\omega}_k + \sum_{k=1}^{2} \dot{e}_k \omega_k \\
&= -\sum_{k=1}^{2} k_{dk} \dot{e}_k^2 - \sum_{k=1}^{2} \dot{e}_k \tilde{\omega}_k
\end{aligned} \tag{5.80}
$$

Therefore, we can obtain

$$
\dot{V} \leq -\sum_{k=1}^{2} k_{dk} \dot{e}_k^2 - \sum_{k=1}^{2} \dot{e}_k \tilde{\omega}_k \leq -\sum_{k=1}^{2} \left(k_{dk} - \frac{1}{2}\right) \dot{e}_k^2 + \frac{1}{2} \sum_{k=1}^{2} \tilde{\omega}_k^2 \tag{5.81}
$$

Since $\frac{1}{2} \sum_{k=1}^{2} \tilde{\omega}_k^2$ is bounded and converges to zero as $t \to \infty$ by Lemma 5.7, when $\|\dot{e}_k\| \geq \sqrt{\frac{\tilde{\Xi}_k^2}{k_{dk} - \frac{1}{2}}}$, we have $\dot{V} < 0$. Therefore, $\|\dot{e}_k\| \leq \sqrt{\frac{\tilde{\Xi}_k^2}{k_{dk} - \frac{1}{2}}}$. Noting $\|\tilde{\Xi}_k^2\| \to 0$, as $t \to \infty$, $\|\dot{e}_k\| \to 0$ can be obtained. Thus, we have $\ddot{e}_k \to 0$ as $t \to \infty$. It follows $e_k \to 0$ as $t \to \infty$.

(ii) Integrating both sides of the above equation gives

$$
\begin{aligned}
V(t) - V(0) &\leq -\int_0^t \sum_{k=1}^{2} \left(k_{dk} - \frac{1}{2}\right) e_k^2 \, ds + \frac{1}{2} \sum_{k=1}^{2} \int_0^t \tilde{\Xi}_k^2 \, dt \\
&\leq -\int_0^t \sum_{k=1}^{2} \left(k_{dk} - \frac{1}{2}\right) e_k^2 \, ds + \frac{1}{2} \sum_{k=1}^{2} \int_0^t \delta^2 e^{-2\sigma t} \, dt \\
&= -\int_0^t \sum_{k=1}^{2} \left(k_{dk} - \frac{1}{2}\right) e_k^2 \, ds + \frac{\delta^2}{4\sigma} \sum_{k=1}^{2} \left(1 - e^{-2\sigma t}\right)
\end{aligned} \tag{5.82}
$$

by noting $\lim_{t \to \infty} 1 - e^{-2\sigma t} = 1$. Thus, V is bounded, it can be obtained that $e_k, \dot{e}_k \in L_\infty$. As we have established $e, \dot{e} \in L_\infty$, we conclude that $\zeta, \dot{\zeta}, \ddot{\zeta} \in L_\infty$.

Therefore, all the signals on the right-hand side of (5.51) are bounded, it is easy to conclude that τ is bounded from (5.76). □

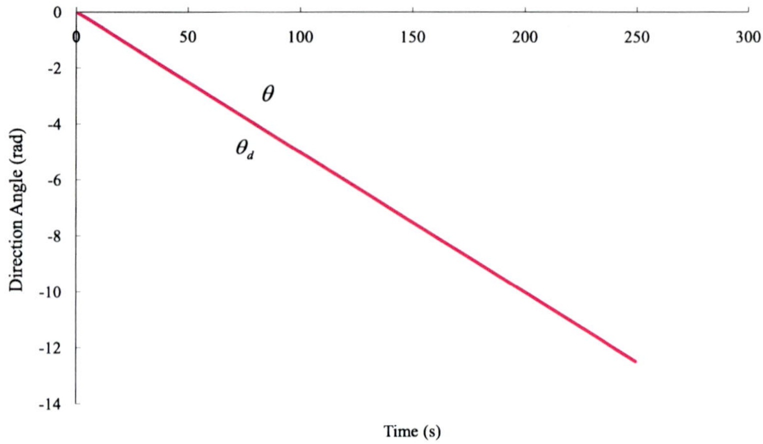

Fig. 5.2 Tracking the direction angle by model-based control

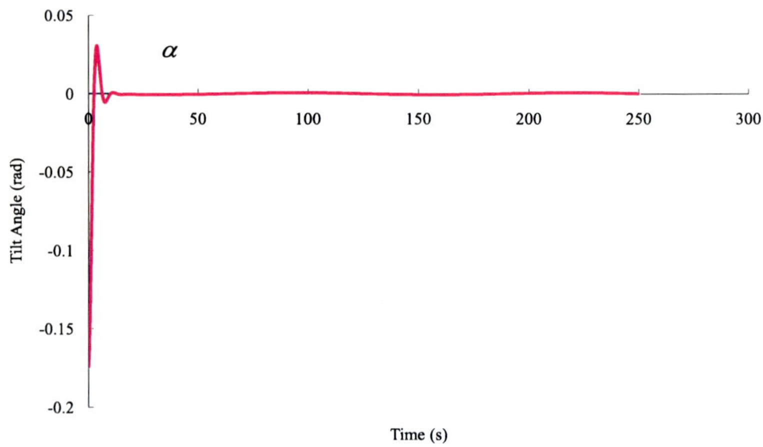

Fig. 5.3 Tracking the desired tilt angle by model-based control

5.7 Simulation Studies

To verify the effectiveness of the proposed control algorithm, let us consider a wheeled inverted pendulum as shown in Fig. 3.1. The details of dynamics can be founded in Chap. 3. In the simulation, we assume the parameters $m = 2.5$ kg, $I_M = 2.5$ kg m^2, $I_p = 1.0$ kg m^2, $I_w = 0.2$ kg m^2, $M_w = 0.2$ kg, $l = 1$ m, $d = 1$ m, $r = 0.5$ m, $\mathcal{B} = \text{diag}(1.0)$, $\zeta(0) = [-0.2, 0, \pi/180]^T$, $\dot{\zeta}(0) = [0.0, 0.1, 0.0]^T$. The desired trajectories are chosen as $\theta_d = 0.2t$ rad, $\alpha_d = 0$ rad, initial velocity is 0.1 m/s. The external disturbances are set as $1.0\sin(t)$ and $1.0\cos(t)$.

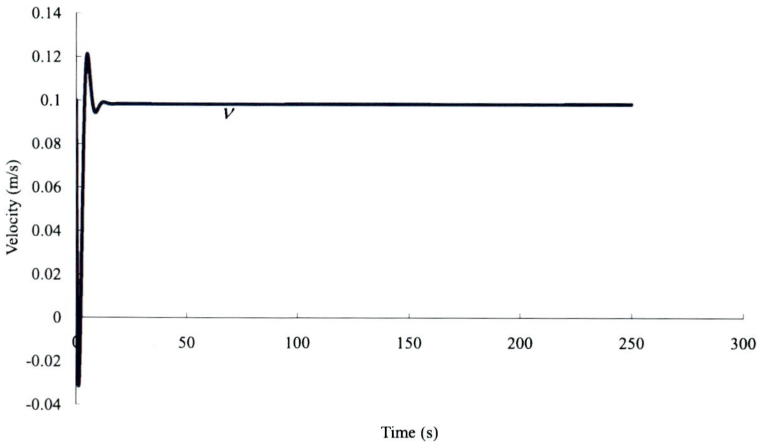

Fig. 5.4 The stable velocity by model-based control

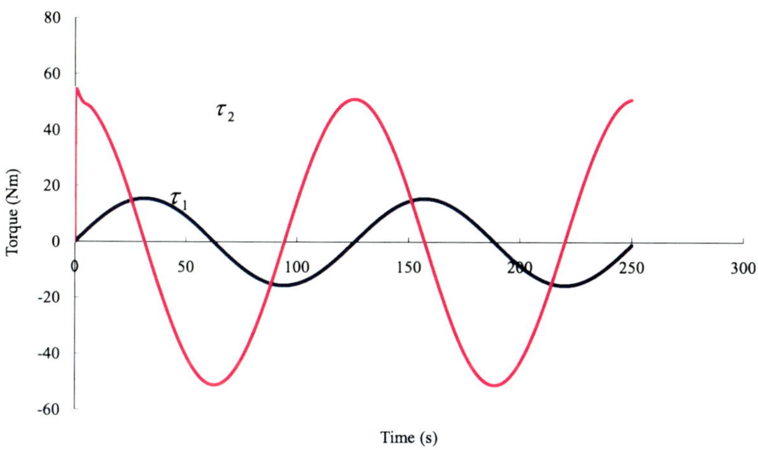

Fig. 5.5 Input torques by model-based control

Without considering the unmodel dynamics, we consider model-based controller as $\tau_1 = -k_{1P} r_1 + m_{11}\ddot{\zeta}_{1r} + v_{11}\dot{\zeta}_{1r} + v_{13}\dot{\zeta}_3 + f_1 + g_1 + d_1$ and $\tau_2 = -k_{3P} r_3 - \mu_3$. The position tracking using the model based control are shown in Figs. 5.2–5.5. The input torques are shown in Fig. 5.5. The final velocity is shown in Fig. 5.4. The tilt angle is shown in Fig. 5.3. From these figures, we can see the model-based control is with good performance.

Considering the unmodeled dynamics, we introduce the effects of 10 % parametric uncertainties in the dynamic model. In the simulation, we found that since the ζ_3 is very sensitive to the model, more than 10 % model uncertainty would cause the system unstable under the model-based controller. The presence of parametric

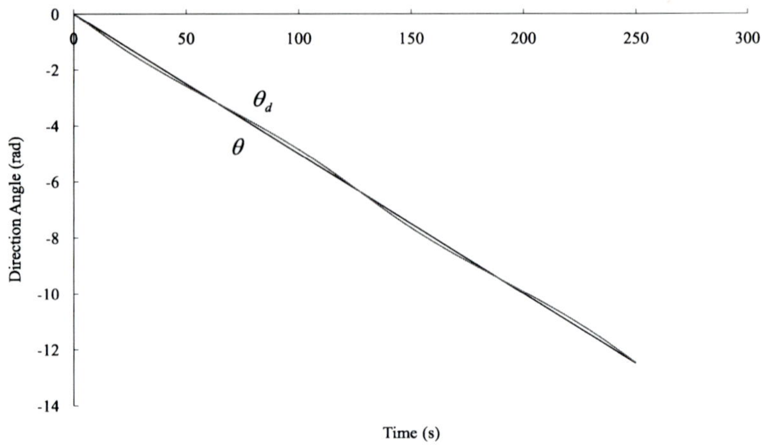

Fig. 5.6 Tracking the direction angle by model-based control with unmodeled dynamics

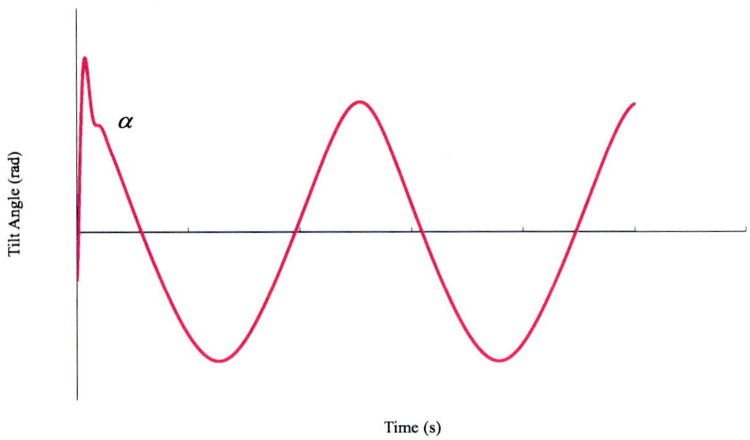

Fig. 5.7 Tracking the desired tilt angle by model-based control with unmodeled dynamics

errors is a common problem for model-based controllers since the identification of dynamic parameters is error-prone. For instance, the controlled conditions in the test facility under which the parameters are identified are often very different from actual conditions, thus rendering the parameters inaccurate for real operating conditions. Neural network control presented in this paper are not susceptible to this problem, since the unknown parameters are learn during the wheeled inverted pendulum operation in actual conditions. The position tracking using the model based control are shown in Figs. 5.6–5.9. The tilt angle is shown in Fig. 5.7. The final velocity is shown in Fig. 5.8, which fluctuates greatly. From these figures, the model-based control under the unmodeled dynamics is not good performance.

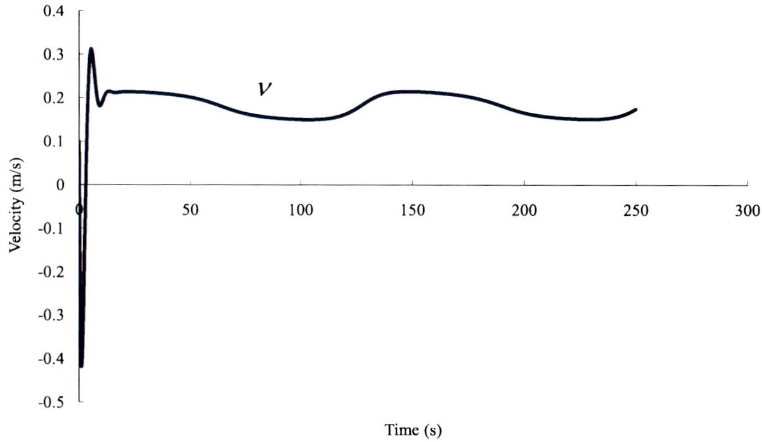

Fig. 5.8 The stable velocity by model-based control with unmodeled dynamics

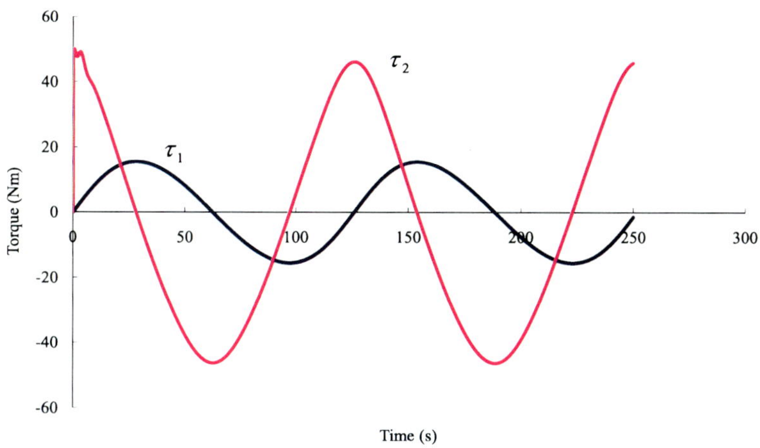

Fig. 5.9 Input torques by model-based control with unmodeled dynamics

For the disturbance observer, we choose $K_p = 3.5$, $K_d = 0.6$. The parameters of the observers are chosen as $\mu_1 = \mu_2 = 6$, $\upsilon_1 = \upsilon_2 = 4$, $\gamma_1 = \gamma_2 = 8$. Initial velocity and angular velocity are taken as 0.1 m/s and 0, respectively; the target posture is taken as velocity and angular velocity are taken as $\alpha(t) = 0$ m/s and $\theta(t) = 1$ rad/s, respectively. The peaks, frequencies of two disturbances are optional. For the purpose of simulation, the parameters of two disturbances are taken as: $d_1 = 0.1 \sin(40\pi t)$ and $d_2 = 0.15 \cos(20\pi t)$. The simulation results for velocity tracking and angular velocity tracking are shown in Figs. 5.10, 5.11, 5.12, 5.13.

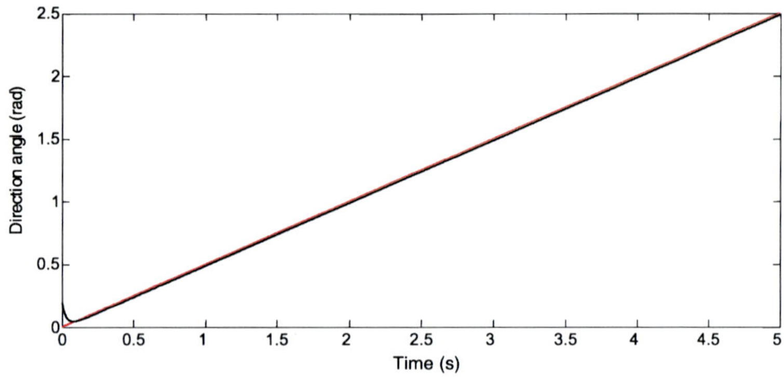

Fig. 5.10 Tracking the direction angle by model-based disturbance rejection control

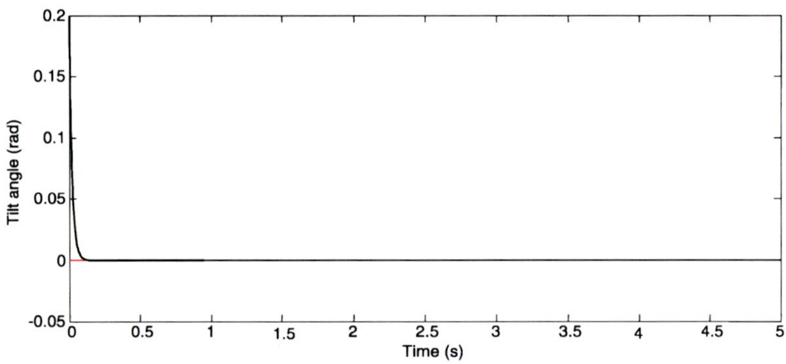

Fig. 5.11 Tracking the desired tilt angle by model-based disturbance rejection control

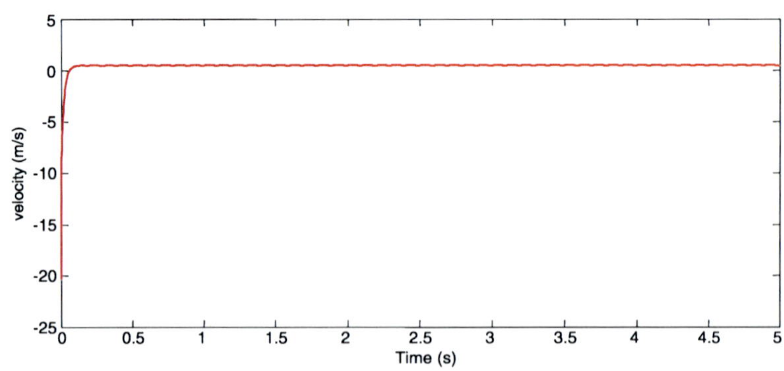

Fig. 5.12 The stable velocity by model-based disturbance rejection control

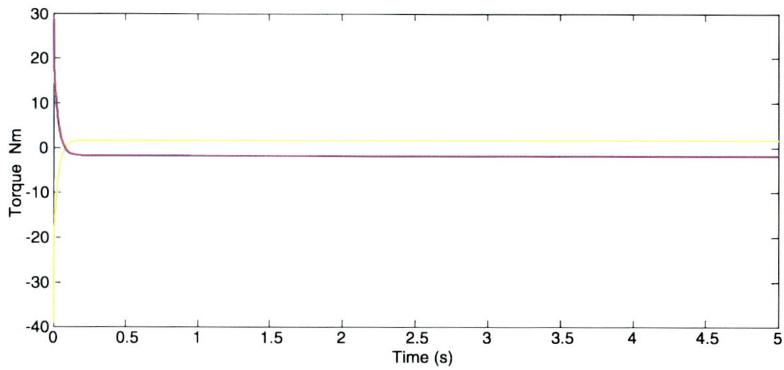

Fig. 5.13 Input torques by model-based disturbance rejection control

5.8 Conclusion

In this chapter, three nonlinear control methods of wheeled inverted pendulum systems are introduced: feedback linearization, model-based control and model-based disturbance rejection control. The former two methods perform suitably well in the ideal case that assumes exact knowledge of system. In practical cases, the system may be subject to the unknown external disturbance, and would lead to errors. An improvement might be to include a disturbance observer as feedforward compensation. The disturbance observer mechanisms that would eliminate any existing external unknown disturbance. The extensive simulation studies have been conducted to verify the effectiveness of the presented approaches.

Chapter 6
Adaptive Control

6.1 Introduction

Developing a highly accurate model could be much more complex than the design of a control mechanism. In almost every case, we have to simplify the model, under some assumptions, such that design of control can be carried out. Consequently, modeling errors always exist and in addition, WIP systems are characterized by unstable balance and unmodeled dynamics, and are subject to time varying external disturbances, which are difficult to model accurately. As a result, traditional model-based control may not be an ideal approach since it works effectively only when the dynamic model is exactly known and accurate.

In this chapter, we will investigate adaptive robust control approach in the presence of system uncertainties and disturbances which could disrupt the function of the traditional model-based feedback control and lead to unbalance of the pendulum. The study will focus on how to reduce the workload on modeling dramatically as well as to compensate for modeling errors. The proposed robust adaptive controls based on physical properties of wheeled inverted pendulums make use of online adaptation mechanism to cancel the unmodeled dynamics. Based on Lyapunov synthesis, the proposed controls ensure that the system outputs track the given bounded reference signals within a small neighborhood of zero, and guarantee semi-global uniform boundedness of all closed loop signals. The effectiveness of the proposed controls is verified through extensive simulations.

6.2 Motion Control

As mentioned, WIP systems have attracted a great deal of attention in recent years [21, 40, 56, 97, 103], while similar systems such as cart and pendulum were studied in the literature just a few years ago [46]. Differences from these systems are that the motion of the inverted pendulum in the present system is not planar and the motors driving the wheels are directly mounted on the body of the pendulum [103].

Z. Li et al., *Advanced Control of Wheeled Inverted Pendulum Systems*,
DOI 10.1007/978-1-4471-2963-9_6, © Springer-Verlag London 2013

The motion of WIP systems is governed by under-actuated configuration, i.e., the number of the control inputs is less than the number of the degrees of freedom to be stabilized [68], which makes it difficult to apply conventional approaches of robotics based on Euler–Lagrange description of the systems. Due to these reasons, increasing efforts have been made towards the design of controls that guarantee stability and robustness for WIP systems.

6.2.1 Adaptive Robust Control Design

In the previous chapter, we have developed a simple model-based controller by canceling the physical model of the system. We also have proved that the developed control laws warranted the stability of the system. However, in reality, physical model of the system cannot be exactly known, i.e., there exist model uncertainties. In addition, external disturbances may also affect the performance of the system. In this section, we take both factors into consideration to develop an adaptive robust control scheme to deal with uncertainties as well as external disturbances.

ζ_1-Subsystem

Assumption 6.1 There exist some finite positive constants $c_{1i} > 0$ $(1 \le i \le 3)$ such that $\forall \zeta_1 \in R$, $\forall \dot{\zeta}_1 \in R$, $|m_{11}| \le c_{11}$, $|v_{11}| \le c_{12} + c_{13}\|\dot{\zeta}\|$, $v_{13} \le c_{14} + c_{15}\|\dot{\zeta}\|$, $|g_1| \le c_{16}$, and $\sup_{t\ge 0}|d_1| \le c_{17}$.

Assumption 6.2 Time varying positive function ς converges to zero as $t \to \infty$ and satisfies

$$\lim_{t\to\infty}\int_0^t \varsigma(s)\,ds = \rho < \infty$$

with finite constant ρ.

Since $\dot{\zeta}_1 = \dot{\zeta}_{1r} + r_1$ and $\ddot{\zeta}_1 = \ddot{\zeta}_{1r} + \dot{r}_1$, Eq. (5.17) becomes

$$m_{11}\dot{r}_1 = \tau_1 - (m_{11}\ddot{\zeta}_{1r} + v_{11}\dot{\zeta}_1 + v_{13}\dot{\zeta}_3 + g_1 + d_1) \tag{6.1}$$

Consider the following control adaptive robust laws:

$$\tau_1 = -k_{p1}r_1 - k_{i1}\int_0^t r_1\,ds - \sum_{i=1}^{7}\frac{r_1\hat{c}_{1i}\Psi_{1i}^2}{\Psi_{1i}|r_1| + \delta_{1i}} \tag{6.2}$$

$$\Phi_1 = \hat{C}_1^T\Psi_1 \tag{6.3}$$

$$\dot{\hat{c}}_{1i} = -\sigma_{1i}\hat{c}_{1i} + \sum_{i=1}^{7}\frac{\gamma_{1i}\Psi_{1i}^2|r_1|^2}{|r_1|\Psi_{1i} + \delta_{1i}} \tag{6.4}$$

Let $\hat{C}_1 = [\hat{c}_{11}, \ldots, \hat{c}_{17}]^T$; $\Psi_1 = [\|\ddot{\zeta}_{1r}\|, 1, \|\zeta\|\|\zeta_1\|, 1, \|\zeta\|\|\zeta_3\|, 1, 1]^T$, where k_{p1} and k_{i1}, are positive constants; $\gamma_{1i} > 0$, $\delta_{1i} > 0$ and $\sigma_{1i} > 0$ satisfy Assumption 5.2:

$$\int_0^\infty \delta_{1i}(s)\,ds = \rho_{1i\delta} < \infty, \qquad \int_0^\infty \sigma_{1i}(s)\,ds = \rho_{1i\sigma} < \infty$$

with the finite constants $\rho_{1i\delta}$ and $\rho_{1i\sigma}$. Note in (6.2), the third term of the right-hand side represents a function adaptive to the state ζ_1 and state errors r_1.

To analyze the closed loop stability for the ζ_1-subsystem, consider the following Lyapunov function:

$$\mathbb{V}_1 = \frac{m_{11}}{2} r_1^2 + \frac{1}{2}\left(\int_0^t r_1\,ds\right) k_{i1} \int_0^t r_1\,ds + \frac{1}{2}\tilde{C}_1^T \Gamma_1^{-1} \tilde{C}_1 \tag{6.5}$$

where $\Gamma_1 = \mathrm{diag}[\gamma_{11}, \ldots, \gamma_{17}]$, $\tilde{C}_1 = C_1 - \hat{C}_1$.

Its time derivative is given by

$$\dot{\mathbb{V}}_1 = r_1\left(m_{11}\dot{r}_1 + k_{i1}\int_0^t r_1\,ds\right) + \tilde{C}_1^T \Gamma_1^{-1} \dot{\tilde{C}}_1 \tag{6.6}$$

Substituting (6.1) into (6.6), we have

$$\begin{aligned} \dot{\mathbb{V}}_1 = {} & r_1\left[\tau_1 - (m_{11}\ddot{\zeta}_{1r} + \upsilon_{11}\dot{\zeta}_1 + \upsilon_{13}\dot{\zeta}_3 + g_1 + d_1) + k_{i1}\int_0^t r_1\,ds\right] \\ & + \tilde{C}_1^T \Gamma_1^{-1} \dot{\tilde{C}}_1 \end{aligned} \tag{6.7}$$

Note that in the above equation, there are parameters of the physical model of the system involved, i.e., $m_{11}\ddot{\zeta}_{1r} + \upsilon_{11}\dot{\zeta}_1 + \upsilon_{13}\dot{\zeta}_3 + g_1 + d_1$ which are assumed unknown because of uncertainties. For simplicity, we let an unknown parameter μ_1 replace them and obtain

$$\begin{aligned} \dot{\mathbb{V}}_1 = {} & -k_{p1}r_1^2 - r_1\mu_1 - r_1\sum_{i=1}^{7}\frac{r_1\hat{c}_{1i}\Psi_{1i}^2}{|r_1|\Psi_{1i} + \delta_{1i}} \\ & + \hat{C}_1^T \Sigma_1 \Gamma_1^{-1}\tilde{C}_1 - \sum_{i=1}^{7}\frac{|r_1|^2\tilde{c}_{1i}\Psi_{1i}^2}{|r_1|\Psi_{1i} + \delta_{1i}} \\ \le {} & -k_{p1}r_1^2 + |r_1|\Phi_1 - \sum_{i=1}^{7}\frac{|r_1|^2 c_{1i}\Psi_{1i}^2}{|r_1|\Psi_{1i} + \delta_{1i}} + \hat{C}_1^T \Sigma_1 \Gamma_1^{-1}\tilde{C}_1 \\ \le {} & -k_{p1}r_1^2 + C_1^T\Delta_1 + \hat{C}_1^T \Sigma_1 \Gamma_1^{-1}\tilde{C}_1 \\ = {} & -k_{p1}r_1^2 + C_1^T\Delta_1 + \hat{C}_1^T \Sigma_1 \Gamma_1^{-1}(C_1 - \hat{C}_1) \\ = {} & -k_{p1}r_1^2 + C_1^T\Delta_1 - \frac{1}{4}C_1^T \Sigma_1 \Gamma_1^{-1} C_1 \\ & + \frac{1}{4}C_1^T \Sigma_1 \Gamma_1^{-1} C_1 + \hat{C}_1^T \Sigma_1 \Gamma_1^{-1} C_1 - \hat{C}_1^T \Sigma_1 \Gamma_1^{-1} \hat{C}_1 \\ = {} & -k_{p1}r_1^2 + C_1^T\Delta_1 - \left(\frac{1}{2}C_1^T - \hat{C}_1^T\right)\Sigma_1 \Gamma_1^{-1}\left(\frac{1}{2}C_1 - \hat{C}_1\right) \end{aligned}$$

$$+\frac{1}{4}C_1^T\Sigma_1\Gamma_1^{-1}C_1$$
$$\leq -k_{p1}r_1^2 + C_1^T\Delta_1 + \frac{1}{4}C_1^T\Sigma_1\Gamma_1^{-1}C_1$$

with $\Sigma_1 = \text{diag}[\sigma_{11},\ldots,\sigma_{17}]$, $\Delta_1 = [\delta_{11},\ldots,\delta_{17}]^T$.

Therefore, $\dot{\mathbb{V}}_1 \leq -k_{p1}r_1^2 + C_1^T\Delta_1 + \frac{1}{4}C_1^T\Sigma_1\Gamma^{-1}C_1$. Since $C_1^T\Delta_1 + \frac{1}{4}C_1^T\Sigma_1\Gamma^{-1}\times C_1$ is bounded, there exists $t > t_1$, $C_1^T\Delta_1 + \frac{1}{4}C_1^T\Sigma_1\Gamma^{-1}C_1 \leq \rho_1$ with the finite constant ρ_1, when $|r_1| \geq \sqrt{\frac{\rho_1}{k_{p1}}}$, then $\dot{\mathbb{V}}_1 \leq 0$. For $|r_1| \geq \sqrt{\frac{\rho_1}{k_{p1}}}$, and r_1 will converge to a compact set denoted by

$$\Omega_1 := \left\{ r_1 : |r_1| \leq \sqrt{\frac{\rho_1}{k_{p1}}} \right\} \tag{6.8}$$

From all the above, r_1 converges to a small set containing the origin as $t \to \infty$. Moreover, $\rho_1 \to 0$ as $t \to \infty$ because of δ_{1i} and σ_{1i}, therefore, Ω_1 converges to the origin, there $r_1 \to 0$.

Integrating both sides of $\dot{\mathbb{V}}_1$ gives

$$\begin{aligned}\mathbb{V}_1(t) - \mathbb{V}_1(0) \leq & -\int_0^t k_{p1}r_1^2\,ds \\ & + \int_0^t \left(C_1^T\Delta_1 + \frac{1}{4}C_1^T\Sigma_1\Gamma^{-1}C_1\right)ds\end{aligned} \tag{6.9}$$

Since C_1 and Γ_1 are constants, $\int_0^\infty \Delta_1\,ds = \rho_{1\delta} = [\rho_{11\delta},\ldots,\rho_{17\delta}]^T$ and $\int_0^\infty \Sigma_1\,ds = \rho_{1\sigma} = [\rho_{11\sigma},\ldots,\rho_{17\sigma}]^T$, we can rewrite (6.9) as

$$\begin{aligned}\mathbb{V}_1(t) - \mathbb{V}_1(0) \leq & -\int_0^t k_{p1}r_1^2\,ds + C_1^T\left(\int_0^t \Delta_1\,ds\right) \\ & + \frac{1}{4}C_1^T\left(\int_0^t \Sigma_1\,ds\right)\Gamma_1^{-1}C_1 \\ \leq & -\int_0^t k_{p1}r_1^2\,ds + C_1^T\rho_{1\delta} + C_1^T\rho_{1\sigma}\Gamma_1^{-1}C_1 \\ < & -\int_0^t k_{p1}r_1^2\,ds + C_1^T\rho_{1\delta} + C_1^T\rho_{1\sigma}\Gamma_1^{-1}C_1 \\ < & \infty\end{aligned} \tag{6.10}$$

Thus $\mathbb{V}_1$ is bounded, which implies that $r_1 \in L_\infty$. From (6.10), we have

$$\int_0^t k_{p1}r_1^2\,ds \leq \mathbb{V}_1(0) - \mathbb{V}_1(t) + C_1^T\rho_{1\delta} + C_1^T\rho_{1\sigma}\Gamma_1^{-1}C_1 \tag{6.11}$$

which leads to $r_1 \in L_2$. From $r_1 = \dot{e}_1 + \Lambda_1 e_1$, it can be obtained that $e_1, \dot{e}_1 \in L_\infty$. As we have established $e_1, \dot{e}_1 \in L_\infty$, from the predefined trajectory boundedness, we conclude that $\zeta_1, \dot{\zeta}_1, \dot{\zeta}_{1r}$, and $\ddot{\zeta}_{1r} \in L_\infty$.

Therefore, all the signals on the right hand side of (6.1) are bounded, and we can conclude that $\dot{r}_1$ and therefore $\ddot{\zeta}_1$ are bounded. From (6.8), when $\rho_1 \to 0$, $r_1 \to 0$

as $t \to \infty$ can be obtained. Consequently, we have $e_1 \to 0, \dot{e}_1 \to 0$ as $t \to \infty$. It follows that $e_1, \dot{e}_1 \to 0$ as $t \to \infty$.

ζ_3-Subsystem Since $\dot{\zeta}_3 = \dot{\zeta}_{3r} + r_3$, $\ddot{\zeta}_3 = \ddot{\zeta}_{3r} + \dot{r}_3$, Eq. (5.19) becomes

$$\begin{aligned} &\frac{m_{22}m_{33} - m_{23}^2(\zeta_3)}{m_{23}(\zeta_3)}\dot{r}_3 - \tilde{\upsilon}_{23}(\zeta, \dot{\zeta})r_3 \\ &\quad = -\tau_2 - \left(\frac{m_{22}m_{33} - m_{23}^2(\zeta_3)}{m_{23}(\zeta_3)}\ddot{\zeta}_{3r} - \hat{\upsilon}_{23}r_3 - \upsilon_{23}(\zeta, \dot{\zeta})\dot{\zeta}_{3r}\right) \\ &\quad\quad - \frac{m_{23}(\zeta_3)}{m_{33}}(\upsilon_{31}\dot{\zeta}_1 + f_3 + g_3) + (f_2 + g_2 + d_2) \end{aligned} \tag{6.12}$$

where we decompose $\upsilon_{23} = \hat{\upsilon}_{23} - \tilde{\upsilon}_{23}$, such that $\frac{d}{dt}(\frac{m_{22}m_{33} - m_{23}^2(\zeta)}{m_{23}(\zeta_3)}) - 2\tilde{\upsilon}_{23} = 0$.

Assumption 6.3 There exist some finite positive constants $c_{3i} > 0$ $(1 \le i \le 7)$ such that $\forall \zeta_2, \zeta_3 \in R$, $\forall \dot{\zeta}_2, \dot{\zeta}_3 \in R$, $|\frac{m_{22}m_{33} - m_{23}^2(\zeta_3)}{m_{23}(\zeta_3)}| \le c_{31}$, $|\upsilon_{23}| < c_{32} + c_{33}\|\dot{\zeta}\|$, $|\frac{m_{23}(\zeta_3)}{m_{33}}\upsilon_{31}| < c_{34} + c_{35}\|\dot{\zeta}\|$, $|f_2(\dot{\zeta}) + \frac{m_{23}(\zeta_3)}{m_{33}}f_3(\dot{\zeta})| < c_{36}\|\dot{\zeta}\|$, $|\hat{\upsilon}_{23}| < c_{37} + c_{38}|\dot{\zeta}|$, $|\frac{m_{23}(\zeta_3)}{m_{33}}g_3 + g_2 + d_2| < c_{39}$.

Consider the following control laws:

$$\tau_3 = k_{p3}r_3 + k_{i3}\int_0^t r_3\,dt + \sum_{i=1}^{9}\frac{r_3\hat{c}_{3i}\Psi_{3i}^2}{\Psi_{3i}|r_3| + \delta_{3i}} \tag{6.13}$$

$$\Phi_3 = \hat{C}_3^T\Psi_3 \tag{6.14}$$

$$\dot{\hat{c}}_{3i} = -\sigma_{3i}\hat{c}_{3i} + \sum_{i=1}^{9}\frac{\gamma_{3i}\Psi_{3i}^2|r_3|^2}{|r_3|\Psi_{3i} + \delta_{3i}} \tag{6.15}$$

Let $\hat{C}_3 = [\hat{c}_{31}, \ldots, \hat{c}_{39}]^T$; $\Psi_3 = [|\ddot{\zeta}_{3r}|, |\dot{\zeta}_{3r}|, \|\dot{\zeta}\||\dot{\zeta}_{3r}|, |\dot{\zeta}|, \|\dot{\zeta}\|\|\dot{\zeta}\|, |\dot{\zeta}|, |r_3|, \|\dot{\zeta}\||r_3|, 1]^T$; where k_{p3} and k_{i3} are positive constants. And $\gamma_{3i} > 0$, $\delta_{3i} > 0$ and $\sigma_{3i} > 0$ satisfy Assumption 5.2:

$$\int_0^\infty \delta_{3i}(s)\,ds = \rho_{3i\delta} < \infty, \qquad \int_0^\infty \sigma_{3i}(s)\,ds = \rho_{3i\sigma} < \infty$$

with the finite constants $\rho_{3i\delta}$ and $\rho_{3i\sigma}$.

To analyze the closed loop stability for the ζ_3-subsystem, consider the following Lyapunov function:

$$\begin{aligned} \mathbb{V}_3 = &\frac{1}{2}\frac{m_{22}m_{33} - m_{23}^2(\zeta_3)}{m_{23}(\zeta_3)}r_3^2 + \frac{1}{2}\left(\int_0^t r_3\,ds\right)k_{i3}\int_0^t r_3\,ds \\ &+ \frac{1}{2}\tilde{C}_3^T\Gamma_3^{-1}\tilde{C}_3 \end{aligned} \tag{6.16}$$

where $\Gamma_3 = \mathrm{diag}[\gamma_{31}, \ldots, \gamma_{39}]$, $\tilde{C}_3 = C_3 - \hat{C}_3$.

Its time derivative is given by

$$\dot{\mathbb{V}}_3 = r_3\left(\frac{m_{22}m_{33} - m_{23}^2(\zeta_3)}{m_{23}(\zeta_3)}\dot{r}_3 + k_{i3}\int_0^t r_3\,ds\right) + \tilde{C}_3{}^T\Gamma_3^{-1}\dot{\tilde{C}}_3 \tag{6.17}$$

where Property 5.4 has been used. Substituting (6.12) into (6.17), we have

$$\begin{aligned}\dot{\mathbb{V}}_3 = r_3\bigg[-\tau_2 - \left(\frac{m_{22}(q)m_{33} - m_{23}^2(\zeta_3)}{m_{23}(\zeta)}\ddot{\zeta}_{3r} - \hat{\upsilon}_{23}r_3 - \upsilon_{23}(\zeta,\dot{\zeta})\dot{\zeta}_{3r}\right) \\ + \frac{m_{23}(\zeta_3)}{m_{33}}(-\upsilon_{31}\dot{\zeta}_1 - f_3 - g_3) + (f_2 + g_2 + d_2)\bigg] \\ + \tilde{C}_3{}^T\Gamma_3^{-1}\dot{\tilde{C}}_3\end{aligned} \tag{6.18}$$

Let the unknown function in (6.18) as

$$\begin{aligned}\mu_3 = \frac{m_{22}m_{33} - m_{23}^2(\zeta_3)}{m_{23}(\zeta_3)}\ddot{\zeta}_{3r} - \hat{\upsilon}_{23}r_3 - \upsilon_{23}(\zeta,\dot{\zeta})\dot{\zeta}_{3r} \\ + \frac{m_{23}}{m_{33}}(-\upsilon_{31}\dot{\zeta}_1 - f_3 - g_3) + (f_2 + g_2 + d_2)\end{aligned}$$

Therefore, we have

$$\begin{aligned}\dot{\mathbb{V}}_3 &= -k_{p3}r_3^2 - r_3\mu_3 - r_3\sum_{i=1}^{9}\frac{r_3\hat{c}_{3i}\Psi_{3i}^2}{|r_3|\Psi_{3i} + \delta_{3i}} \\ &\quad + \hat{C}_3^T\Sigma_3\Gamma_3^{-1}\tilde{C}_3 - \sum_{i=1}^{9}\frac{|r_3|^2\tilde{c}_{3i}\Psi_{3i}^2}{|r_3|\Psi_{3i} + \delta_{3i}} \\ &\le -k_{p3}r_3^2 + |r_3|\Phi_3 - \sum_{i=1}^{9}\frac{|r_3|^2 c_{3i}\Psi_{3i}^2}{|r_3|\Psi_{3i} + \delta_{3i}} + \hat{C}_3^T\Sigma_3\Gamma_3^{-1}\tilde{C}_3 \\ &\le -k_p r_3^2 + C_3^T\Delta_3 + \hat{C}_3^T\Sigma_3\Gamma_3^{-1}\tilde{C}_3 \\ &= -k_{p3}r_3^2 + C_3^T\Delta_3 + \hat{C}_3^T\Sigma_3\Gamma_3^{-1}(C_3 - \hat{C}_3) \\ &= -k_{p3}r_3^2 + C_3^T\Delta_3 - \frac{1}{4}C_3^T\Sigma_3\Gamma_3^{-1}C_3 \\ &\quad + \frac{1}{4}C_3^T\Sigma_3\Gamma_3^{-1}C_3 + \hat{C}_3^T\Sigma_3\Gamma_3^{-1}C_3 - \hat{C}_3^T\Sigma_3\Gamma_3^{-1}\hat{C}_3 \\ &= -k_{p3}r_3^2 + C_3^T\Delta_3 - \left(\frac{1}{2}C_3^T - \hat{C}_3^T\right)\Sigma_3\Gamma_3^{-1}\left(\frac{1}{2}C_3 - \hat{C}_3\right) \\ &\quad + \frac{1}{4}C_3^T\Sigma_3\Gamma_3^{-1}C_3 \\ &\le -k_{p3}r_3^2 + C_3^T\Delta_3 + \frac{1}{4}C_3^T\Sigma_3\Gamma_3^{-1}C_3\end{aligned}$$

with $\Sigma_3 = \mathrm{diag}[\sigma_{31},\ldots,\sigma_{39}]$, $\Delta_3 = [\delta_{31},\ldots,\delta_{39}]^T$.

Therefore,

$$\dot{\mathbb{V}}_3 \le -k_{p3} r_3^2 + C_3^T \Delta_3 + \frac{1}{4} C_3^T \Sigma_3 \Gamma_3^{-1} C_3 \tag{6.19}$$

Since $C_3^T \Delta_3 + \frac{1}{4} C_3^T \Sigma_3 \Gamma_3^{-1} C_3$ is bounded, there exists $t > t_3$, $C_3^T \Delta_3 + \frac{1}{4} C_3^T \Sigma_3 \Gamma_3^{-1} C_3 \le \rho_3$ with the finite constant ρ_3, when $|r_3| \ge \sqrt{\frac{\rho_3}{k_{p3}}}$, then $\dot{\mathbb{V}}_3 \le 0$. For $|r_3| \ge \sqrt{\frac{\rho_3}{k_{p3}}}$, and r_3 will converge to a compact set denoted by

$$\Omega_3 := \left\{ r_3 : |r_3| \le \sqrt{\frac{\rho_3}{k_{p3}}} \right\} \tag{6.20}$$

From all above, r_3 converges to a small set containing the origin as $t \to \infty$. Moreover, $\rho_3 \to 0$ as $t \to \infty$ because of δ_{3i} and σ_{3i}, therefore, Ω_3 converges to the origin, there $r_3 \to 0$.

Integrating both sides of $\dot{\mathbb{V}}_3$ gives

$$\begin{aligned} \mathbb{V}_3(t) - \mathbb{V}_3(0) \le & -\int_0^t k_{p3} r_3^2 \, ds \\ & + \int_0^t \left(C_3^T \Delta_3 + \frac{1}{4} C_3^T \Sigma_3 \Gamma_3^{-1} C_3 \right) ds \end{aligned} \tag{6.21}$$

Since C_3 and Γ_3 are constant, $\int_0^\infty \Delta_3 \, ds = \rho_{3\delta} = [\rho_{31\delta}, \ldots, \rho_{37\delta}]^T$ and $\int_0^\infty \Sigma_3 \, ds = \rho_{3\sigma} = [\rho_{31\sigma}, \ldots, \rho_{37\sigma}]^T$, we can rewrite (6.21) as

$$\begin{aligned} \mathbb{V}_3(t) - \mathbb{V}_3(0) & \le -\int_0^t k_{p3} r_3^2 \, ds + C_3^T \left(\int_0^t \Delta_3 \, ds \right) \\ & \quad + \frac{1}{4} C_3^T \left(\int_0^t \Sigma_3 \, ds \right) \Gamma_3^{-1} C_3 \\ & \le -\int_0^t k_{p3} r_3^2 \, ds + C_3^T \rho_{3\delta} + C_3^T \rho_{3\sigma} \Gamma_3^{-1} C_3 \\ & < -\int_0^t k_{p3} r_3^2 \, ds + C_3^T \rho_{3\delta} + C_3^T \rho_{3\sigma} \Gamma_3^{-1} C_3 \\ & < \infty \end{aligned} \tag{6.22}$$

Thus $\mathbb{V}_3$ is bounded, which implies that $r_3 \in L_\infty$. From (6.22), we have

$$\int_0^t k_{p3} r_3^2 \, ds \le \mathbb{V}_3(0) - \mathbb{V}_3(t) + C_3^T \rho_{3\delta} + C_3^T \rho_{3\sigma} \Gamma_3^{-1} C_3 \tag{6.23}$$

which leads to $r_3 \in L_2$. From $r_3 = \dot{e}_3 + \Lambda_3 e_3$, it can be obtained that $e_3, \dot{e}_3 \in L_\infty$. As we have established $e_3, \dot{e}_3 \in L_\infty$, from Assumption 5.2, we conclude that ζ_3, $\dot{\zeta}_3$, $\dot{\zeta}_{3r}$, and $\ddot{\zeta}_{3r} \in L_\infty$.

Therefore, all the signals on the right-hand side of (6.12) are bounded, and we can conclude that $\dot{r}_3$ and therefore $\ddot{\zeta}_3$ are bounded. Thus, from (6.20), we know $\rho_3 \to 0$, $r_3 \to 0$ as $t \to \infty$ can be obtained. Consequently, we have $e_3 \to 0$, $\dot{e}_3 \to 0$ as $t \to \infty$.

Remark 6.1 Compared with the linear controls proposed by [56, 59], and the control proposed on the velocity level [103], the proposed controls (6.2) and (6.13) are extended further to the dynamic level based on the nonlinear parameter uncertainty and unknown external disturbances.

ζ_2-Subsystem For system (5.17)–(5.19) under the control laws (6.2) and (6.13), the ζ_2-subsystem (5.18) can be rewritten as

$$\dot{\varphi} = f(\xi, \varphi, u) \tag{6.24}$$

where $\varphi = [\zeta_2, \dot{\zeta}_2]^T$, $\xi = [\zeta_1, \zeta_3, \dot{\zeta}_1, \dot{\zeta}_3]^T$, $u = [\tau_1, \tau_2]^T$. Then the zero dynamics can be addressed as [42]

$$\dot{\varphi} = f\big(0, \varphi, u^*(0, \varphi)\big) \tag{6.25}$$

where $u^* = [\tau_1^*, \tau_2^*]^T$, where τ_1^*, τ_2^* are the inputs when $\xi = 0$.

Assumption 6.4 The system (5.17), (5.18), (5.19) is hyperbolically minimum-phase, i.e., zero dynamics (6.25) is exponentially stable. In addition, assume that the control input u is designed as a function of the states (ξ, φ) and the reference signal satisfies the predefined trajectory boundedness, and the function $f(\xi, \varphi, u)$ is Lipschitz in ξ, i.e., there exists Lipschitz constants L_ξ and L_f for $f(\xi, \varphi, u)$ such that

$$\big\| f(\xi, \varphi, u) - f(0, \varphi, u_\varphi)\big\| \le L_\xi \|\xi\| + L_f \tag{6.26}$$

where $u_\varphi = u^*(0, \varphi)$.

Under Assumptions 6.1 and 6.3, by the converse theorem of Lyapunov [60], there exists a Lyapunov function $V_0(\varphi)$ which satisfies [42]

$$\sigma_2 \|\varphi\|^2 \le V_0(\varphi) \le \sigma_1 \|\varphi\|^2 \tag{6.27}$$

$$\frac{\partial V_0}{\partial \varphi} f(0, \varphi, u_\varphi) \le -\lambda_a \|\varphi\|^2 \tag{6.28}$$

$$\left\| \frac{\partial V_0}{\partial \varphi} \right\| \le \lambda_b \|\varphi\| \tag{6.29}$$

where σ_1, σ_2, λ_a and λ_b are positive constants.

From the previous stability analysis about the ζ_1-subsystem (5.14) and the ζ_3-subsystem (5.19), we know that ζ_1, ζ_3, $\dot{\zeta}_1$, and $\dot{\zeta}_3$ are bounded. Accordingly, ξ is bounded. We denote the upper bounded of ξ as

$$\|\xi\| \le \|\xi\|_{\max} \tag{6.30}$$

where $\|\xi\|_{\max}$ is a positive constant.

Lemma 6.2 *For the internal dynamics $\dot{\varphi} = f(\xi, \varphi, u)$ of the system, if Assumptions 5.2 and 6.4 are satisfied, then there exist positive constants L_φ and T_0, such that*

$$\big\|\varphi(t)\big\| \le L_\varphi, \quad \forall t > T_0 \tag{6.31}$$

Proof According to Assumption 6.4, there exists a Lyapunov function $V_0(\varphi)$. Differentiating $V_0(\varphi)$ along (5.17), (5.18), (5.19) yields

$$\begin{aligned}\dot{V}_0(\varphi) &= \frac{\partial V_0}{\partial \varphi} f(\xi, \varphi, u) = \frac{\partial V_0}{\partial \varphi} f(0, \varphi, u_\varphi) \\ &\quad + \frac{\partial V_0}{\partial \varphi}\left[f(\xi, \varphi, u) - f(0, \varphi, u_\varphi)\right]\end{aligned} \tag{6.32}$$

Noting (6.26)–(6.29) and (6.32) can be written as

$$\dot{V}_0(\varphi) \leq -\lambda_a\|\varphi\|^2 + \lambda_b\|\varphi\|\left(L_\xi\|\xi\| + L_f\right) \tag{6.33}$$

Noting (6.30), we have

$$\dot{V}_0(\varphi) \leq -\lambda_a\|\varphi\|^2 + \lambda_b\|\varphi\|\left(L_\xi\|\xi\|_{\max} + L_f\right) \tag{6.34}$$

Therefore, $\dot{V}_0(\varphi) \leq 0$, whenever

$$\|\varphi\| \geq \frac{\lambda_b}{\lambda_a}\|\varphi\|\left(L_\xi\|\xi\|_{\max} + L_f\right) \tag{6.35}$$

By letting $L_\varphi = \frac{\lambda_b}{\lambda_a}\|\varphi\|(L_\xi\|\xi\|_{\max} + L_f)$, we conclude that there exists a positive constant T_0, such that (6.31) holds. □

Theorem 6.3 *Consider the system* (5.17)–(5.19) *with Assumptions* 5.2, 6.1 *and* 6.3, *under the action of the control laws* (6.2) *and* (6.13) *and adaptation laws* (6.4) *and* (6.15). *For the compact set* Ω_{10}, *where* $(\zeta_1(0), \dot{\zeta}_1(0), \hat{C}_1(0)) \in \Omega_{10}$, *the compact set* Ω_{30}, *where* $(\zeta_3(0), \dot{\zeta}_3(0), \hat{C}_3(0)) \in \Omega_{30}$, *and the compact set* Ω_{20}, *where* $(\zeta_2(0), \dot{\zeta}_2(0)) \in \Omega_{20}$, *the tracking errors* r_1 *and* r_3 *converge to the compact sets* Ω_1 *and* Ω_3, *respectively, defined by* (6.8) *and* (6.20), *and all the signals in the closed loop system are bounded.*

Proof From the results of (6.10) and (6.22), it is clear that the tracking errors r_1 and r_3 converge to the compact sets Ω_1 and Ω_3, respectively, defined by (6.8) and (6.20). In addition, the signals $\tilde{C}_1$ are $\tilde{C}_3$ are bounded. From Lemma 2.48, we can know e_1, $\dot{e}_1$, e_3, and $\dot{e}_3$ are also bounded. From the boundedness of ζ_{1d} and ζ_{3d} in Assumption 5.2, we know that ζ_1 and ζ_3 are bounded. Since $\dot{\zeta}_{1d}$ and $\dot{\zeta}_{3d}$ are also bounded, it follows that $\dot{\zeta}_1$ and $\dot{\zeta}_3$ are bounded. With C_1 and C_3 as constants, we know that $\hat{C}_1$ and $\hat{C}_3$ are also bounded. From Lemma 6.2, we know that ζ_2 and $\dot{\zeta}_2$ are bounded, therefore, we can obtain the ζ_2-subsystem (5.18) is stable. This completes the proof. □

6.2.2 Zero-Dynamics Stability Analysis

First, we briefly mention the method used to analyze the zero-dynamics stability. By substituting the two control inputs τ_1 and τ_3 into the zero-dynamics and after some simplifications, an equation in terms of $\dot{\zeta}_2, \ddot{\zeta}_2$ was obtained. The analysis of the stability of the zero-dynamics ζ_2-subsystem is complicated by the presence of dynamic modeling errors $\tilde{C}_1$, $\tilde{C}_3$, which cannot simply be assumed as zeros. However, according to Lemma 6.2, we know that the zero-dynamics in (5.18) is stable in theory, provided that Assumption 6.4 is true. In this chapter, we will justify, through the simulation results, that the zero dynamics are indeed stable.

6.2.3 Simulation Studies

Let us consider a mobile wheeled inverted pendulum as shown in Fig. 3.1. In the simulation, we assume the parameters $M = 5.0$ kg, $m = 2.5$ kg, $I_M = 1.0$ kg m^2, $I_p = 1.0$ kg m^2, $I_w = 0.2$ kg m^2, $M_w = 0.2$ kg, $l = 1$ m, $d = 0.5$ m, $r = 0.5$ m, $\mathcal{B} = 0.1$, $\zeta(0) = [-0.2, 0, \pi/180]^T$, and $\dot{\zeta}(0) = [0.0, 0.1, 0.0]^T$. The disturbances from environments on the system are introduced as $1.0\sin(t)$ and $1.0\cos(t)$ in the simulation model. The desired trajectories are chosen as $\theta_d = 0.2t$ rad and $\alpha_d = 0$ rad, and the initial velocity is 0.1 m/s.

In order to validate the better performance of the proposed control, in the same conditions, we compare (i) the model based control, which assume 10 % model uncertainty, (ii) robust control with 30–50 % model uncertainty under the conditions that the constants C_1 and C_3 can be obtained easily, therefore, the adaptive update laws in (6.4) and (6.15) are not included in the implemented control, and (iii) the proposed control 30–50 % model uncertainty without the conditions of (ii). In the robust control, we choose $C_1 = [10.0, \ldots, 10.0]^T$, $C_3 = [10.0, \ldots, 10.0]^T$. In the adaptive robust control, by Theorem 6.3, the control gains are selected as $k_{p1} = 1.0$, $k_{p3} = 80.0$, $k_{i1} = k_{i3} = 0.0$, $\hat{C}_1(0) = [10.0, \ldots, 10.0]^T$, $\hat{C}_3(0) = [10.0, \ldots, 10.0]^T$, $\delta_i = \sigma_{ji} = 1/(1+t)^2$, and $\gamma_{ji} = 0.05$.

The tracking performance of the three controls are illustrated in Figs. 6.1–6.8. The direction angle tracking by the three controls are shown in Figs. 6.1, 6.2, 6.3, the input torques are shown in Figs. 6.5, 6.6, 6.7, and the comparison of stable velocity is shown in Fig. 6.8. The balance control performance comparison is shown in Fig. 6.4. As these figures show, we can obtain good performance by the proposed control with the initial disturbances boundedness from the environment given, even if the nominal parameters of the system have 30–50 % model uncertainty. Moreover, by the model based control, in the same conditions, more than 10 % uncertainty would cause the system divergent. Therefore, we know the proposed control scheme could achieve better tracking performance over the model-based control and robust control. The better tracking performances is largely due to the "adaptive" mechanism. Although the parametric uncertainties and the external disturbances are both introduced into the simulation model, the motion/balance control performance of system, under the proposed control, is not degraded. The simulation results demonstrate the effectiveness of the proposed adaptive control in the presence of unknown nonlinear dynamic system and environments. Different motion/balance tracking performance can be achieved by adjusting parameter adaptation gains and control gains.

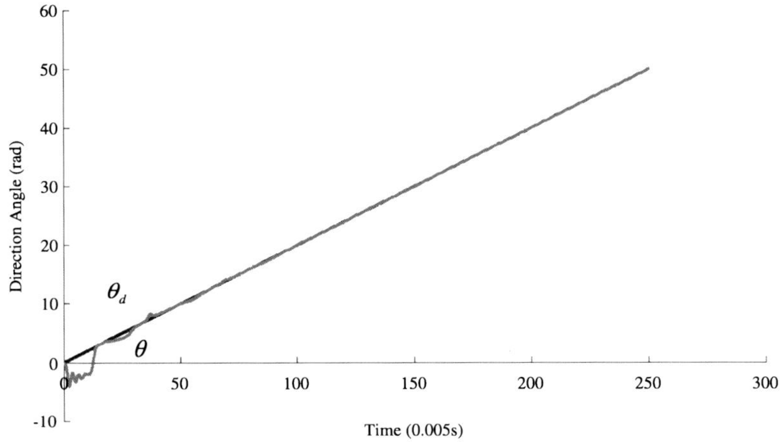

Fig. 6.1 Tracking the direction angle by the adaptive robust control

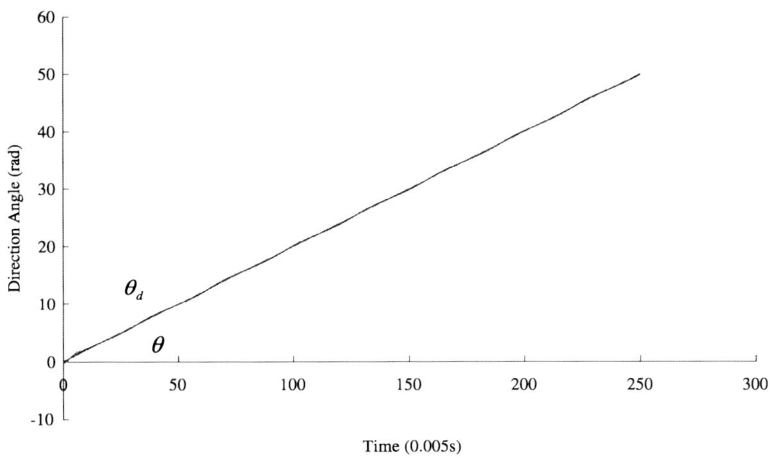

Fig. 6.2 Tracking the desired direction angle by the robust control

6.3 Hybrid Force and Motion Control

Previous works for the WIP systems usually eliminate nonholonomic constraint force in order to make the control design easier, under the assumption that the friction force from the ground is large enough as needed. Nevertheless, such assumption is unfeasible in the practical applications. In this section, adaptive robust motion/force control for wheeled inverted pendulums is investigated with parametric and functional uncertainties.

Although wheeled inverted pendulums systems are intrinsically nonlinear and their dynamics are described by nonlinear differential equations, if the system operates around an operating point, and the signals involved are small, we can obtain a

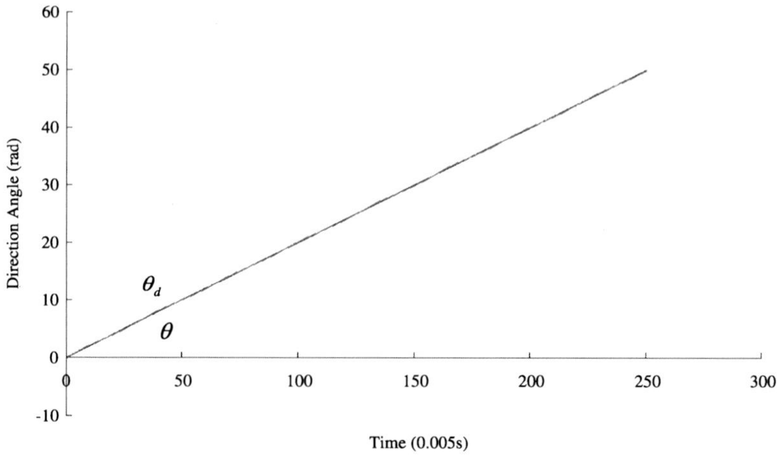

Fig. 6.3 Tracking the desired direction angle by the model based control

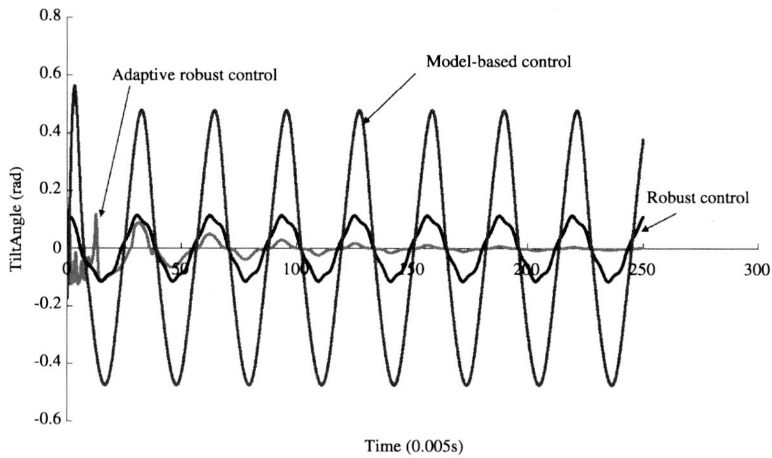

Fig. 6.4 Tracking the desired tilt angle by the adaptive robust control, robust control, and model-based control

linear model approximating the nonlinear system in the region of operation. In [59], motion control was proposed using linear state-space model. In [56], dynamics was derived using a Newtonian approach and the control was designed based on the dynamic equations linearized around an operating point. In [106], dynamic equations of the inverted pendulum were studied involving pitch and rotation angles of the two wheels as the variables of interest, and in [107] a linear controller was designed for stabilization considering robustness as a condition. In [19], a linear stabilizing controller was derived by a planar model without considering yaw. In [75], the exact dynamics of two-wheeled inverted pendulum was investigated, and linear feedback control was developed on the dynamic model. In [103], a two-level velocity con-

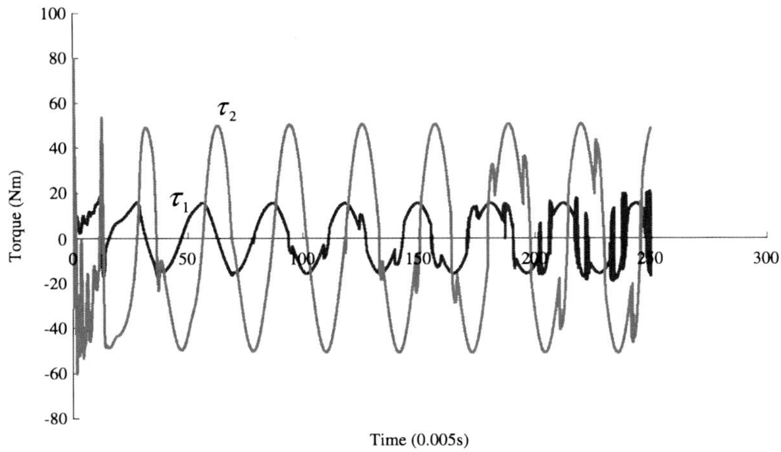

Fig. 6.5 Input torques by the adaptive robust control

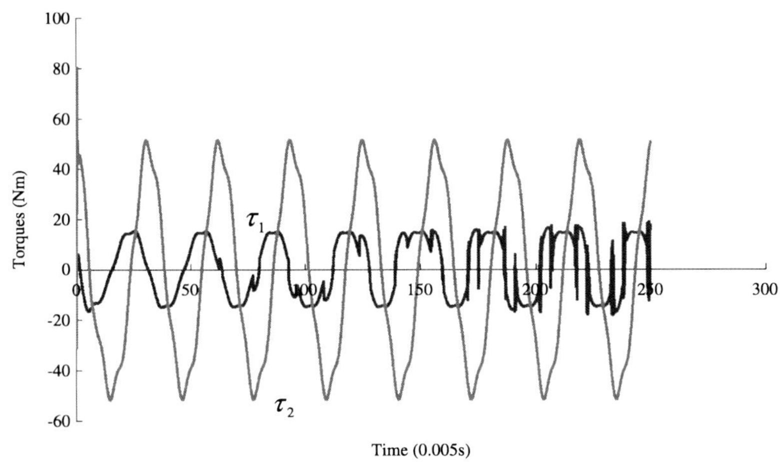

Fig. 6.6 Input torques by the robust control

troller via partial feedback linearization and a stabilizing position controller were derived.

Based on the idea of linearization, model-based approach is generally utilized in dynamic control. If accurate knowledge of the dynamic model is available, the model-based control can provide an effective solution to the problem. However, wheeled inverted pendulum control is characterized by unstable balance and unmodeled dynamics, and subject to time varying external disturbances, in the form of parametric and functional uncertainties, which are generally difficult to model accurately. Therefore, traditional model-based control may not be the ideal approach since it generally works best only when the dynamic model is known exactly. The

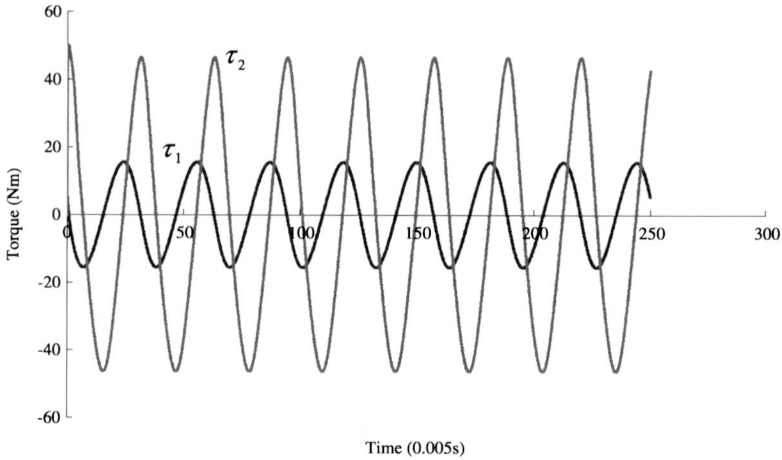

Fig. 6.7 Input torques by the model based control

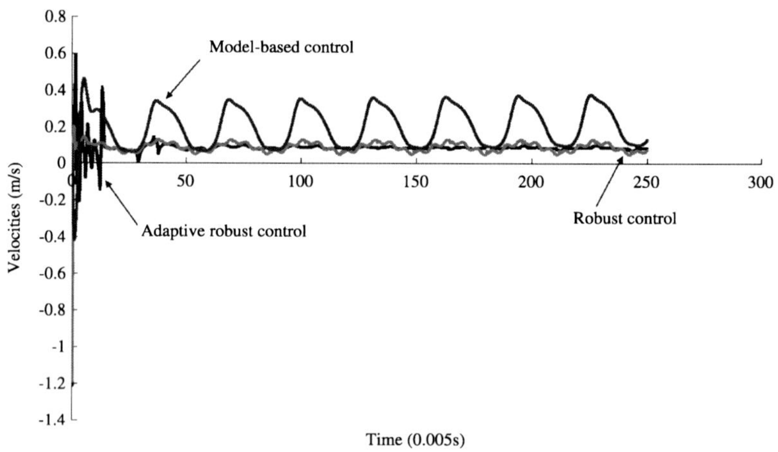

Fig. 6.8 The stable velocity by the adaptive robust control, robust control and model-based control

presence of uncertainties and disturbances would disrupt the function of the traditional model-based feedback control and lead to the unstable balance.

In this chapter, by discovering and utilizing the unique physical property of the WIP systems, we separate the zero-dynamics subsystem to simplify the model. Then, we propose the robust adaptive hybrid motion and force control for the WIP systems. Since the system except the zero-dynamics subsystem is still MIMO nonlinear system, we propose adaptive robust controls to accommodate the presence of parametric and functional uncertainties in the dynamics of WIP systems.

The proposed robust adaptive controls based on physical properties of WIP systems make use of online adaptation mechanism to cancel the unmodeled dynamics.

Based on Lyapunov synthesis, the proposed controls ensure that the system outputs track the given bounded reference signals within a small neighborhood of zero, and guarantee semi-global uniform boundedness of all closed loop signals. The effectiveness of the proposed controls is verified through extensive simulations.

6.3.1 Preliminaries and Dynamics Transformation

Consider the nonholonomic constraint described in (5.4), since $J_\upsilon(q) \in \mathbb{R}^{(n_\upsilon - m)\times n}$, we introduce $J_\alpha \in \mathbb{R}^{n_\alpha \times n}$, and $J = [J_\upsilon, J_\alpha]^T \in \mathbb{R}^{(n-m)\times n}$ with $n_\upsilon + n_\alpha = n$, such that it is possible to find a $(m + n_\alpha)$ rank matrix $R(q) \in \mathbb{R}^{n\times(m+n_\alpha)}$ formed by a set of smooth and linearly independent vector fields spanning the null space of $J(q)$, i.e.,

$$R^T(q)J^T(q) = 0 \tag{6.36}$$

Denote $R(q) = [r_1(q), \ldots, r_m(q), r_{m+1}(q), \ldots, r_{m+n_\alpha}]$ and define an auxiliary time function $\dot{z}(t) \in \mathbb{R}^{m+n_\alpha}$, and $\dot{z}(t) = [\dot{z}_1(t), \ldots, \dot{z}_m(t), \dot{z}_{m+1}, \ldots, \dot{z}_{m+n_\alpha}]^T$ such that

$$\begin{aligned}\dot{q} = R(q)\dot{z}(t) &= r_1(q)\dot{z}_1(t) + \cdots + r_m(q)\dot{z}_m(t)\\ &\quad + r_{m+1}(q)\dot{z}_{m+1}(t) + \cdots + r_{m+n_\alpha}(q)\dot{z}_{m+n_\alpha}(t)\end{aligned} \tag{6.37}$$

Equation (6.37) is the kinematic model for the WIP systems. Usually, $\dot{z}(t)$ has physical meaning, consisting of the angular velocity ω, the linear velocity υ, and the tilt angle α, i.e., $\dot{z}(t) = [\upsilon\ \omega\ \dot{\alpha}]^T$. Equation (6.37) describes the kinematic relationship between the motion vector q and the velocity vector $\dot{z}(t)$.

Differentiating (6.37) yields

$$\ddot{q} = \dot{R}(q)\dot{z} + R(q)\ddot{z} \tag{6.38}$$

From (6.37), $\dot{z}$ can be obtained from q and $\dot{q}$ as

$$\dot{z} = \left[R^T(q)R(q)\right]^{-1} R^T(q)\dot{q} \tag{6.39}$$

The WIP dynamics equation which satisfies the nonholonomic constraint (5.4), can be rewritten in terms of the internal state variable $\dot{z}$ as

$$M(q)R(q)\ddot{z} + V^*\dot{z} + G(q) + F = B(q)\tau + J^T(q)\lambda \tag{6.40}$$

with $V^* = [M(q)\dot{R}(q) + V(q,\dot{q})R(q)]$. Combining (6.37) and (6.38) with system dynamics described in (5.1), we have the transformed nonholonomic system

$$M_1(q)\ddot{z} + V_1(q,\dot{q})\dot{z} + G_1(q) + F_1 = R^T u \tag{6.41}$$

where $M_1(q) = R^T M(q)R$, $V_1(q,\dot{q}) = R^T[M(q)\dot{R} + V(q,\dot{q})R]$, $G_1(q) = R^T G(q)$, $u = B(q)\tau$, $F_1 = R^T F$. which is more appropriate for the controller design as the constraint λ has been eliminated from the dynamic equation.

The force multipliers λ can be obtained by (6.40)

$$\lambda = Z\big(\big[M(q)\dot{R}(q) + V(q,\dot{q})R(q)\big]\dot{z} + G + F - u\big) \tag{6.42}$$

where $Z = (JM^{-1}J^T)^{-1}JM^{-1}$. Consider the control input u in (6.41) decoupled into the motion control τ_a and the force control τ_b in the following form:

$$u = \tau_a - J^T\tau_b \tag{6.43}$$

Then, (6.41) and (6.42) can be changed to

$$M_1\ddot{z} + V_1\dot{z} + G_1 + F_1 = R^T\tau_a \tag{6.44}$$

$$\lambda = Z\big(\big[M(q)\dot{R}(q) + V(q,\dot{q})R(q)\big]\dot{z} + G + F - \tau_a\big) + \tau_b \tag{6.45}$$

Considering (6.41), $\dot{z}(t)$ has physical meaning, for example, consisting of the linear velocity v and the angular velocity ω, and the tilt angle α, i.e., $\dot{z}(t) = [\dot{z}_1\,\dot{z}_2\,\dot{z}_3]^T = [v\,\omega\,\dot{\alpha}]^T$, we denote the variable z_3 for the inverted pendulum, and select z_1 coupled with z_3 from the remaining variables such that the dimension of z_1 and z_3 is same, and the left variables are denoted with z_2, we have

$$M_1 = \begin{bmatrix} m_{11} & m_{12} & m_{13} \\ m_{21} & m_{22} & m_{23} \\ m_{31} & m_{32} & m_{33} \end{bmatrix}, \quad z = \begin{bmatrix} z_1 \\ z_2 \\ z_3 \end{bmatrix}, \quad V_1\dot{z} = \begin{bmatrix} v_1 \\ v_2 \\ v_3 \end{bmatrix}$$
$$G_1 = \begin{bmatrix} g_1 \\ g_2 \\ g_3 \end{bmatrix}, \quad F_1 = \begin{bmatrix} d_1 \\ d_2 \\ d_3 \end{bmatrix}, \quad R^T\tau_a = \begin{bmatrix} \tau_1 \\ \tau_2 \\ 0 \end{bmatrix} \tag{6.46}$$

where $m_{11}, \ldots, m_{33}, v_1, v_2, v_3, g_1, g_2, g_3, d_1, d_2, d_3$, are unknown functions.

From (5.1) and (6.46), we know the system is under-actuated, control design directly based on (5.1) is difficult because there is zero-dynamics within the system. We can observe that z_1 and z_3 are coupled. Making use of the physical properties just mentioned plus a simple manipulation of (6.46), we can obtain three subsystems: z_1-subsystem, z_2-subsystem, and z_3-subsystem, respectively, as follows

$$m_{11}\ddot{z}_1 = \tau_1 - v_1 - g_1 - d_1 - m_{12}\ddot{z}_2 - m_{13}\ddot{z}_3 \tag{6.47}$$

$$\begin{aligned} &\big(m_{22} - m_{21}m_{11}^{-1}m_{12}\big)\ddot{z}_2 + \big(m_{23} - m_{21}m_{11}^{-1}m_{13}\big)\ddot{z}_3 \\ &\quad + v_2 + g_2 + d_2 - m_{21}m_{11}^{-1}v_1 - m_{21}m_{11}^{-1}g_1 - m_{21}m_{11}^{-1}d_1 \\ &= \tau_2 - m_{21}m_{11}^{-1}\tau_1 \end{aligned} \tag{6.48}$$

$$\begin{aligned} &\big(m_{32} - m_{31}m_{11}^{-1}m_{12}\big)\ddot{z}_2 + \big(m_{33} - m_{31}m_{11}^{-1}m_{13}\big)\ddot{z}_3 \\ &\quad + v_3 + g_3 + d_3 - m_{31}m_{11}^{-1}v_1 - m_{31}m_{11}^{-1}g_1 - m_{31}m_{11}^{-1}d_1 \\ &= -m_{31}m_{11}^{-1}\tau_1 \end{aligned} \tag{6.49}$$

Let $a = m_{22} - m_{21}m_{11}^{-1}m_{12}$, $b = m_{23} - m_{21}m_{11}^{-1}m_{13}$, $l = m_{32} - m_{31}m_{11}^{-1}m_{12}$, $j = m_{33} - m_{31}m_{11}^{-1}m_{13}$, $e = (m_{22} - m_{21}m_{11}^{-1}m_{12})\dot{z}_2 + (m_{23} - m_{21}m_{11}^{-1}m_{13})\dot{z}_3$, $f = (m_{32} - m_{31}m_{11}^{-1}m_{12})\dot{z}_2 + (m_{33} - m_{31}m_{11}^{-1}m_{13})\dot{z}_3$, $h = (m_{21} - m_{21}m_{11}^{-1}m_{11})\dot{z}_1 + g_2 + d_2 - m_{21}m_{11}^{-1}g_1 - m_{21}m_{11}^{-1}d_1$, $k = (m_{31} - m_{31}m_{11}^{-1}m_{11})\dot{z}_1 + g_3 + d_3 -$

$m_{31}m_{11}^{-1}g_1 - m_{31}m_{11}^{-1}d_1$. Then, let $\xi = [z_3, z_2]^T$, we can rewrite (6.48) and (6.49) as

$$\mathcal{M}\ddot{\xi} + \mathcal{V}\dot{\xi} + \mathcal{D} = \mathcal{B}_1 \Lambda \mathcal{U} \tag{6.50}$$

where

$$\begin{aligned}
\mathcal{M} &= \begin{bmatrix} j & l \\ b & a \end{bmatrix} \\
\mathcal{D} &= \begin{bmatrix} k \\ h \end{bmatrix} \\
\mathcal{B}_1 &= \begin{bmatrix} m_{31}m_{11}^{-1} & 0 \\ m_{21}m_{11}^{-1} & 1 \end{bmatrix} \\
\mathcal{V} &= \begin{bmatrix} m_{33} - m_{31}m_{11}^{-1}m_{13} & m_{32} - m_{31}m_{11}^{-1}m_{12} \\ m_{23} - m_{21}m_{11}^{-1}m_{13} & m_{22} - m_{21}m_{11}^{-1}m_{12} \end{bmatrix} \\
\Lambda &= \mathrm{diag}[-1, 1], \qquad \mathcal{U} = [\tau_1, \tau_2]^T
\end{aligned} \tag{6.51}$$

Property 6.4 The inertia matrix $\mathcal{M}$ is symmetric and positive definite.

Property 6.5 The matrix $\dot{\mathcal{M}} - 2\mathcal{V}$ is skew-symmetric.

Property 6.6 The eigenvalues of the inertia matrix $\mathcal{B}_1$ are positive.

Remark 6.7 There exist the minimum and maximum eigenvalues $\lambda_{\min}(\mathcal{B}_1)$ and $\lambda_{\max}(\mathcal{B}_1)$, such that $\forall x \in \mathbb{R}^{(m+n_\alpha)}$, $x^T\lambda_{\min}(\mathcal{B}_1)Ix \le x^T\mathcal{B}_1 x \le x^T\lambda_{\max}(\mathcal{B}_1)Ix$ with the unit matrix I, and the known positive parameter b satisfying $0 < b < \lambda_{\min}(\mathcal{B}_1)$, that is, $x^T bIx \le x^T\lambda_{\min}(\mathcal{B}_1)Ix$.

Remark 6.8 The dynamics model in (6.46) is the general form for the mechanical system, but for the WIP systems, the detained dynamical model is presented in Chap. 3.

By appropriate selection of a set of vector $\dot{z}(t) \in \mathbb{R}^{m+n_\alpha}$, since the control input is only m dimension, and $m > n_\alpha$, only m variables of $z(t)$ can be controlled, the control objective can be specified as: design a controller that ensures the tracking errors of z_i $(1 \le i \le m)$ from their respective desired trajectories $z_{id}(t)$ to be within a small neighborhood of zero, i.e.,

$$\left|z_i(t) - z_{id}(t)\right| \le \epsilon_i, \quad i = 2, 3 \tag{6.52}$$

where $\epsilon_i > 0$. Ideally, ϵ_i should be the threshold of measurable noise, while the constraint force error $(\lambda - \lambda_d)$ is bounded in a certain region. At the same time, n_α variables of $z(t)$ are to be kept bounded. The variables z_i $(1 \le i \le m)$, can be thought as output equation of the system.

Assumption 6.5 The desired reference trajectory $z_{id}(t)$ $(1 \le i \le m)$ is assumed to be bounded and uniformly continuous, and has bounded and uniformly continuous derivatives up to the second order. The desired $\lambda_d(t)$ is bounded and uniformly continuous.

Remark 6.9 Since we can plan and design the desired trajectory for $z_{id}(t)$ before implementing control, it is reasonable and feasible that we give the trajectory satisfying Assumption 6.5.

Lemma 6.10 *For* $x \ge 0$ *and* $\rho = 1 + \frac{1}{(1+t)^2} \ge 1$ *with* $t > 0$, *we have* $\ln(\cosh(x)) + \rho \ge x$.

Proof If $x \ge 0$, we have $\int_0^x \frac{2}{e^{2\delta}+1} d\delta < \int_0^x \frac{2}{e^{2\delta}} d\delta = 1 - e^{-2x} < 1 + \frac{1}{(1+t)^2}$. Therefore, $\ln(\cosh(x)) + \rho \ge \ln(\cosh(x)) + \int_0^x \frac{2}{e^{2\delta}+1} d\delta$. Let $f(x) = \ln(\cosh(x)) + \int_0^x \frac{2}{e^{2\delta}+1} d\delta - x$, we have $\dot{f}(x) = \tanh(x) + \frac{2}{e^{2x}+1} - 1 = \frac{e^x - e^{-x}}{e^x + e^{-x}} + \frac{2}{e^{2x}+1} - 1 = 0$. From the Mean Value theorem, we have $f(x) - f(0) = \dot{f}(x)(x - 0)$. Since $f(0) = 0$, we have $f(x) = 0$, that is, $\ln(\cosh(x)) + \int_0^x \frac{2}{e^{2\delta}+1} d\delta = x$, then, we have $\ln(\cosh(x)) + \rho \ge x$. This completes the proof. □

6.3.2 Motion Control of z_2 and z_3-Subsystems

Let us define the following notations as $e_\xi = \xi - \xi_d$, $e_\lambda = \lambda - \lambda_d$, $\dot{\xi}_r = \dot{\xi}_d - \rho_1 e_\xi$, $s = \dot{e}_\xi + \rho_1 e_\xi$, where $\dot{\xi}_d = [\dot{z}_{3d}, \dot{z}_{2d}]^T$, $\dot{\xi}_r = [\dot{z}_{3r}, \dot{z}_{2r}]^T$ is the reference signal described in internal state space, and ρ_1 is positive diagonal. Apparently, we have

$$\dot{\xi} = \dot{\xi}_r + s \tag{6.53}$$

From the dynamic equation (6.50) together with (6.53), we have

$$\mathcal{M}\dot{s} = -\mathcal{V}s - \mathcal{M}\ddot{\xi}_r - \mathcal{V}\dot{\xi}_r - \mathcal{D} + \mathcal{B}_1 \Lambda \mathcal{U} \tag{6.54}$$

Let $\mathcal{M}_0$, $\mathcal{V}_0$, $\mathcal{D}_0$ and $\mathcal{B}_{10}$ be nominal parameter vector which gives the corresponding nominal function $\mathcal{M}_0\ddot{\xi}_r + \mathcal{V}_0\dot{\xi}_r + \mathcal{D}_0$ and $(\mathcal{B}_{10})^{-1}$, respectively. There exist some finite positive constants $c_i > 0$ $(1 \le i \le 6)$, such that $\forall q, \dot{q} \in \mathbb{R}^n$, $\|\mathcal{M} - \mathcal{M}_0\| \le c_1$, $\|\mathcal{V} - \mathcal{V}_0\| \le c_2 + c_3\|\dot{q}\|$, $\|\mathcal{D} - \mathcal{D}_0\| \le c_4 + c_5\|\dot{q}\|$, $\|\mathcal{B}_1 - \mathcal{B}_{10}\| \le c_6$.

The proposed control for the system is given as

$$\Lambda \mathcal{U} = u_1 + u_2 \tag{6.55}$$

where u_1 is the nominal control

$$u_1 = -\mathcal{B}_{10}^{-1} K_p s + \mathcal{B}_{10}^{-1}(\mathcal{M}_0\ddot{\xi}_r + \mathcal{V}_0\dot{\xi}_r + \mathcal{D}_0) \tag{6.56}$$

where K_p is a diagonal positive constant, and u_2 is designed to compensate for the parameter errors and the function approximation errors arising from approximating the unknown function as

$$u_2 = u_{21} + u_{22} + u_{23} + u_{24} + u_{25} + u_{26} \tag{6.57}$$

$$u_{21} = -\frac{1}{b}\operatorname{sgn}(s)\big(\ln\big(\cosh(\hat{\Phi}_1)\big) + \rho\big) \tag{6.58}$$

$$u_{22} = -\frac{1}{b}\operatorname{sgn}(s)\big(\ln\big(\cosh(\hat{\Phi}_2)\big) + \rho\big) \tag{6.59}$$

$$u_{23} = -\frac{1}{b}\operatorname{sgn}(s)\big(\ln\big(\cosh(\hat{\Phi}_3)\big) + \rho\big) \tag{6.60}$$

$$u_{24} = -\frac{1}{b}\operatorname{sgn}(s)\big(\ln\big(\cosh(\hat{\Phi}_4)\big) + \rho\big) \tag{6.61}$$

$$u_{25} = -\frac{1}{b}\operatorname{sgn}(s)\big(\ln\big(\cosh(\hat{\Phi}_5)\big) + \rho\big) \tag{6.62}$$

$$u_{26} = -\frac{1}{b}\operatorname{sgn}(s)\big(\ln\big(\cosh(\hat{\Phi}_6)\big) + \rho\big) \tag{6.63}$$

with $\operatorname{sgn}(s) = \frac{s}{\|s\|}$, which are adaptively tuned according to

$$\dot{\hat{c}}_1 = -\alpha_1\hat{c}_1 + \gamma_1\|s\|\|\ddot{\xi}_r\|, \quad \hat{c}_1(0) > 0 \tag{6.64}$$

$$\dot{\hat{c}}_2 = -\alpha_2\hat{c}_2 + \gamma_2\|s\|\|\dot{\xi}_r|, \quad \hat{c}_2(0) > 0 \tag{6.65}$$

$$\dot{\hat{c}}_3 = -\alpha_3\hat{c}_3 + \gamma_3\|s\|\|\dot{\xi}_r\|\|\dot{q}\|, \quad \hat{c}_3(0) > 0 \tag{6.66}$$

$$\dot{\hat{c}}_4 = -\alpha_4\hat{c}_4 + \gamma_4\|s\|, \quad \hat{c}_4(0) > 0 \tag{6.67}$$

$$\dot{\hat{c}}_5 = -\alpha_5\hat{c}_5 + \gamma_5\|s\|\|\dot{q}\|, \quad \hat{c}_5(0) > 0 \tag{6.68}$$

$$\dot{\hat{c}}_6 = -\alpha_6\hat{c}_6 + \gamma_6\|s\|\|u_1\|, \quad \hat{c}_6(0) > 0 \tag{6.69}$$

with $\hat{\Phi}_1 = \hat{c}_1\|\ddot{\xi}_r\|$, $\hat{\Phi}_2 = \hat{c}_2\|\dot{\xi}_r\|$, $\hat{\Phi}_3 = \hat{c}_3\|\dot{q}\|\|\dot{\xi}_r\|$, $\hat{\Phi}_4 = \hat{c}_4$, $\hat{\Phi}_5 = \hat{c}_5\|\dot{q}\|$, $\hat{\Phi}_6 = \hat{c}_6\|u_1\|$, and $\alpha_i > 0$ being design parameters and satisfying

$$\lim_{t\to\infty} \alpha_i = 0 \tag{6.70}$$

$$\int_0^\infty \alpha_i(t)\,dt = \varrho_\alpha < \infty \tag{6.71}$$

with finite constant ϱ_α, and $\gamma_i > 0$ is design parameter.

Remark 6.11 As discussed in Chap. 3, the WIP systems subject to kinematic constraints and dynamic constraints are apparently different from generalized nonholonomic systems, the control (6.57) is proposed based on the obtained reduced model (6.50), for the generalized nonholonomic system, if we can obtain the corresponding reduced model similar as (6.50), the control (6.57) is also applicable.

To analyze closed loop stability for the z_2 and z_3 subsystems, consider the following Lyapunov function candidate

$$\mathbf{V}_1 = \frac{1}{2}s^T\mathcal{M}s + \frac{1}{2}\sum_{i=1}^{6}\tilde{c}_i\gamma_i^{-1}\tilde{c}_i \tag{6.72}$$

where $\hat{c}_i = c_i - \tilde{c}_i$, $i = 1, \ldots, 6$, therefore, $\dot{\hat{c}}_i = -\dot{\tilde{c}}_i$. Its time derivative is given by

$$\dot{\mathbf{V}}_1 = s^T\left(\mathcal{M}\dot{s} + \frac{1}{2}\dot{\mathcal{M}}s\right) + \sum_{i=1}^{6} \tilde{c}_i \gamma_i^{-1} \dot{\tilde{c}}_i \tag{6.73}$$

Considering Property 6.5, and substituting (6.54) into (6.73), we have

$$\dot{\mathbf{V}}_1 = s^T\left[\mathcal{B}_1 \Lambda U - (\mathcal{M}\ddot{\xi}_r + \mathcal{V}\dot{\xi}_r + \mathcal{D})\right] + \sum_{i=1}^{6} \tilde{c}_i \gamma_i^{-1} \dot{\tilde{c}}_i \tag{6.74}$$

Integrating (6.56) and (6.57) into (6.74), we have

$$\begin{aligned}
\dot{\mathbf{V}}_1 &= s^T\left[(\mathcal{B}_1 - \mathcal{B}_{10})u_1 + \mathcal{B}_1 u_2 + \mathcal{B}_{10}u_1 - \mathcal{M}\ddot{\xi}_r - \mathcal{V}\dot{\xi}_r - \mathcal{D}\right] \\
&\quad + \sum_{i=1}^{6} \tilde{c}_i \gamma_i^{-1} \dot{\tilde{c}}_i \\
&= s^T\left[(\mathcal{B}_1 - \mathcal{B}_{10})u_1 + \mathcal{B}_1 \sum_{i=1}^{6} u_{2i} - K_p s - (\mathcal{M} - \mathcal{M}_0)\ddot{\xi}_r \right.\\
&\quad \left. - (\mathcal{V} - \mathcal{V}_0)\dot{\xi}_r - (\mathcal{D} - \mathcal{D}_0)\right] + \sum_{i=1}^{6} \tilde{c}_i \gamma_i^{-1} \dot{\tilde{c}}_i \\
&= -s^T K_p s + s^T\left[\mathcal{B}_1 u_{21} - (\mathcal{M} - \mathcal{M}_0)\ddot{\xi}_r\right] \\
&\quad + s^T\left[\mathcal{B}_1(u_{22} + u_{23}) - (\mathcal{V} - \mathcal{V}_0)\dot{\xi}_r\right] \\
&\quad + s^T\left[\mathcal{B}_1(u_{24} + u_{25}) - (\mathcal{D} - \mathcal{D}_0)\right] \\
&\quad + s^T\left[\mathcal{B}_1 u_{26} + (\mathcal{B}_1 - \mathcal{B}_{10})u_1\right] + \sum_{i=1}^{6} \frac{1}{\gamma_i} \tilde{c}_i \dot{\tilde{c}}_i
\end{aligned} \tag{6.75}$$

Considering (6.58) and (6.64), and Lemma 6.10, the second right-hand term of (6.75) is bounded by

$$\begin{aligned}
& s^T\left[\mathcal{B}_1 u_{21} - (\mathcal{M} - \mathcal{M}_0)\ddot{\xi}_r\right] + \frac{1}{\gamma_1}\tilde{c}_1 \dot{\tilde{c}}_1 \\
&\quad \le c_1 \|s\| \|\ddot{\xi}_r\| - \|s\| \left(\ln\left(\cosh(\hat{\Phi}_1)\right) + \rho\right) + \frac{1}{\gamma_1}\tilde{c}_1 \dot{\tilde{c}}_1 \\
&\quad \le c_1 \|s\| \|\ddot{\xi}_r\| - \hat{c}_1 \|s\| \|\ddot{\xi}_r\| + \tilde{c}_1 \left[\frac{\alpha_1}{\gamma_1}\hat{c}_1 - \|s\| \|\ddot{\xi}_r\|\right] \\
&\quad = \frac{\alpha_1}{\gamma_1}\tilde{c}_1 \hat{c}_1 = -\frac{\alpha_1}{\gamma_1}\left(\hat{c}_1 - \frac{1}{2}c_1\right)^2 + \frac{\alpha_1}{4\gamma_1}c_1^2
\end{aligned} \tag{6.76}$$

Considering (6.59), (6.60), and (6.65), (6.66), and Lemma 6.10, the third right-hand term of (6.75) is bounded by

$$
\begin{aligned}
& s^T\big[\mathcal{B}_1(u_{22}+u_{23})-(\mathcal{V}-\mathcal{V}_0)\dot{\xi}_r\big]+\sum_{i=2}^{3}\frac{1}{\gamma_i}\tilde{c}_i\dot{\tilde{c}}_i \\
& \le c_2\|s\|\|\dot{\xi}_r\|+c_3\|s\|\|\dot{q}\|\|\dot{\xi}_r\|-\sum_{i=2}^{3}\|s\|\big(\ln\big(\cosh(\hat{\Phi}_i)\big)+\rho\big) \\
& \quad +\sum_{i=2}^{3}\frac{\alpha_i}{\gamma_i}\tilde{c}_i\dot{\tilde{c}}_i \\
& \le c_2\|s\|\|\dot{\xi}_r\|+c_3\|s\|\|\dot{q}\|\|\dot{\xi}_r\|-\hat{c}_2\|s\|\|\dot{\xi}_r\|-\hat{c}_3\|s\|\|\dot{q}\|\|\dot{\xi}_r\| \\
& \quad +\tilde{c}_2\left[\frac{\alpha_2}{\gamma_2}\hat{c}_2-\|s\|\|\dot{\xi}_r\|\right]+\tilde{c}_3\left[\frac{\alpha_3}{\gamma_3}\hat{c}_3-\|s\|\|\dot{q}\|\|\dot{\xi}_r\|\right] \\
& =\sum_{i=2}^{3}\frac{\alpha_i}{\gamma_i}\tilde{c}_i\hat{c}_i=-\sum_{i=2}^{3}\frac{\alpha_i}{\gamma_i}\left(\hat{c}_i-\frac{1}{2}c_i\right)^2+\frac{\alpha_i}{4\gamma_i}c_i^2
\end{aligned}
\tag{6.77}
$$

Similarly, considering (6.61), (6.62) and (6.67), (6.68) and Lemma 6.10, the fourth right-hand term of (6.75) is bounded by

$$
\begin{aligned}
& s^T\big[\mathcal{B}_1(u_{24}+u_{25})-(\mathcal{D}-\mathcal{D}_0)\big]+\sum_{i=4}^{5}\frac{1}{\gamma_i}\tilde{c}_i\dot{\tilde{c}}_i \\
& \le c_4\|s\|+c_5\|s\|\|\dot{q}\|\|\dot{\xi}\|-\sum_{i=4}^{5}\|s\|\big(\ln\big(\cosh(\hat{\Phi}_i)\big)+\rho\big) \\
& \quad +\sum_{i=4}^{5}\frac{1}{\gamma_i}\tilde{c}_i\dot{\tilde{c}}_i \\
& \le c_4\|s\|+c_5\|s\|\|\dot{q}\|-\hat{c}_4\|s\|-\hat{c}_5\|s\|\|\dot{q}\| \\
& \quad +\tilde{c}_4\left[\frac{\alpha_4}{\gamma_4}\hat{c}_4-\|s\|\right]+\tilde{c}_5\left[\frac{\alpha_5}{\gamma_5}\hat{c}_5-\|s\|\|\dot{q}\|\right] \\
& =\sum_{i=4}^{5}\frac{\alpha_i}{\gamma_i}\tilde{c}_i\hat{c}_i=-\sum_{i=4}^{5}\frac{\alpha_i}{\gamma_i}\left(\hat{c}_i-\frac{1}{2}c_i\right)^2+\frac{\alpha_i}{4\gamma_i}c_i^2
\end{aligned}
\tag{6.78}
$$

Similarly, considering (6.63) and (6.69) and Lemma 6.10, the fifth right-hand term of (6.75) is bounded by

$$
\begin{aligned}
& s^T\big[\mathcal{B}_1u_{26}+(\mathcal{B}_1-\mathcal{B}_{10})u_1\big]+\frac{1}{\gamma_6}\tilde{c}_6\dot{\tilde{c}}_6 \\
& \le c_6\|s\|\|u_1\|-\hat{c}_6\|s\|\|u_1\|+\tilde{c}_6\left[\frac{\alpha_6}{\gamma_6}\hat{c}_6-\|s\|\|u_1\|\right] \\
& =\frac{\alpha_6}{\gamma_6}\tilde{c}_6\hat{c}_6=-\frac{\alpha_6}{\gamma_6}\left(\hat{c}_6-\frac{1}{2}c_6\right)^2+\frac{\alpha_6}{4\gamma_6}c_6^2
\end{aligned}
\tag{6.79}
$$

Combining (6.76)–(6.79), we could obtain

$$\dot{\mathbf{V}}_1 \leq -\lambda_{\min}(K_p)\|s\|^2 - \sum_{i=1}^{6} \frac{\alpha_i}{\gamma_i}\left(\hat{c}_i - \frac{1}{2}c_i\right)^2 + \sum_{i=1}^{6} \frac{\alpha_i}{4\gamma_i}c_i^2 \tag{6.80}$$

Therefore,

$$\dot{\mathbf{V}}_1 \leq -\lambda_{\min}(K_p)\|s\|^2 + \sum_{i=1}^{6} \frac{\alpha_i}{4\gamma_i}c_i^2 \tag{6.81}$$

Considering (6.70) and (6.71), $\sum_{i=1}^{6} \frac{\alpha_i}{4\gamma_i}c_i^2$ is bounded, there exists $T > t_1$, $\sum_{i=1}^{6} \frac{\alpha_i}{4\gamma_i}c_i^2 \leq \varrho_1$ with the finite constant ϱ_1, when $\|s\| \geq \sqrt{\frac{\varrho_1}{\lambda_{\min}(K_p)}}$, then $\dot{\mathbf{V}}_1 \leq 0$. From all the above, s converges to a small set Ω: $\|s\| \leq \sqrt{\frac{\varrho_1}{\lambda_{\min}(K_p)}}$ containing the origin as $T \to \infty$. Integrating both sides of (6.81) gives $\mathbf{V}_1(t) - \mathbf{V}_1(0) \leq -\int_0^T \lambda_{\min}(K_p)\|s\|^2\,dt + \int_0^T \sum_{i=1}^{6} \frac{\alpha_i}{4\gamma_i}c_i^2\,dt$. Since c_i and γ_i are constants, moreover, $\int_0^\infty \alpha_i\,dt = \varrho_{i\alpha}$, we can rewrite (6.81) as $\mathbf{V}_1(t) - \mathbf{V}_1(0) \leq -\int_0^T \lambda_{\min}(K_p) \times \|s\|^2\,dt + \sum_{i=1}^{6} \frac{\varrho_{i\alpha}}{4\gamma_i}c_i^2 < \infty$. Thus $\mathbf{V}_1$ is bounded, which implies that $s \in L_\infty$. From the above equation, we have $\int_0^T \lambda_{\min}(K_p)\|s\|^2 dt \leq \mathbf{V}_1(0) - \mathbf{V}_1(t) + \sum_{i=1}^{6} \frac{\varrho_{i\alpha}}{4\gamma_i}c_i^2$, which leads to $s \in L_2$. From $s = \dot{e}_\xi + \rho_1 e_\xi$, it can be obtained that $e_\xi, \dot{e}_\xi \in L_\infty$. As we have established $e_\xi, \dot{e}_\xi \in L_\infty$, from Assumption 5.2, we conclude that $\xi, \dot{\xi}_r \in L_\infty$.

Therefore, all the signals on the right-hand side of (6.54) are bounded, and we can conclude that $\dot{s}$ and therefore $\ddot{\xi}$ is bounded. Thus, $s \to 0$ as $t \to \infty$ can be obtained. Consequently, we have $e_\xi \to 0$, $\dot{e}_\xi \to 0$ as $t \to \infty$. It follows that $e_\xi, \dot{e}_\xi \to 0$ as $t \to \infty$.

6.3.3 Stability Analysis of the z_1-Subsystem

Finally, for system (6.48)–(6.49) under control laws (6.55), apparently, the z_1-subsystem (6.47) can be rewritten as

$$\dot{\varphi} = f(\nu, \varphi, \mathcal{U}) \tag{6.82}$$

where $\varphi = [z_1, \dot{z}_1]^T$, $\nu = [\xi^T, \dot{\xi}^T]^T = [z_3, z_2, \dot{z}_3, \dot{z}_2]^T$.

Assumption 6.6 From (6.48) and (6.49), the reference signal satisfies Assumption 5.2, and the following functions are Lipschitz in γ, i.e., there exists Lipschitz positive constants $L_{1\gamma}$, $L_{2\gamma}$, and L_{1f}, L_{2f}, such that

$$\|v_1 + g_1 + d_1\| \leq L_{1\gamma}\|\gamma\| + L_{1f} \tag{6.83}$$

$$\|f + k\| \leq L_{2\gamma}\|\gamma\| + L_{2f} \tag{6.84}$$

where $\gamma = [z_3, z_2, \dot{z}_3, \dot{z}_2]^T$, moreover, from the stability analysis of ζ_2 and ζ_3 subsystems, γ converges to a small neighborhood of $\gamma_d = [z_{3d}, z_{2d}, \dot{z}_{3d}, \dot{z}_{2d}]^T$.

Remark 6.12 Under the stability of z_2 and z_3 subsystems, and considering (6.52), we have $\|\gamma - \gamma_d\| \le \varsigma_1$, it is easy to obtain $\|\gamma\| \le \|\gamma_d\| + \varsigma_1$, and similarly, let $\mu = [\ddot{z}_3, \ddot{z}_2]^T$, and $\mu_d = [\ddot{z}_{3d}, \ddot{z}_{2d}]^T$, $\|\mu\| \le \|\mu_d\| + \varsigma_2$, where ς_1 and ς_2 are small bounded errors.

Lemma 6.13 *The z_1-subsystem* (6.47), *if z_2-subsystem and z_3-subsystem are stable, is globally asymptotically stable, too.*

Proof From (6.47), (6.48), and (6.49), we choose the following function as

$$\mathbf{V}_2 = \mathbf{V}_1 + \ln\big(\cosh(\dot{z}_1)\big) \tag{6.85}$$

Differentiating (6.85) along (6.47) gives

$$\begin{aligned}\dot{\mathbf{V}}_2 &= \dot{\mathbf{V}}_1 + \tanh(\dot{z}_1)\ddot{z}_1 \\ &= \dot{\mathbf{V}}_1 + \tanh(\dot{z}_1)m_{11}^{-1}(\tau_1 - v_1 - g_1 - d_1 \\ &\quad - m_{12}\ddot{z}_2 - m_{13}\ddot{z}_3)\end{aligned} \tag{6.86}$$

From (6.49), we have

$$\tau_1 = -m_{11}m_{31}^{-1}(l\ddot{z}_2 + j\ddot{z}_3 + f + k) \tag{6.87}$$

Integrating (6.87) into (6.86), we have

$$\begin{aligned}\dot{\mathbf{V}}_2 &= \dot{\mathbf{V}}_1 + \tanh(\dot{z}_1)\big(-m_{31}^{-1}(l\ddot{z}_2 + j\ddot{z}_3 + f + k) \\ &\quad + m_{11}^{-1}(-v_1 - g_1 - d_1 - m_{12}\ddot{z}_2 - m_{13}\ddot{z}_3)\big) \\ &= \dot{\mathbf{V}}_1 - \begin{bmatrix} \tanh(\dot{z}_1)(m_{31}^{-1}l + m_{11}^{-1}m_{12}) \\ \tanh(\dot{z}_1)(m_{31}^{-1}j + m_{11}^{-1}m_{13}) \end{bmatrix}^T \begin{bmatrix} \ddot{z}_2 \\ \ddot{z}_3 \end{bmatrix} \\ &\quad - \tanh(\dot{z}_1)(f + k) \\ &\quad - \tanh(\dot{z}_1)m_{11}^{-1}(v_1 + g_1 + d_1)\end{aligned}$$

Since $\|\tanh(\dot{z}_1)\| \le 1$, m_{31}^{-1}, m_{12} and m_{13} are all bounded. Let

$$\left\| \begin{bmatrix} m_{31}^{-1}l + m_{11}^{-1}m_{12} \\ m_{31}^{-1}j + m_{11}^{-1}m_{13} \end{bmatrix}^T \right\| \le \eta_1 \tag{6.88}$$

and $\|m_{11}^{-1}\| \le \eta_2$, where η_1 and η_2 are bounded constants. Considering Assumption 6.6 and Remark 6.12, we have

$$\begin{aligned}\dot{\mathbf{V}}_2 &\le -\frac{1}{2}\lambda_{\min}(K_p)\|s\|^2 + \sum_{i=1}^{6}\frac{\alpha_i}{4\gamma_i}c_i^2 \\ &\quad + \eta_1\big(\|\mu_d\| + \varsigma_2\big) + L_{2\gamma}\big(\|\gamma_d\| + \varsigma_2\big) + L_{2f} \\ &\quad + \eta_2\big(L_{1\gamma}\big(\|\gamma_d\| + \varsigma_1\big) + L_{1f}\big)\end{aligned} \tag{6.89}$$

Let $\Theta = \sum_{i=1}^{6} \frac{\alpha_i}{4\gamma_i}\rho_i^2 + \eta_1(\|\mu_d\| + \varsigma_2) + L_{2\gamma}(\|\gamma_d\| + \varsigma_2) + L_{2f} + \eta_2(L_{1\gamma}(\|\gamma_d\| + \varsigma_1) + L_{1f})$ and it is apparently bounded positive, we have $\dot{\mathbf{V}}_2 \leq 0$, when $\|s\| \geq \sqrt{\frac{2\Theta}{\lambda_{\min}(K_p)}}$, we can choose the properly K_p such that s can be arbitrarily small. Therefore, we can obtain the internal dynamics is stable with respect to the output $\dot{z}_1$. Therefore, the z_1-subsystem (6.47) is globally asymptotically stable. □

Theorem 6.14 *Consider the system* (6.48)–(6.49) *with Assumption* 6.5, *under the action of control laws* (6.55). *For compact set* Ω, *where* $(z_2(0), z_3(0), \dot{z}_2(0), \dot{z}_3(0)) \in \Omega$, *the tracking errors* s *converges to the compact set* Ω *defined in* (6.52), *and all the signals in the closed loop system are bounded.*

Proof From the results (6.52), it is clear that the tracking error s converges to the compact set Ω defined by (6.52). From Lemma 2.48, we can know e_ξ, $\dot{e}_\xi$ are also bounded. From the boundedness of z_{2d}, z_{3d} in Assumption 6.5, we know that z_2, z_3 are bounded. Since $\dot{z}_{2d}, \dot{z}_{3d}$ are also bounded, it follows that $\dot{z}_2, \dot{z}_3$ are bounded. From Lemma 6.13, we know that the z_1-subsystem (6.48) is stable, and $z_1, \dot{z}_1$ are bounded. This completes the proof. □

6.3.4 Force Control

The force control input τ_b is designed as

$$\tau_b = \hat{c}_1 Z R^{+^T} \ddot{z}_d + \lambda_d - K_f e_\lambda \tag{6.90}$$

where $e_\lambda = \lambda - \lambda_d$, K_f is a constant matrix of proportional control feedback gains. Substituting the control (6.55) and (6.90) into the reduced order dynamics (6.45) yields

$$\begin{aligned}(I + K_f)e_\lambda &= Z\big[\big(M(q)\dot{R}(q) + C(q,\dot{q})R(q)\dot{z}\big) + G + F - \tau_a\big] + \tau_b \\ &= -ZR^{+^T} M_1 \ddot{z} + \hat{c}_1 Z R^{+^T} M_1 \ddot{z}_d \end{aligned} \tag{6.91}$$

Since $\ddot{z}_2 \to \ddot{z}_{2d}$, $\ddot{z}_3 \to \ddot{z}_{3d}$, $\ddot{z}_1$ is bounded from Theorem 6.14, $-ZR^{+^T} M_1 \ddot{z} + \hat{c}_1 Z R^{+^T} M_1 \ddot{z}_d$ is also bounded, therefore, the size of e_λ can be adjusted by choosing the proper gain matrix K_f. Since s, z, $\dot{z}$, ξ_r, $\dot{\xi}_r$, and e_λ are all bounded, it is easy to conclude that τ is bounded from (6.55) and (6.90).

6.3.5 Simulation Studies

Let us consider a WIP as shown in Fig. 3.1. In the simulation, we assume the parameters $I_w = 0.5\ \mathrm{kg\,m^2}$, $M_w = 0.2$ kg, $I_m = 1.0\ \mathrm{kg\,m^2}$, $m = 10.0$ kg, $l = 1.0$ m, $d = 1.0$ m, $r = 0.5$ m, $\kappa_1 = \kappa_2 = \kappa_3 = 1.0$, $z(0) = [0, -0.2, \pi/180]^T$,

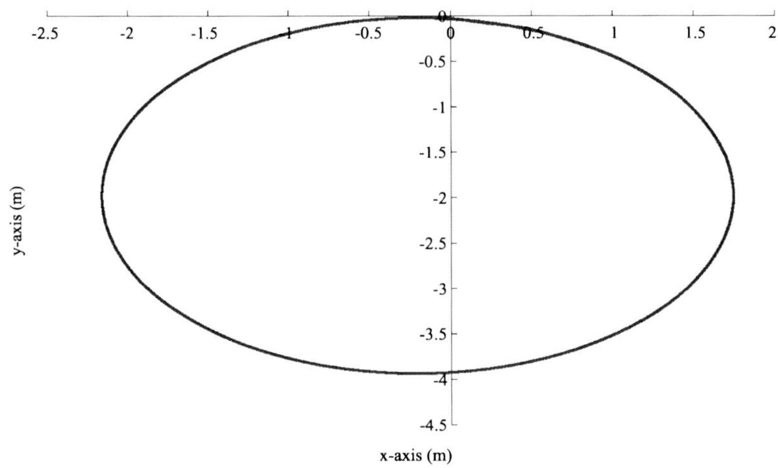

Fig. 6.9 The produced trajectory

$\dot{z}(0) = [0.1, 0.0, 0.0]^T$. The desired trajectories are chosen as $\theta_d = -0.05t$ rad, $\alpha_d = 0.0$ rad, initial velocity is 0.1 m/s. The external disturbances are set as $1.0\sin(t)$ and $1.0\cos(t)$. The desired nonholonomic constraint force is set as $10N$. The system state is observed through the noisy linear measurement channel, zero-mean Gaussian noises are added to the state information. All noises are assumed to be mutually independent. The noises have variances corresponding to a 5 % noise to signal radio. The design parameters of the above controllers are: $K_p = \text{diag}[10.0, 800.0]$, $\rho_1 = \text{diag}[1.0, 5.0]$, $\gamma_i = 0.5$, $\alpha_i = \delta = 1/(t+1)^2$, $B_{10} = [\{1.0, 0\}; \{1.0, 1.0\}]$, $b = 0.9$, $[\hat{c}_1(0), \ldots, \hat{c}_6(0)]^T = [1.0, 1.0, 10.0, 20.0, 10.0, 10.0]^T$, $K_f = 10.0$.

The trajectory tracking by the proposed control approach is shown in Fig. 6.9, and direction angle, the tilt angles for the dynamic balance and the stable velocities are shown in Figs. 6.10, 6.11, 6.12 and 6.13, respectively. The input torques are shown respectively in Fig. 6.14. The nonholonomic constraint force error with the desired force is shown in Fig. 6.15. Therefore, the slipping and slippage can not happen since the friction force converges within the region of $10N$. From these figures, even if without the prior knowledge of the system, we can obtain good performance by the proposed adaptive robust control. Robust adaptive approach is tolerant of modeling errors because accurate modeling of WIP systems dynamics is difficult, and time-consuming and uncertain. The presence of parametric errors is a common problem since the identification of dynamic parameters is error-prone. Robust adaptive control presented in this chapter is not susceptible to this problem, since the unknown parameters are learned during the WIP systems operation in actual conditions. Although the parametric uncertainties and the external disturbances are both introduced into the simulation model, the motion/force control performance of system, under the proposed control, is not degraded. The simulation results demonstrate the effectiveness of the proposed adap-

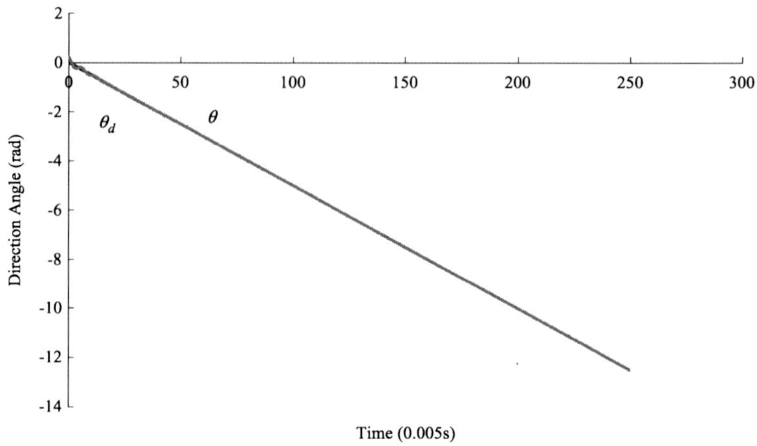

Fig. 6.10 Tracking the direction angle

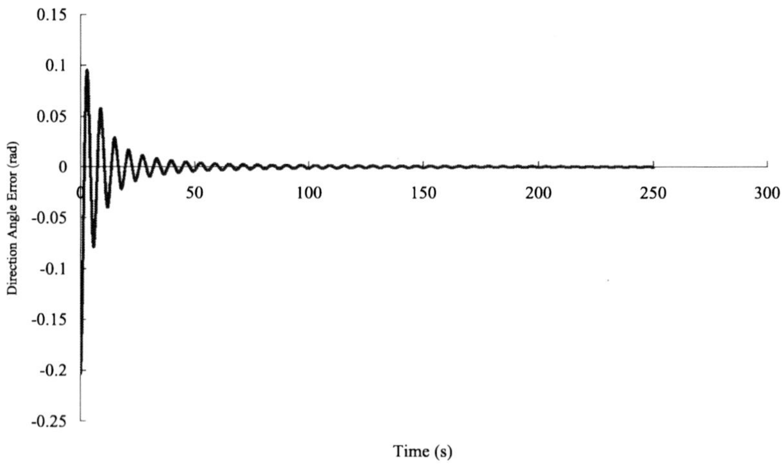

Fig. 6.11 The direction angle error

tive control in the presence of unknown nonlinear dynamic system and environment.

6.3.6 Conclusions

In this chapter, firstly, adaptive robust control design is carried out for dynamic balance and motion tracking of wheeled inverted pendulum. The proposed control considers the presence of unmodeled dynamics, or parametric/functional uncertainties. Comparison simulation results have been conducted to demonstrate that the

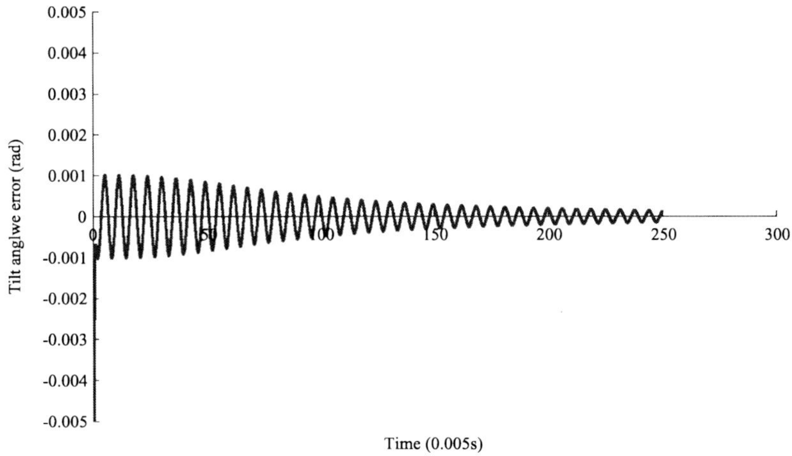

Fig. 6.12 The tilt angle tracking error

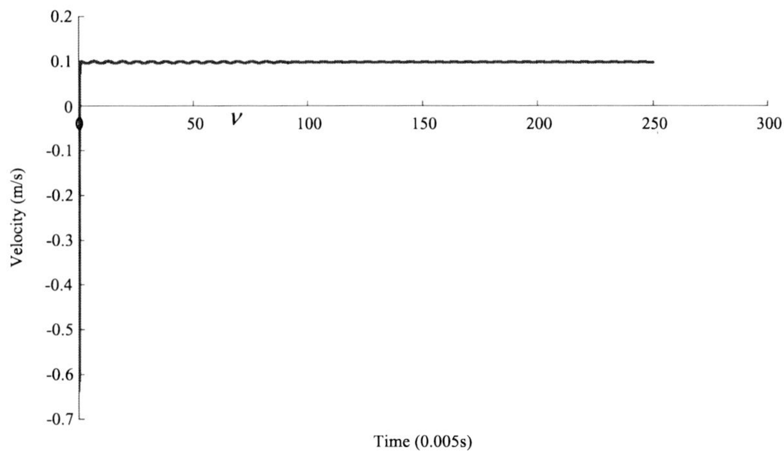

Fig. 6.13 The stable velocity

adaptive robust control is more capable at tracking reference signals satisfactorily, with all the closed loop signals uniformly bounded.

Moreover, considering the nonholonomic force control, robust adaptive motion/force control design is investigated for WIP systems, in the presence of unmodeled dynamics, parametric/functional uncertainties and the nonholonomic constraint force. The control is mathematically shown to guarantee semi-globally uniformly bounded stability, and the steady state compact sets to which the closed loop error signals converge are derived. The size of compact sets for motion and force can be made small through appropriate choice of control design parameters. Simulation

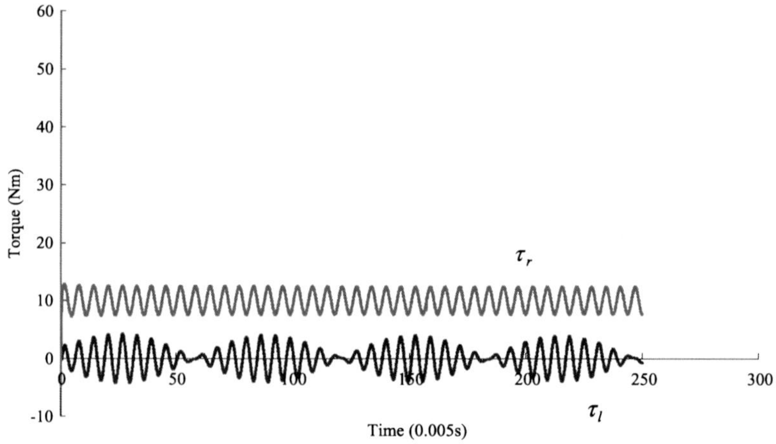

Fig. 6.14 Input torques

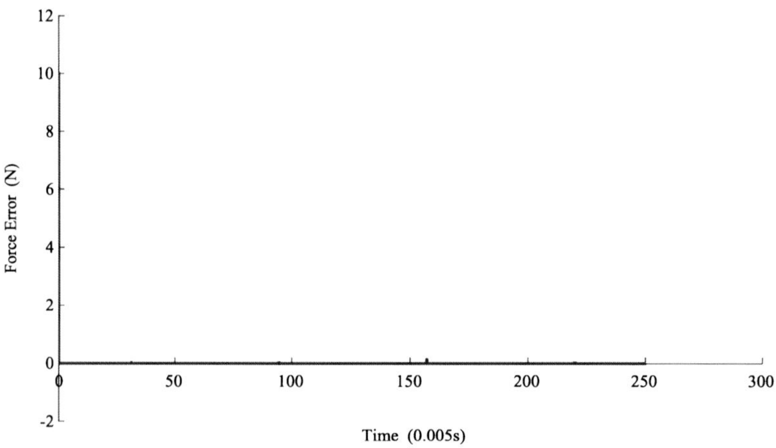

Fig. 6.15 The nonholonomic constraint force error with $\lambda_d = 10N$

results demonstrate that the system is able to track reference signals satisfactorily, with all closed loop signals uniformly bounded.

Chapter 7
Intelligent Control

7.1 Introduction

In this chapter, three model-free intelligent control approaches are proposed for the dynamic balance and motion control of wheeled inverted pendulums (WIP). Firstly, LS-SVM (Least Squares Support Vector Machine) approximation is considered for wheeled inverted pendulums (WIP) in the presence of parametric and functional dynamics uncertainties. Based on Lyapunov synthesis, the proposed control mechanisms use the advantage of LS-SVM combined with on-line parameters estimation strategy in order to have an efficient approximation. Under the controller designed, we can ensure that the outputs of the system track the given bounded reference signals within a small neighborhood of zero, and guarantee semi-global uniform boundedness of all the closed loop signals. Simulation results are presented to verify the effectiveness of the proposed control.

Second, adaptive fuzzy logic control of dynamic balance and motion is investigated for wheeled inverted pendulums with parametric and functional uncertainties. The proposed adaptive fuzzy logic control based on physical properties of wheeled inverted pendulums makes use of a fuzzy logic engine and a systematic online adaptation mechanism to approximate the unknown dynamics. Based on Lyapunov synthesis, the fuzzy control ensures that the system outputs track the given bounded reference signals to within a small neighborhood of zero, and guarantees semi-global uniform boundedness of all closed loop signals. The effectiveness of the proposed control is verified through extensive simulations.

Third, we propose a novel autonomous vehicle shown in Fig. 7.1, WIP-car based on the widely used WIP model. This novel transportation system is composed of a wheeled inverted pendulum system, a driven chair, an acceleration pedal and a deceleration pedal, which are used to drive the chair forward or backward such that the car can be accelerated or decelerated. Since the human within the WIP-car may be difficult to predict and the human's dynamics parameters do not obtain beforehand, output feedback adaptive neural network (NN) control incorporating a linear dynamic compensator is developed for approximating the inverse dynamics so as to achieve stable dynamic balance and tracking desired trajectories. Simulation

Z. Li et al., *Advanced Control of Wheeled Inverted Pendulum Systems*,
DOI 10.1007/978-1-4471-2963-9_7,

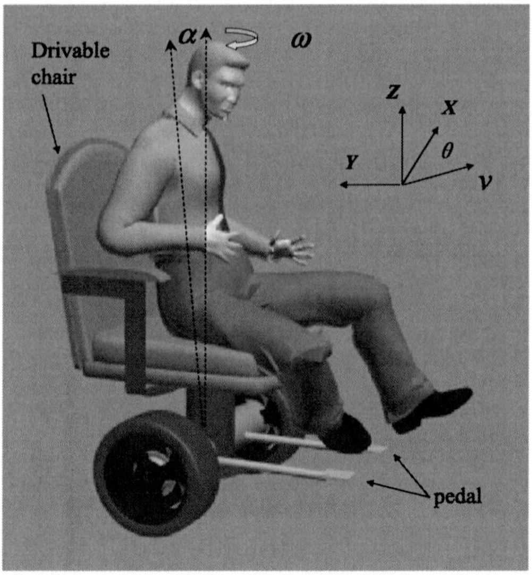

Fig. 7.1 A transportation system based on wheeled inverted pendulum

results demonstrate that the system is able to track reference signals satisfactorily and keep the stable balance.

7.2 SVM Control

As mentioned, most of the previously cited works require reasonably precise knowledge of the dynamic models in order to achieve satisfactory performance. But in practice, since the human within the WIP-car may be difficult to predict and the human's dynamics parameters do not obtain beforehand, it is not possible to obtain a perfect, or even reasonably accurate, dynamic model of WIP-car system. In the dynamic balance and motion control of WIP systems, an important concern is how to deal with unknown perturbations to the nominal model, in the form of parametric and functional uncertainties, unmodeled dynamics, and disturbances from the environment. Wheeled inverted pendulum control is characterized by unstable balance and unmodeled dynamics, and subject to time varying external disturbances, which are generally difficult to model accurately.

Therefore, traditional model-based control may not be the ideal approach since it generally works best when the dynamic model is known exactly. The presence of uncertainties and disturbances could disrupt the function of the traditional model-based feedback control and leads to the unstable balance. How to handle with the parametric and functional uncertainties, unmodeled dynamics, and disturbances from the environment is one of the important issues in the control of wheeled inverted pendulum.

Least squares support vector machine (LS-SVM) has been proposed for solving nonlinear function estimation problems [116, 118]. LS-SVM takes equality instead of inequality constraints of SVM in the problem formulation such that LS-SVM is easy to train. The previous works about SVM learning approaches have been proposed in [93, 122, 131, 134] for modeling nonlinear system and SVM-based nonlinear controls, however, those works lack the definite stability proof of the closed-loop system using SVM approaches. In [18], the control design on the SVM is developed to achieve accurate haptic display, the approximation model of friction is established off-line through SVM learning for online feed forward friction compensation. However, in practical control applications, it is desirable to have systematic methods to ensure on-line stability, robustness, and performance of the overall system in the unknown environments beforehand. Although the on-line adaptive fuzzy approximation is a well-known technique, for the real-time mechanical system, it is difficult to choose a satisfying fuzzy rule number, the performance would be degraded with less rule number, on the other hand, the "explosion" of rule number would bring great trouble to the limit computation resource. However, Support Vector Machines approach could avoid this problem. While to our best knowledge, there are few works dealing with SVM-based control proposed for the wheeled inverted pendulum up to now.

The previous works, which utilized the SVMs approximation for system modeling, assume beforehand that the state variables of the system are bounded in a compact set since the approximated data can be provided off-line, while for the stability of system by on-line SVM approximation, the system errors can not be predicted. It is not reasonable to assume that the bounds within the compact set beforehand. The direct SVMs approximation is not valid, while many applications using off-line SVM approximation (all data obtained is bounded) limit the extension of SVMs for the control of nonlinear dynamic systems, for example wheeled inverted pendulums.

In this section, by discovering and utilizing the unique physical property of the wheeled inverted pendulum, we could decouple the system to simplify the model, such that we can design the control easily. Moreover, we make full use of the physical properties of the wheeled inverted pendulum and then design on-line LS-SVM based control by sliding window to accommodate the presence of parametric and functional uncertainties in the dynamics of wheeled inverted pendulums. The developed SVM-based control combing the physical properties of the system is original.

7.2.1 Preliminaries

From Chap. 5, consider the following wheeled inverted pendulum dynamics described by Lagrangian formulation:

$$\begin{aligned}&\begin{bmatrix} D_v & D_{v\alpha} \\ D_{\alpha v} & D_\alpha \end{bmatrix}\begin{bmatrix} \ddot{q}_v \\ \ddot{\alpha} \end{bmatrix}+\begin{bmatrix} C_v & C_{v\alpha} \\ C_{\alpha v} & C_\alpha \end{bmatrix}\begin{bmatrix} \dot{q}_v \\ \dot{\alpha} \end{bmatrix}\\ &\quad+\begin{bmatrix} F_v \\ F_\alpha \end{bmatrix}+\begin{bmatrix} G_v \\ G_\alpha \end{bmatrix}+\begin{bmatrix} d_v \\ d_\alpha \end{bmatrix}\\ &\quad=\begin{bmatrix} B_v\tau_v \\ 0 \end{bmatrix}+\begin{bmatrix} J_v^T\lambda \\ 0 \end{bmatrix} \end{aligned} \tag{7.1}$$

Assumption 7.1 The nonholonomic constraints on the wheeled inverted pendulum is known and not to be violated.

Remark 7.1 In actual implementation, we can adopt the methods of producing enough friction between the wheels of the mobile platform and the ground such that the assumption of nonholonomic constraints holds.

The vehicle is subjected to nonholonomic constraints J_υ, which can be expressed as

$$J_\upsilon \dot{q}_\upsilon = 0 \tag{7.2}$$

We could find a set of smooth and linearly independent vector fields $S_1(q)$ and $S_2(q)$ constituting the matrix $S = [S_1(q), S_2(q)] \in R^{3\times 2}$ with full rank $S^T S$, which in the local coordinates satisfy the following relation

$$S^T J_\upsilon^T = 0 \tag{7.3}$$

Constraint equation (7.2) implies the existence of vector $\dot{\chi} = [\omega, \upsilon]^T \in \mathbb{R}^2$ with ω representing the angular velocity of the platform denoted in Fig. 3.1 and υ denoting the forward velocity of mobile platform such that

$$\dot{q}_\upsilon = S(q)\dot{\chi} \tag{7.4}$$

Consider the derivative of (7.4), define new variables $\dot{\zeta} = [\dot{\chi}, \dot{\alpha}]^T = [\dot{\zeta}_1, \dot{\zeta}_2, \dot{\zeta}_3]^T = [\omega, \upsilon, \dot{\alpha}]^T$, and multiply diag$[S^T, I]$ by both sides of (7.1) to eliminate J_υ^T, the dynamics of wheeled inverted pendulum can be expressed as

$$\begin{aligned}
&\begin{bmatrix} S^T D_\upsilon S & S^T D_{\upsilon\alpha} \\ D_{\alpha\upsilon} S & D_\alpha \end{bmatrix} \begin{bmatrix} \ddot{\chi} \\ \ddot{\alpha} \end{bmatrix} + \begin{bmatrix} S^T D_\upsilon \dot{S} + S^T C_\upsilon S & S^T C_{\upsilon\alpha} \\ D_{\alpha\upsilon}\dot{S} + C_{\alpha\upsilon} S & C_\alpha \end{bmatrix} \begin{bmatrix} \dot{\chi} \\ \dot{\alpha} \end{bmatrix} \\
&\quad + \begin{bmatrix} S^T F_\upsilon \\ F_\alpha \end{bmatrix} + \begin{bmatrix} S^T G_\upsilon \\ G_\alpha \end{bmatrix} + \begin{bmatrix} S^T d_\upsilon \\ d_\alpha \end{bmatrix} \\
&= \begin{bmatrix} S^T B_\upsilon \tau_\upsilon \\ 0 \end{bmatrix}
\end{aligned} \tag{7.5}$$

Assumption 7.2 [23, 37] The desired trajectories $\zeta_{1d}(t)$ and $\zeta_{3d}(t)$ and their time derivatives up to the 3rd order are continuously differentiable and bounded for all $t \geq 0$.

Remark 7.2 Since we can plan and design the desired trajectory for $\zeta_{1d}(t)$ and $\zeta_{3d}(t)$ before implementing control, it is reasonable and feasible that we give the trajectory satisfying Assumption 7.2.

The objective of the control can be specified as: given desired trajectories $\zeta_{1d}(t)$ and $\zeta_{3d}(t)$, we are to determine a control law such that for any $(\zeta_j(0), \dot{\zeta}_j(0)) \in \Omega$, $\zeta_j, \dot{\zeta}_j$ converge to a manifold Ω_d specified as Ω where

$$\Omega_d = \{(\zeta_j, \dot{\zeta}_j) \mid |\zeta_j - \zeta_{jd}| \leq \epsilon_{j1}, \ |\dot{\zeta}_j - \dot{\zeta}_{jd}| \leq \epsilon_{j2}\} \tag{7.6}$$

where $\epsilon_{ji} > 0$, $i = 1, 2$, $j = 1, 3$. Ideally, ϵ_{ji} should be the threshold of tolerable noise. At the same time, all the closed loop signals are to be kept bounded. The variables ζ_1 and ζ_3 can be thought as output equation of the system, the choice of ζ is, as an example, illustrated in the next section.

Assumption 7.3 $F_1(\dot{\zeta}) = [f_1, f_2, f_3]^T$ is an independent vector, for simplification and convenience, such as $F_1(\dot{\zeta}) = \rho\dot{\zeta}$ with the friction coefficient ρ, which is diagonal positive definite.

Remark 7.3 The friction model can be found in [73], and F_1 is a function including $\dot{\zeta}$, for simplification, we give the above assumption.

7.2.2 Reduced Dynamics and Physical Properties

By exploiting the physical properties of the WIP embedded in the dynamics of (7.5) are given by

$$D_1 = \begin{bmatrix} S^T D_v S & S^T D_{v\alpha} \\ D_{\alpha v} S & D_\alpha \end{bmatrix} = \begin{bmatrix} D_{11}(\zeta_3) & 0 & 0 \\ 0 & D_{22} & D_{23}(\zeta_3) \\ 0 & D_{32}(\zeta_3) & D_{33} \end{bmatrix} \tag{7.7}$$

$$C_1 = \begin{bmatrix} S^T D_v \dot{S} + S^T C_v S & S^T C_{v\alpha} \\ D_{\alpha v}\dot{S} + C_{\alpha v} S & C_\alpha \end{bmatrix} = \begin{bmatrix} C_{11} & 0 & C_{13} \\ 0 & 0 & C_{23} \\ C_{31} & 0 & 0 \end{bmatrix} \tag{7.8}$$

$$\begin{bmatrix} S^T F_v \\ F_\alpha \end{bmatrix} = \begin{bmatrix} f_1(\dot{\zeta}_1) \\ f_2(\dot{\zeta}_2) \\ f_3(\dot{\zeta}_3) \end{bmatrix}, \qquad \begin{bmatrix} S^T d_v \\ d_\alpha \end{bmatrix} = \begin{bmatrix} d_1 & d_2 & d_3 \end{bmatrix}^T$$

$$\begin{bmatrix} S^T B_v \tau_v \\ 0 \end{bmatrix} = \begin{bmatrix} \tau_1 \\ \tau_2 \\ 0 \end{bmatrix}, \qquad \begin{bmatrix} S^T G_v \\ G_\alpha \end{bmatrix} = \begin{bmatrix} 0 \\ 0 \\ g_3(\zeta_3) \end{bmatrix}$$

where D_{22}, D_{33} are unknown constants, $D_{11}(\zeta_3)$, $D_{23}(\zeta_3)$, $D_{32}(\zeta_3)$, C_{11}, C_{13}, C_{23}, C_{31}, $f_1(\dot{\zeta}_1)$, $f_2(\dot{\zeta}_2)$, $f_3(\dot{\zeta}_3)$, $g_3(\zeta_3)$, and d_1, d_2, d_3 are unknown functions.

From (7.5), (7.7) and (7.8), we can obtain three subsystems: ζ_1-subsystem, ζ_2-subsystem, and ζ_3-subsystem, respectively, as follows

$$D_{11}(\zeta_3)\ddot{\zeta}_1 + C_{11}\dot{\zeta}_1 + C_{13}\dot{\zeta}_3 + f_1 + d_1 = \tau_1 \tag{7.9}$$

$$\frac{D_{22}D_{33} - D_{23}^2(\zeta_3)}{D_{33}}\ddot{\zeta}_2 - \frac{D_{23}(\zeta_3)}{D_{33}}\big(C_{31}\dot{\zeta}_1 + g_3(\zeta_3) + f_3(\dot{\zeta}_3) + d_3\big) + C_{23}\dot{\zeta}_3 + f_2(\dot{\zeta}_2) + d_2 - \tau_2 = 0 \tag{7.10}$$

$$\frac{D_{22}D_{33} - D_{23}^2(\zeta_3)}{D_{23}(\zeta_3)}\ddot{\zeta}_3 + \frac{D_{22}}{D_{23}(\zeta_3)}\big(C_{31}\dot{\zeta}_1 + g_3(\zeta_3) + f_3(\dot{\zeta}_3) + d_3\big) - C_{23}\dot{\zeta}_3 - f_2(\dot{\zeta}_2) - d_2 = -\tau_2 \tag{7.11}$$

Define the tracking errors and the filtered tracking errors as

$$e_j = \zeta_j - \zeta_{jd} \tag{7.12}$$

$$r_j = \dot{e}_j + \Lambda_j e_j \tag{7.13}$$

where Λ_j is a positive number, and $j = 1, 3$. Therefore, we could study the stability of e_j and $\dot{e}_j$ by the properties of r_j. Moreover, it is easy to have the following computable signals:

$$\dot{\zeta}_{jr} = \dot{\zeta}_{jd} - \Lambda_j e_j \tag{7.14}$$

$$\ddot{\zeta}_{jr} = \ddot{\zeta}_{jd} - \Lambda_j \dot{e}_j \tag{7.15}$$

7.2.3 LS-SVM Based Model Learning

Because the dynamics uncertainties of the system are usually hard to measure, such as friction forces and disturbances to the system, we need to estimate those uncertainties to design an effective controller. In this chapter, we make use of Support Vector Machine (SVM) for on-line nonlinear system identification.

LS-SVM

For SVM regression, the basic idea is to map the data to a higher dimensional feature space $\mathcal{H}$ (Reproducing kernel Hilbert space), via a nonlinear mapping ϕ, and then to do the linear regression in this space. Therefore, given a training set of l training samples $(x_1, y_1), \ldots, (x_l, y_l) \in \mathbb{R}^n \times \mathbb{R}$, we introduce a nonlinear mapping $\phi(\cdot) : \mathbb{R}^n \longrightarrow \mathcal{H} \in \mathbb{R}^h$, which maps the training samples to a new data set $(\phi_1(x), y_1), \ldots, (\phi_l(x), y_l)$. In ϵ-insensitive Support Vector Regression the goal is to estimate the following function

$$\hat{f}(x) = \langle \varpi, \phi(x) \rangle + b; \quad \omega \in \mathbb{R}^h,\ b \in \mathbb{R} \tag{7.16}$$

where ϖ and b are the coefficients, which are estimated by the risk function

$$R = \min_{\varpi, b, \mathcal{E}} \left(\frac{1}{2} \|\varpi\|^2 + c \frac{1}{2} \sum_{i=1}^{l} \mathcal{E}_i^2 \right), \quad \text{s.t.} \quad y_i - \hat{f}(x_i) = \mathcal{E}_i \tag{7.17}$$

where l is the number of the training samples and the constant $c > 0$ measure the trade-off between complexity and losses.

Construct a Lagrangian to solve the optimization problem:

$$\max_a \min_{\varpi, b} \left\{ L = \frac{1}{2} \varpi^T \varpi + \frac{1}{2} c \sum_{i=1}^{l} \mathcal{E}_i^2 - \sum_{i=1}^{l} a_i \left\{ y_i - \left[\varpi^T \phi(x_i) + b \right] - \mathcal{E}_i \right\} \right\} \tag{7.18}$$

According to Karush–Kuhn–Tucker optimization condition, we can seek the optimal solution and transform this optimization problem into a matrix function as:

$$\begin{bmatrix} 0 & 1 & \cdots & 1 \\ 1 & K(x_1, x_1) + \frac{1}{c} & \cdots & K(x_1, x_l) \\ \vdots & \vdots & \ddots & \vdots \\ 1 & K(x_l, x_1) & \cdots & K(x_l, x_l) + \frac{1}{c} \end{bmatrix} \begin{bmatrix} b \\ a_1 \\ \vdots \\ a_l \end{bmatrix} = [0 \quad y_1 \quad \ldots \quad y_l]^T \tag{7.19}$$

where $K(x_i, x_j) = \phi(x_i)^T \phi(x_j)$, $i, j = 1, \ldots, l$, it is a kernel function, which satisfies the Mercer's theorem.

The obtained nonlinear model is:

$$\hat{f}(x) = \sum_{i=1}^{l} a_i K(x_i, x_j) + b \tag{7.20}$$

where x_i and a_i are obtained from solving the set of linear (7.19) and x_j is the jth actual state vector, if the on-line sliding windows is introduced as next section, x_j is the actual state vector at time j. The data x_i is used as support vector data for the control signal [116].

On-line Training with Time Window

In order to have a proper performance of LS-SVM, we need to select as many samples as possible for training, however, the dimension of SVM will greatly increase in the process of on-line training. Based on the aim of designing a control which depends on the current state of the nonlinear dynamic system, the training data collected earlier might not suit for real time system, the large data set might lead to time consuming calculation. Therefore, sliding time window is constructed by l with selected sampling time interval, then sample data is collected orderly from current to past. Moreover, a new data sample is collected while the oldest data being dropped. We assume that the nearest data can more properly describe the feature of the system than the oldest data.

Theorem 7.4 *For any given continuous real function $f(x)$ on a compact set $U \in \mathbb{R}^n$ and arbitrary $\varepsilon > 0$, there exists an LS-SVM approximation $\hat{f}(x)$ formed by* (7.20) *such that*

$$\sup_{x \in U} \left| \hat{f}(x) - f(x) \right| < \varepsilon \tag{7.21}$$

Proof See [123]. □

Theorem 7.5 *For any given continuous real function $f(x)$ on a compact set, for a large enough length of sliding time window l combined with properly selected* *sampling time interval and any given* $\epsilon > 0$*, there exists an LS-SVM approximation function* $\hat{f}(x)$ *formed by* (7.20) *such that*

$$\sup_{t\in[T,T-l+1]} \left|\hat{f}(x) - f(x)\right| \leq \varepsilon \tag{7.22}$$

Proof Let $\{a_1, a_2, \ldots, a_l\}$ be the estimated weights of LS-SVM by minimizing $L(\cdot)$ given by (7.18). Let the input data be $(x_{T-l+1}, y_{T-l+1}), \ldots, (x_{T-1}, y_{T-1}), (x_T, y_T)$, here, T denotes the current time. It is easy to know that the length l and the selected sampling time interval determines the size of data samples. From (7.18), we know the regularization constant c determines the trade-off between the empirical risk and the model complexity. From Theorem 7.4 and [119], when the time interval is properly selected and l is sufficiently large, we can obtain that for any given $\epsilon > 0$, there exists an LS-SVM approximation function $\hat{f}(x)$ formed by (7.20) such that

$$\sup_{t\in[T,T-l+1]} \left|\hat{f}(x) - f(x)\right| \leq \varepsilon \tag{7.23}$$

□

Remark 7.6 For SVMs approximation, we know the computation cost of SVMs is directly proportional to the number of training data pair. However, in order to guarantee high accuracy, we need to choose a large sliding window presented in Theorem 7.5, for example, in the simulation presented later, we choose the sliding windows containing 200 sample data, the control performance is good. Since SVM system approximation ability only holds on a compact set, it is more rigorous to assume the states are within a sufficiently large compact set beforehand. Otherwise, the SVM system approximation will be violated. After the derivation, we can find know that if we choose the sliding window to be large enough for SVM system approximation to cover the sufficiently large compact set, the states are bounded in the set by themselves.

7.2.4 LS-SVM Based Control Design

In reality, dynamics of the system cannot be exactly known. In addition, external disturbances may also affect the performance of the system. In this section, we take both factors into consideration to develop an LS-SVM based control with adaptive law to deal with uncertainties as well as external disturbances.

ζ_1 and ζ_3-Subsystems Since $\dot{\zeta}_j = \dot{\zeta}_{jr} + r_j$ and $\ddot{\zeta}_j = \ddot{\zeta}_{jr} + \dot{r}_j$, $j = 1, 3$, let

$$\mu_1 = C_{11}\dot{\zeta}_{1r} + C_{13}\dot{\zeta}_3 + f_1(\dot{\zeta}_1) + d_1 + D_{11}(\zeta_3)\ddot{\zeta}_{1r} \tag{7.24}$$

$$\begin{aligned}\mu_3 = {} & \frac{D_{22}D_{33} - D_{23}^2(\zeta_3)}{D_{23}(\zeta_3)}\ddot{\zeta}_{3r} - \hat{C}_{23}r_3 - C_{23}\dot{\zeta}_{3r} \\ & + \frac{D_{22}}{D_{23}(\zeta_3)}(C_{31}\dot{\zeta}_1 + f_3 + g_3 + d_3) \\ & - (f_2 + d_2)\end{aligned} \tag{7.25}$$

where we decompose $C_{23} = \hat{C}_{23} - \tilde{C}_{23}$ such that

$$\frac{d}{dt}\left(\frac{D_{22}D_{33} - D_{23}^2(\zeta_3)}{D_{23}(\zeta_3)}\right) - 2\tilde{C}_{23} = 0 \tag{7.26}$$

Equations (7.9) and (7.11) become

$$D_{11}(\zeta_3)\dot{r}_1 + C_{11}r_1 = \tau_1 - \mu_1 \tag{7.27}$$

$$\frac{D_{22}D_{33} - D_{23}^2(\zeta_3)}{D_{23}(\zeta_3)}\dot{r}_3 + \tilde{C}_{23}r_3 = -\tau_2 - \mu_3 \tag{7.28}$$

Define now control new inputs as

$$u_1 = \tau_1 - \mu_1 \tag{7.29}$$

$$u_2 = -\tau_2 - \mu_3 \tag{7.30}$$

where u_1 and u_2 are auxiliary control inputs to be optimized later, therefore, the close loop system for ζ_1 and ζ_3 coupled system becomes

$$\mathcal{D}\dot{r} + \mathcal{C}r = \mathcal{U} \tag{7.31}$$

where

$$\mathcal{D} = \begin{bmatrix} D_{11}(\zeta_3) & 0 \\ 0 & \frac{D_{22}D_{33} - D_{23}^2(\zeta_3)}{D_{23}(\zeta_3)} \end{bmatrix}$$

$$\mathcal{C} = \begin{bmatrix} C_{11} & 0 \\ 0 & \tilde{C}_{23} \end{bmatrix}$$

$$r = \begin{bmatrix} r_1 & r_3 \end{bmatrix}^T, \qquad \mathcal{U} = \begin{bmatrix} u_1 & u_2 \end{bmatrix}^T$$

The following properties are useful for the stability.

Property 7.7 The inertia matrix $\mathcal{D}$ is positive definite and symmetric, from which we know that the terms $D_{11}(\zeta_3)$ and $\frac{D_{22}D_{33} - D_{23}^2(\zeta_3)}{D_{33}}$ are positive.

Property 7.8 The matrix $\dot{\mathcal{D}} - 2\mathcal{C}$ is skew-symmetric, and then we could obtain $x^T(\dot{\mathcal{D}} - 2\mathcal{C})x = 0$, where $x \in \mathbb{R}^2$.

From (7.13), we obtain the error dynamics as

$$\dot{e} = -\Lambda e + r \tag{7.32}$$

where $e = [e_1, e_3]^T$, $\Lambda = \text{diag}[\Lambda_j]$, and $r = [r_1, r_3]^T$.

We could build up the following augmented system as

$$\dot{X} = \begin{bmatrix} \dot{e} \\ \dot{r} \end{bmatrix} = \begin{bmatrix} -\Lambda & I \\ 0 & -\mathcal{D}^{-1}\mathcal{C} \end{bmatrix}\begin{bmatrix} e \\ r \end{bmatrix} + \begin{bmatrix} 0 \\ \mathcal{D}^{-1} \end{bmatrix}\mathcal{U} \tag{7.33}$$

which can be described by brief form

$$\dot{X} = A(\zeta, \dot{\zeta})X + B(\zeta)\mathcal{U} \tag{7.34}$$

with $A \in \mathbb{R}^{4\times 4}$, $B \in \mathbb{R}^{4\times 2}$, and $X \in \mathbb{R}^{4\times 1}$.

Lemma 7.9 *Consider the dynamics described by* (7.33). *Given a weighted matrix* $Q = Q^T > 0$, *if there exists a symmetric positive definite matrix* $K = K^T > 0$ *satisfying the following algebraic Riccati-like equation*

$$PA + A^T P - PBR^{-1}B^T P + \dot{P} + Q = 0 \tag{7.35}$$

for a gain matrix $R^{-1} = Q_{22}$ *and positive definite matrix* $Q = \begin{bmatrix} Q_{11} & Q_{12} \\ Q_{12}^T & Q_{22} \end{bmatrix}$, *and* $Q_{12} + Q_{21}^T < 0$, $P = \mathrm{diag}[K, \mathcal{D}]$. *The feedback control*

$$\mathcal{U} = -\frac{1}{2}R^{-1}B^T PX = -\frac{1}{2}R^{-1}r \tag{7.36}$$

guarantees that all the variables of the close loop system are bounded and the tracking performance is achieved.

Proof Let us choose a Lyapunov function as

$$\mathbb{V}_1 = \frac{1}{2}X^T PX \tag{7.37}$$

Taking the time derivative of V along the dynamics (7.33), we obtain

$$\begin{aligned}
\dot{\mathbb{V}}_1 &= \frac{1}{2}\big(\dot{X}^T PX + X^T \dot{P}X + X^T P\dot{X}\big) \\
&= \frac{1}{2}\big(X^T A^T(\zeta, \dot{\zeta}) + \mathcal{U}^T B^T(\zeta)\big)PX + \frac{1}{2}X^T \dot{P}X \\
&\quad + \frac{1}{2}X^T\big(PA(\zeta, \dot{\zeta})X + PB(\zeta)\mathcal{U}\big) \\
&= \frac{1}{2}X^T A^T(\zeta, \dot{\zeta})PX + \frac{1}{2}\mathcal{U}^T B^T(\zeta)PX + \frac{1}{2}X^T \dot{P}X \\
&\quad + \frac{1}{2}X^T PA(\zeta, \dot{\zeta})X + \frac{1}{2}X^T PB(\zeta)\mathcal{U} \\
&= \frac{1}{2}X^T\big(A^T(\zeta, \dot{\zeta})P + \dot{P} + PA(\zeta, \dot{\zeta})\big)X + X^T PB(\zeta)\mathcal{U}
\end{aligned}$$

Consider (7.35) and (7.36), we have

$$\dot{\mathbb{V}} = -\frac{1}{2}X^T \begin{bmatrix} Q_{11} & Q_{12} \\ Q_{12}^T & Q_{22} \end{bmatrix} X \le 0 \tag{7.38}$$

Therefore, using the feedback control (7.36) results in the controller nonlinear system

$$\dot{X} = \left(A(\zeta, \dot{\zeta}) - \frac{1}{2}B(\zeta)R^{-1}B^T(\zeta)P\right)X \tag{7.39}$$

being globally exponentially stable about the origin in $\mathbb{R}^2$, that is, e and r are bounded. □

Let $\hat{F}_1$ and $\hat{F}_3$ denote SVM estimation from (7.20) with on-line LS-SVM method, due to the approximation property of SVM and Theorem 7.9, such that

$$|\hat{F}_1 - \mu_1| \leq \varepsilon_1 \tag{7.40}$$

$$|\hat{F}_3 - \mu_3| \leq \varepsilon_3 \tag{7.41}$$

where ε_1 and ε_3 are bounded errors by Theorem 7.5. For simplicity, we assume a known upper limit

$$\|\varepsilon\| \leq \varepsilon_m \tag{7.42}$$

Since the control objective is to make $r_1 = 0$ and $r_3 = 0$, that is, $\mu_1 = \tau_1$ and $\mu_3 = \tau_2$, the effectiveness of τ_1 and τ_2 are to converge $r_1 \to 0$ and $r_3 \to 0$, therefore, $\tau_1 \to \mu_1$, $\tau_2 \to -\mu_3$.

To estimate μ_j with SVM method, we select the sample pairs as $(x_1, y_1) = (r_1, \tau_1)$ and $(x_3, y_3) = (r_3, -\tau_2)$, $\phi_j(x) = [\phi_j^{(1)}(x), \phi_j^{(2)}(x), \ldots, \phi_j^{(h)}(x)]^T$, where $\phi_j(x)$ is determined by the kernel function $K_j(x, x)$, then $\mu_j = \langle w_j, \phi_j(x)\rangle + b_j = \hat{F}_j(x) + \varepsilon_j = \sum_{i=1}^{l} a_i^{(1)} K_1(x, x_j) + b_j + \varepsilon_j$.

Remark 7.10 From Lemma 7.9, we know r_1, e_1, r_3 and e_3 are bounded uniformly in time, therefore, Theorem 7.5 can be applied.

Therefore, the external torques are given as

$$\tau_1 = \hat{F}_1 + u_1 + \tau_{r1} \tag{7.43}$$

$$\tau_2 = -\hat{F}_3 - u_2 - \tau_{r2} \tag{7.44}$$

where τ_{rj} is a robustifying vector defined later.

Then, (7.27) and (7.28) can be rewritten as

$$D_{11}(\zeta_3)\dot{r}_1 + C_{11}r = u_1 + \tau_{r1} + \hat{F}_1 - \mu_1 \tag{7.45}$$

$$\frac{D_{22}D_{33} - D_{23}^2(\zeta_3)}{D_{23}(\zeta_3)}\dot{r}_3 + \tilde{C}_{23}r_3 = u_2 + \tau_{r2} + \hat{F}_3 - \mu_3 \tag{7.46}$$

The state space description (7.34) can be given by

$$\dot{X} = A(\zeta, \dot{\zeta})X + B(\zeta)(\mathcal{U} + \hat{F} + \tau_r - \mu) \tag{7.47}$$

where $\tau_r = [\tau_{r1}, \tau_{r2}]^T$.

Considering the feedback control (7.36), we have

$$\dot{X} = \left(A - \frac{1}{2}BR^{-1}B^T P\right)X + B(\hat{F} + \tau_r - \mu) \tag{7.48}$$

Theorem 7.11 *Consider the $\mathcal{U}$ provided by* (7.36) *with the robust term given by*

$$\tau_r = -\frac{r\varepsilon_m^2}{\|r\|\varepsilon_m + \delta(t)} \tag{7.49}$$

with the positive ε_m defined in (7.42) *and r defined by* (7.13), *$\delta(t)$ is a time varying positive function converging to zero as $t \to \infty$, such that $\int_0^t \delta(\omega)\,d\omega = a < \infty$ with bounded constant a, then X converges to a set containing the origin with a rate at least as fast as $e^{-\nu t}$ with $\nu = \lambda_{\min}(Q) > 0$.*

Proof Considering (7.37), we have

$$\dot{\mathbb{V}}_1 = X^T\left(P\dot{X} + \frac{1}{2}\dot{P}\right)X \tag{7.50}$$

Introducing (7.48) into (7.50) yields

$$\begin{aligned}\dot{\mathbb{V}}_1 &= X^T PAX - \frac{1}{2}X^T PBR^{-1}B^T PX + \frac{1}{2}X^T\dot{P}X \\ &\quad + X^T PB(\hat{F} + \tau_r - \mu)\end{aligned} \tag{7.51}$$

Using $X^T PAX = \frac{1}{2}X^T(A^T P + PA)X$, integrating (7.35), the time derivative of Lyapunov function becomes

$$\dot{\mathbb{V}}_1 = -\frac{1}{2}X^T QX + X^T PB(\hat{F} + \tau_r - \mu) \tag{7.52}$$

Since $B = [0\ \mathcal{D}^{-1}]^T$ from (7.33), $X^T PB = [e^T\ r^T]\mathrm{diag}[K, \mathcal{D}][0\ \mathcal{D}^{-1}]^T = r^T$, therefore, we have

$$\begin{aligned}\dot{\mathbb{V}}_1 &\le -\frac{1}{2}\|X\|^2\lambda_{\min}(Q) + \|r\|\varepsilon + X^T PB\tau_r \\ &\le -\frac{1}{2}\|X\|^2\lambda_{\min}(Q) + \|r\|\varepsilon_m - \frac{\|r\|^2\varepsilon_m^2}{\|r\|\varepsilon_m + \delta(t)} \\ &\le -\frac{1}{2}\|X\|^2\lambda_{\min}(Q) + \frac{\|r\|\varepsilon_m\delta(t)}{\|r\|\varepsilon_m + \delta(t)} \\ &\le -\frac{1}{2}\|X\|^2\lambda_{\min}(Q) + \delta(t)\end{aligned} \tag{7.53}$$

Therefore, we arrive at $\dot{V} \le -\nu V + \delta(t)$. Thus, r converges to a set containing the origin with a rate at least as fast as $e^{-\nu t}$.

Integrating both sides of the above equation gives

$$\mathbb{V}_1(t) - \mathbb{V}_1(0) \le -\int_0^t \left(\frac{1}{2}\|X\|^2\lambda_{\min}(Q)\right)ds + a \tag{7.54}$$

Thus V is bounded, which implies that $X \in L_\infty$. From (7.54), we have

$$\int_0^t \left(\frac{1}{2}\|X\|^2\lambda_{\min}(Q)\right)ds \le \mathbb{V}_1(0) - \mathbb{V}_1(t) + a \tag{7.55}$$

which leads to $X \in L_2$. From $r = \dot{e} + \Lambda e$, it can be obtained that $e, \dot{e} \in L_\infty$. As we have established $e, \dot{e} \in L_\infty$, from Assumption 7.2, we conclude that $\dot{\zeta}_1, \dot{\zeta}_3, \dot{\zeta}_{1r}, \dot{\zeta}_{3r} \in L_\infty$. □

ζ_2-Subsystem For system (7.9)–(7.11) under the control laws (7.43) and (7.44), the ζ_2-subsystem (7.10) can be also rewritten as

$$\dot{\varphi} = f(\gamma, \varphi, u) \tag{7.56}$$

where $\varphi = [\zeta_2, \dot{\zeta}_2]^T$, $\gamma = [\zeta_1, \zeta_3, \dot{\zeta}_1, \dot{\zeta}_3]^T$, $u = [\tau_1, \tau_2]^T$.

Assumption 7.4 From (7.9) and (7.11), the reference signal satisfies Assumption 7.2, and the following function is Lipschitz in γ, i.e., there exists Lipschitz positive constants L_γ and L_f such that

$$\left\| C_{31}\dot{\zeta}_1 + g_3 + f_3(\dot{\zeta}_3) + d_3 \right\| \leq L_\gamma \|\gamma\| + L_f \tag{7.57}$$

Moreover, from the stability analysis of ζ_1 and ζ_3 subsystems, γ converges to a small neighborhood of $\gamma_d = [\zeta_{1d}, \zeta_{3d}, \dot{\zeta}_{1d}, \dot{\zeta}_{3d}]^T$.

Remark 7.12 In (7.57), we can see $C_{31}\dot{\zeta}_1$ is a function of γ, therefore, it is easy to obtain that Assumption 7.4 is satisfied.

Remark 7.13 Since the ζ_1 and ζ_3 subsystems satisfy the stability under the proposed controls (7.43) and (7.44), and considering (7.6), let $\|\gamma - \gamma_d\| \leq \rho_2$, it is easy to obtain $\|\gamma\| \leq \|\gamma_d\| + \rho_2$, and from (7.15), $\ddot{\zeta}_{jr} \to \ddot{\zeta}_{jd}$, $\ddot{\zeta}_j \to \ddot{\zeta}_{jd}$, since the close-loop signals are bounded, let $\|\ddot{\zeta}\| \leq \|\ddot{\zeta}_d\| + \rho_3$, where ρ_2 and ρ_3 are small bounded errors.

Lemma 7.14 *The ζ_2-subsystem* (7.10), *if ζ_1-subsystem and ζ_3-subsystem are stable, is globally asymptotically stable, too.*

Proof From (7.9), (7.10), and (7.11), we choose the Lyapunov candidate as

$$\mathbb{V}_2 = \mathbb{V}_1 + \ln\bigl(\cosh(\dot{\zeta}_2)\bigr) \tag{7.58}$$

Differentiating (7.58) along (7.10) gives

$$\begin{aligned} \dot{\mathbb{V}}_2 &= \dot{\mathbb{V}}_1 + \tanh(\dot{\zeta}_2)\ddot{\zeta}_2 \\ &= \dot{\mathbb{V}}_1 + \tanh(\dot{\zeta}_2)\frac{D_{33}}{D_{22}D_{33} - D_{23}^2(\zeta_3)}\left[\frac{D_{23}(\zeta_3)}{D_{33}}\bigl(C_{31}\dot{\zeta}_1 \right. \\ &\quad \left. + g_3(\zeta_3) + f_3(\dot{\zeta}) + d_3\bigr) - C_{23}\dot{\zeta}_3 - f_2(\dot{\zeta}) - d_2 + \tau_2\right] \end{aligned} \tag{7.59}$$

From (7.11), we have

$$\tau_2 - C_{23}(\zeta,\dot{\zeta})\dot{\zeta}_3 - f_2 - d_2 = -\frac{D_{22}}{D_{23}(\zeta_3)}\big(C_{31}\dot{\zeta}_1 + g_3 + f_3(\dot{\zeta}_3) + d_3\big) - \frac{D_{22}D_{33} - D_{23}^2(\zeta_3)}{D_{23}(\zeta_3)}\ddot{\zeta}_3 \tag{7.60}$$

Integrating (7.60) into (7.59), we have

$$\begin{aligned}\dot{\mathbb{V}}_2 &= \dot{\mathbb{V}}_1 + \tanh(\dot{\zeta}_2)\frac{D_{33}}{D_{22}D_{33} - D_{23}^2(\zeta_3)}\frac{D_{23}^2(\zeta_3) - D_{22}D_{33}}{D_{33}D_{23}(\zeta_3)} \\ &\quad \big(C_{31}\dot{\zeta}_1 + g_3 + f_3(\dot{\zeta}_3) + d_3\big) \\ &\quad - \tanh(\dot{\zeta}_2)\frac{D_{33}}{D_{22}D_{33} - D_{23}^2(\zeta_3)}\frac{D_{22}D_{33} - D_{23}^2(\zeta_3)}{D_{23}(\zeta_3)}\ddot{\zeta}_3 \\ &= \dot{\mathbb{V}}_1 - \frac{\tanh(\dot{\zeta}_2)}{D_{23}(\zeta_3)}\big(C_{31}\dot{\zeta}_1 + g_3 + f_3(\dot{\zeta}_3) + d_3 + D_{33}\ddot{\zeta}_3\big)\end{aligned}$$

Since $\|\tanh(\dot{\zeta}_2)\| \leq 1$, $\|D_{23}(\zeta_3)\|$ is bounded, let $\|\frac{\tanh(\dot{\zeta}_2)}{D_{23}(\zeta_3)}\| \leq \rho_1$, where ρ_1 is a bounded constant, from Assumption 7.4, we have $\|C_{31}\dot{\zeta}_1 + g_3 + f_3(\dot{\zeta}_3) + d_3\| \leq L_\gamma\|\gamma_d\| + L_\gamma\rho_2 + L_f$, similarly, since $\ddot{\zeta}_3 = \ddot{\zeta}_{3r} + \dot{r}_3$, considering Assumption 7.2, r_3 is bounded from ζ_1 and ζ_3-subsystems, and $\ddot{\zeta}_{3r} \to \ddot{\zeta}_{3d}$, therefore, we have $\|D_{33}\ddot{\zeta}_3\| \leq D_{33}\|\ddot{\zeta}_d\| + \rho_3$, therefore, we have

$$\begin{aligned}\dot{\mathbb{V}}_2 &\leq -\frac{1}{2}\|X\|^2\lambda_{\min}(Q) + \delta \\ &\quad + \rho_1\big(L_\gamma\|\gamma_d\| + L_\gamma\rho_2 + L_f + D_{33}\|\ddot{\zeta}_d\| + \rho_3\big)\end{aligned} \tag{7.61}$$

Let $\pi = \delta + \rho_1(L_\gamma\|\gamma_d\| + L_\gamma\rho_2 + L_f + D_{33}\|\ddot{\zeta}_d\| + \rho_3)$ and it is apparently bounded positive, we have $\dot{\mathbb{V}}_2 \leq 0$, when $|X| \geq \sqrt{\pi/(2\lambda_{\min}(Q))}$, we can choose the proper Q and R such that the X can be arbitrarily small. Therefore, we can obtain the internal dynamics being stable with respect to the output $\dot{\zeta}_2$. Therefore, the ζ_2-subsystem (7.10) is exponentially stable. □

Theorem 7.15 *Consider the system* (7.9)–(7.11) *with Lemma* 7.14, *under the action of control laws* (7.43) *and* (7.44). *For each compact set* Ω_{10}, *where* $(\zeta_1(0), \dot{\zeta}_1(0)) \in \Omega_{10}$, *each compact set* Ω_{30}, *where* $(\zeta_3(0), \dot{\zeta}_3(0)) \in \Omega_{30}$, *the tracking errors* r_1 *and* r_3 *converge to a set containing the origin with a rate at least as fast as* $e^{-\nu t}$, *and all the signals in the closed loop system are bounded.*

Proof From the results (7.54) in Theorem 7.11, it is clear that the tracking errors r_1 and r_3 converge to a set containing the origin with a rate at least as fast as $e^{-\nu t}$. From Lemma 2.48, we can know e_1, $\dot{e}_1$, e_3, $\dot{e}_3$ are also bounded. From the boundedness of ζ_{1d}, ζ_{3d} in Assumption 7.2, we know that ζ_1, ζ_3 are bounded. Since $\dot{\zeta}_{1d}$, $\dot{\zeta}_{3d}$ are also bounded, it follows that $\dot{\zeta}_1$, $\dot{\zeta}_3$ are bounded. From Lemma 7.14, we know that the ζ_2-subsystem (7.11) is stable, and ζ_2, $\dot{\zeta}_2$ are bounded. □

7.2.5 Simulation Studies

Let us consider a wheeled inverted pendulum as shown in Fig. 3.1. In the simulation, we assume the parameters $M = 10.0$ kg, $J = 1.0$ kg m^2, $J_w = 4.0$ kg m^2, $J_a = 2.0$ kg m^2, $m = 2.0$ kg, $L = 1$ m, $D = 1.0$ m, $R = 0.5$ m, $\rho = \mathrm{diag}[0.1]$, $\zeta(0) = [-0.2, 0, \pi/18]^T$, and $\dot{\zeta}(0) = [0.0, 0.1, 0.0]^T$. The disturbances from environments on the system are introduced as $1.0\sin(t)$ and $1.0\cos(t)$ in the simulation model. The desired trajectories are chosen as $\theta_d = 0.5t$ rad and $\alpha_d = 0$ rad, and the initial velocity is 0.1 m/s. The design parameters of the above two LS-SVM controllers are: $Q_{11} = \mathrm{diag}[80, 120]$, $Q_{12} = Q_{21}^T = \mathrm{diag}[-4, -6]$, $Q_{22} = \mathrm{diag}[40, 300]$, $\Lambda = \mathrm{diag}[10, 10]$, $K = \mathrm{diag}[4, 6]$, $\varepsilon_m = 10$, $\delta = 1/(1+t)^2$, 200 training data are sampled for the sliding window during $[0, 5]$ s. The training data pairs are constructed. We use the radial basis function (RBF) kernels for SVM: $K(x_i, x_j) = \exp\{-\frac{\|x_i - x_j\|^2}{2\sigma^2}\}$. The parameter of RBF kernels is $\sigma = 0.25$, the parameters of LS-SVM are selected as $c_1 = 20$, $c_2 = 40$, $\epsilon = 0.02$. In the same conditions, we compare (i) the model based control, we assume 10 % model uncertainty, the model-based controller is considered as $\tau_1 = D_{11}\ddot{\zeta}_{1d} + C_{11}\dot{\zeta}_1 + C_{13}\dot{\zeta}_3 - k_{11}(\dot{\zeta}_1 - \dot{\zeta}_{1d}) - k_{12}(\zeta_1 - \zeta_{1d})$ and $\tau_2 = -\frac{D_{22}D_{33} - D_{23}^2(\zeta_3)}{D_{23}(\zeta_3)}\ddot{\zeta}_{3d} - k_{31}(\dot{\zeta}_3 - \dot{\zeta}_{3d}) - k_{32}(\zeta_3 - \zeta_{3d}) - C_{23}\dot{\zeta}_3 + \frac{D_{22}}{D_{23}(\zeta_3)}(C_{31}\dot{\zeta}_1 + f_3 + g_3 + d_3) - (f_2 + g_2 + d_2)$ with $k_{11} = k_{12} = k_{31} = k_{32} = 10.0$; (ii) the robust control with 30–50 % model uncertainty and the known bound of model uncertainty, we choose the control law proposed as $\tau_1 = -k_{p1}r_1 - k_{i1}\int_0^t r_1\,ds - \sum_{i=1}^3 \frac{r_1 c_{1i}\Psi_{1i}^2}{\Psi_{1i}|r_1| + \delta_{1i}}$, where $\Phi_1 = \vartheta_1{}^T\Psi_1$, $\vartheta_1 = [\vartheta_{11}\ \vartheta_{12}\ \vartheta_{13}]^T$; $\Psi_1 = [|\ddot{\zeta}_{1r}|, 1, 1]^T$, and $\tau_2 = k_{p3}r_3 + k_{i3}\int_0^t r_3\,dt + \sum_{i=1}^8 \frac{r_3\vartheta_{3i}\Psi_{3i}^2}{\Psi_{3i}|r_3| + \delta_{3i}}$, where $\Phi_3 = C_3{}^T\Psi_3$, $\vartheta_3 = [\vartheta_{31}, \ldots, \vartheta_{38}]^T$; $\Psi_3 = [|\ddot{\zeta}_{3r}|, |\dot{\zeta}_{3r}|, -|\dot{\zeta}|, -1, -1, |\dot{\zeta}|, 1, 1]^T$, the control gains are selected as $k_{p1} = 10.0$, $k_{p3} = 60.0$, $k_{i1} = k_{i3} = 0.0$, $\vartheta_1 = [10.0, 10.0, 10.0]^T$, $\vartheta_3 = [10.0, \ldots, 10.0]^T$, $\delta_{1i} = \delta_{3i} = 1/(1+t)^2$.

The tracking performances by comparison simulations are illustrated in Figs. 7.2–7.6. The comparison balance performance is shown in Fig. 7.3. The direction angle and angle velocity tracking by three approaches are shown in Fig. 7.2, the input torques by three approaches are shown in Fig. 7.4, and the stable velocities by three approaches are shown in Fig. 7.5, and the trajectories of the system by three approaches are shown in Fig. 7.6. As these figures show, the tilt angle by LS-SVM approach is kept in the smallest set around the vertical position compared with the other two approaches, therefore, the balance is more stable, and the forward velocity also converges to a stable value compared with the other two approaches, while the forward velocities and the produced trajectories by the other two approaches apparently diverge and are clearly shown in Fig. 7.6. The performance is achieved under the initial disturbances boundedness from the environment given, even the model parameters of the system are unknown beforehand. Therefore, we know the proposed control scheme could achieve better tracking performance, although the parametric uncertainties and the external disturbances are both introduced into the simulation model, the motion/balance control performance of system, under the

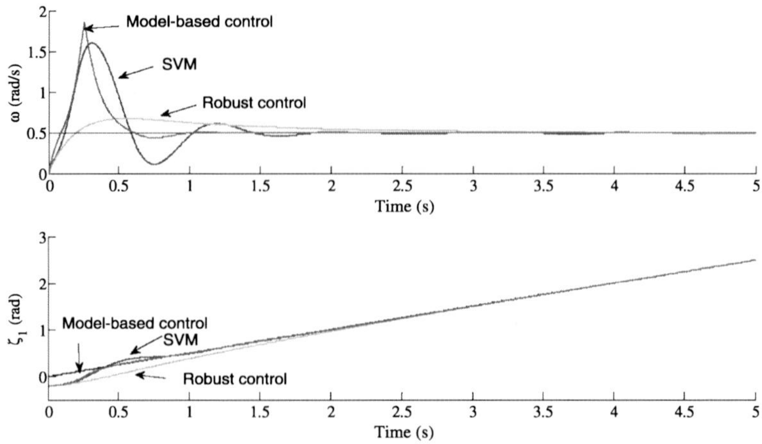

Fig. 7.2 Tracking the direction angle and rotation velocity by the LS-SVM based control

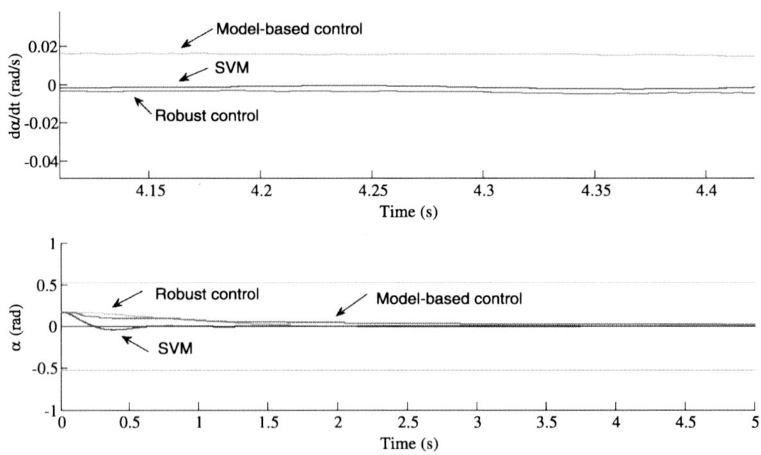

Fig. 7.3 Tracking the desired tilt angle by the LS-SVM based control

proposed control, is not degraded. The simulation results demonstrate the effectiveness of the proposed LS-SVM based control in the presence of unknown nonlinear dynamic system and environments. Different motion/balance tracking performance can be achieved by adjusting control gains.

The performance of model-based approach is sensitive to the accuracy of the dynamic model in Fig. 7.3 and Fig. 7.5, in the model-based control, we only introduce the effects of 10 % parametric uncertainties in the dynamic model, since ζ_3 is very sensitive to the model, more than 10 % model uncertainty would cause the system unstable. For the robust control, we assume to obtain the bound of the dynamic parameter beforehand, however, it is unrealistic to obtain them in the actual application. From Figs. 7.3 and 7.5, the tracking performance is no better than the

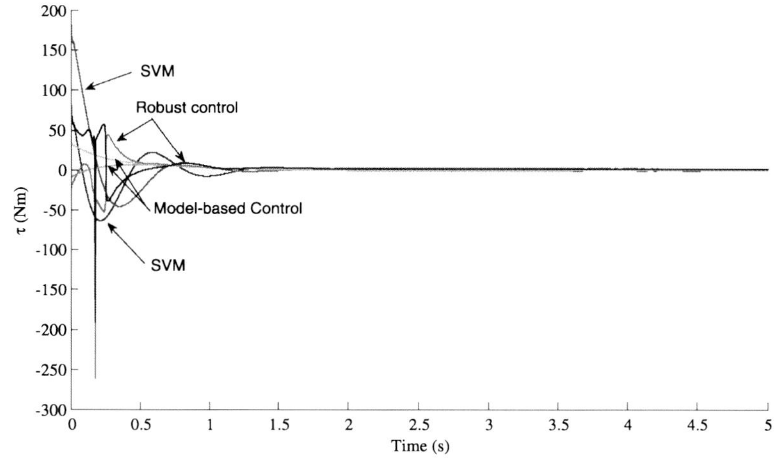

Fig. 7.4 Input torques

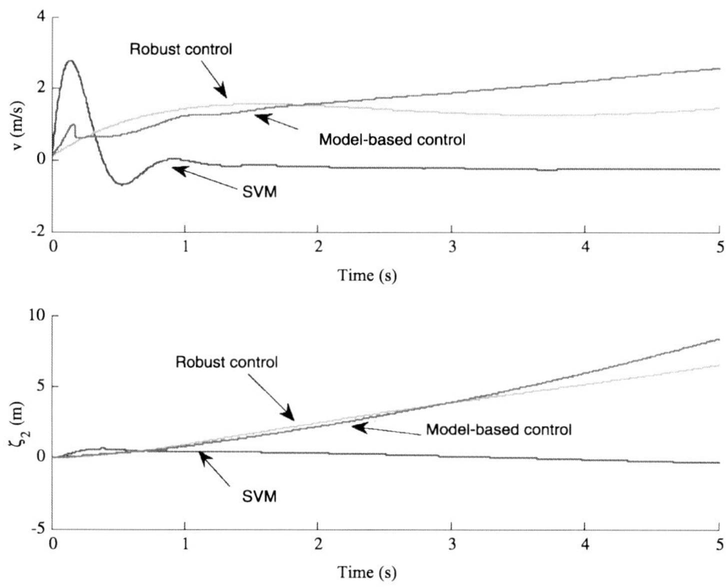

Fig. 7.5 The forward velocity by the LS-SVM based control

LS-SVM. In contrast, LS-SVM control approach is tolerant of modeling errors, and can be viewed as a key advantage over model-based and robust control of wheeled inverted pendulums, for which accurate modeling of wheeled inverted pendulums dynamics is difficult, time-consuming and uncertain. The presence of parametric errors is a common problem for model-based and robust controllers since the identification of dynamic parameters is error-prone. For instance, the controlled condi-

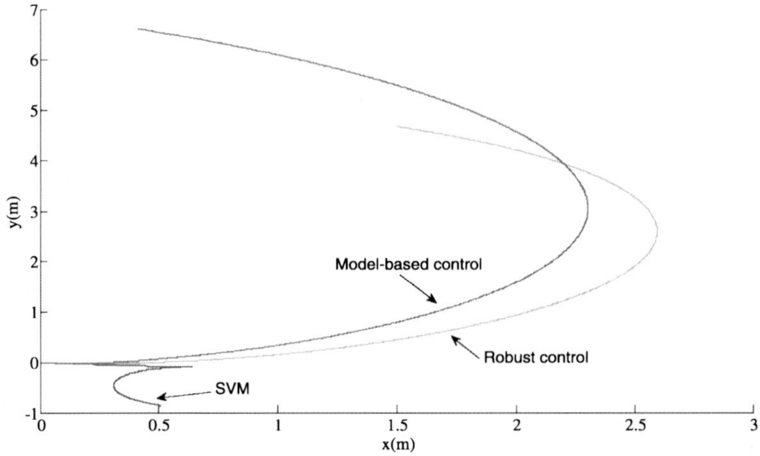

Fig. 7.6 The trajectory of wheeled inverted pendulum

tions in the test facility under which the parameters are identified are often very different from actual conditions, thus rendering the parameters inaccuracy for real operating conditions. LS-SVM control presented in this chapter is not susceptible to this problem, since the unknown parameters are learned during the wheeled inverted pendulum operation in actual conditions. Therefore, we know the proposed control scheme could achieve better tracking performance over the model-based control and the robust control. The better tracking performances are largely due to the "learning" mechanism. Although the parametric uncertainties and the external disturbances are both introduced into the simulation model, the motion/balance control performance of system, under the proposed control, is not degraded. The simulation results demonstrate the effectiveness of the proposed adaptive control in the presence of unknown nonlinear dynamic system and environments. Different motion/balance tracking performance can be achieved by adjusting control gains.

From these figures, the simulation results show the proposed control is with the better performance and is more realistic in practice, which validates the effectiveness of the control law in Theorem 7.15.

7.3 Fuzzy Control

In this section, we study model free intelligent control based on fuzzy logic systems. Fuzzy logic systems (FLS) have been credited in robotics control and applications as powerful tools capable of providing robust controllers for mathematically ill-defined systems that may be subjected to structured and unstructured uncertainties [86, 87]. The universal approximation has been another main driving force behind the increasing popularity of fuzzy logic controllers as it shows that fuzzy systems are theoretically capable of uniformly approximating any continuous real function to any degree of accuracy [61, 87, 124]. In the class of approximators which are linear in parameters, FLS is much closer in spirit to human thinking and natural language [124], and preferred by control engineering practitioners. In practical control applications, it is desirable to have systematic methods of ensuring stability, robustness, and performance of the overall system. Several fuzzy control approaches have been proposed based on Lyapunov synthesis [61, 87], which can guarantee the stability of the closed-loop system.

Developing a highly accurate model could be much more complex than the design of a control mechanism. In almost every case, we have to simplify the model, under some assumptions, such that design of control can be carried out. In this case, modeling errors always exist, which undermines the method of model-based control. By adaptive fuzzy logic control, we attempt to reduce the workload on modeling dramatically and compensate for modeling errors. In this chapter, we present adaptive fuzzy logic control to accommodate the presence of parametric and functional uncertainties in the dynamics of wheeled inverted pendulums.

7.3.1 Preliminaries

Assumption 7.5 The body of the wheeled pendulums is symmetric, and the pendulums is mounted on the middle point between the two wheel centers.

Remark 7.16 Since the angular displacements of the wheels could be represented by the coordination x and y as clearly presented in [13], in this chapter, we use x and y as the generalized coordinates.

Assumption 7.6 $F_1 = [f_1, f_2, f_3]^T$ are the friction forces acting independently in each vector, and the friction force is the function of the velocity, for simplification and convenience, such as $F_1 = \mathcal{B}\dot{\zeta}$ with the friction coefficient $\mathcal{B}$, which is diagonal positive definite.

7.3.2 Functional Universal Approximation Using FLSs

Consider an n_i-inputs, single-output fuzzy logic system [124] with the product-inference rule, singleton fuzzifier, center average defuzzifier, and Gaussian membership function given by n_j fuzzy IF–THEN rules

$$\mathbb{R}^{(j)}: \text{IF } x_1 \text{ is } A_1^j \text{ and } \ldots \text{ and } x_{ni} \text{ is } A_{ni}^j \text{ THEN } y \text{ is } B^j,\quad j = 1, \ldots, n_j \tag{7.62}$$

where $\mathbb{R}^j$ denotes the jth rule, $1 \le j \le n_j$, $(x_1, x_2, \ldots, x_{n_i})^T \in U \subset \mathbb{R}^{n_i}$ and $y \in \mathbb{R}$ are the linguistic variables associated with the inputs and output of the fuzzy logic system, respectively, A_i^j and B^j denote the fuzzy sets in $\mathbb{U}$ and $\mathbb{R}$. The fuzzy logic system performs a nonlinear mapping from $\mathbb{U}$ to $\mathbb{R}$. Based on the generalizations of implications in multi-valued logic, many fuzzy implication rules were proposed in the fuzzy logic literature [61, 124]. In this chapter, we use the strategy of singleton defuzzification, the output of product inference and center-average defuzzification, the fuzzy logic system is

$$y(x)=\frac{\sum_{j=1}^{m} y_j(\prod_{i=1}^{n}\mu_{A_i^j}(x_i))}{\sum_{j=1}^{m}(\prod_{i=1}^{m}\mu_{A_i^j}(x_i))} \tag{7.63}$$

where $x=[x_1,x_2,\dots,x_m]^T$, $\mu_{A_i^j}(x_i)$ is the membership function of linguistic variable x_i with

$$\mu_{A_i^j}(x_i)=\exp\left[-\frac{(x_i-c_{ij})^2}{\sigma_{ij}^2}\right] \tag{7.64}$$

For clarity, let us define the weight vector and fuzzy basis function vector respectively, as

$$B=[y_1,y_2,\dots,y_m]^T$$
$$S(x,c,\sigma)=[s_1,s_2,\dots,s_m]^T$$

where $s_l=\prod_{i=1}^{m}\mu_{A_i^l}(x_i)/[\sum_{k=1}^{m}\prod_{i=1}^{m}\mu_{A_i^k}(x_i)]$, $c=[c_1^T,c_2^T,\dots,c_m^T]^T$ and $\sigma=[\sigma_1^T,\sigma_2^T,\dots,\sigma_m^T]^T$. If the membership functions (i.e., c_{ij}, σ_{ij}) are fixed, the normalized firing strength of jth rule S is function of only x. Then Eq. (7.63) can be represented as

$$y=B^TS(x,c,\sigma) \tag{7.65}$$

It has been proven that fuzzy logic system (7.65) is capable of uniformly approximating any given real continuous function over a compact set to any degree of accuracy.

Thus, for the unknown nonlinear functions $f_i(x_i)$, $i=1,\dots,n$, we have the following approximation over the compact sets Ξ_i

$$f_i(x_i)=B_i^{*T}S(x_i)+\epsilon_i(x_i),\quad \forall x_i\in\Xi_i\subset\mathbb{R}^i \tag{7.66}$$

where $S(x_i)$ is the fuzzy basis vector, $\epsilon_i(x_i)$ is the approximation error and B_i^* is an unknown constant parameter vector.

Theorem 7.17 *For any given real continuous function $g(x)$ on the compact set $\mathbb{U}\in\mathbb{R}^n$ and arbitrary $\epsilon>0$, there exists a function $f(x)$ in the form of* (7.65) *such that* $\sup_{x\in\mathbb{U}}\|g(x)-f(x)\|\le\epsilon$ [124].

Remark 7.18 The optimal weight vector B_i^* in (7.66) is an artificial quantity required only for analytical purposes. Typically, B_i^* is chosen as the value of B_i that minimizes $\epsilon_i(x_i)$ for all $x_i\in\Xi_i$ where $\Xi_i\subset\mathbb{R}^i$ is a compact set, i.e.,

$$B_i^*:=\arg\min_{c\in\mathbb{R}^n}\left\{\sup_{x\in\Xi_i}\left|f_i(x_i)-B_i^TS(x_i)\right|\right\}$$

Assumption 7.7 On a compact region $\Xi_i \in \mathbb{R}^i$

$$\left|\epsilon_i(x_i)\right| \le \epsilon_i^* \tag{7.67}$$

where $\epsilon_i^* > 0$ is unknown bound.

From the above analysis, we see that the system uncertainties are converted to the estimation of unknown parameters B_i^* and unknown bounds ϵ_i^*.

7.3.3 Adaptive Fuzzy Control

ζ_1**-Subsystem** Consider the definition in Chap. 5:

$$\dot{\zeta}_1 = \dot{\zeta}_{1r} + r_1$$
$$\ddot{\zeta}_1 = \ddot{\zeta}_{1r} + \dot{r}_1$$

then Eq. (5.17) becomes

$$\begin{aligned} & m_{11}(\zeta_3)\dot{r}_1 + v_{11}r_1 \\ & \quad = \tau_1 - \big(m_{11}(\zeta_3)\ddot{\zeta}_{1r} + v_{11}\dot{\zeta}_{1r} + v_{13}\dot{\zeta}_3 + f_1 + g_1 + d_1\big) \end{aligned} \tag{7.68}$$

The unknown continuous function $m_{11}(\zeta_3)\ddot{\zeta}_{1r} + v_{11}\dot{\zeta}_{1r} + v_{13}\dot{\zeta}_3 + f_1 + g_1 + d_1$ in (7.68) can be approximated by FLSs to arbitrary any accuracy as

$$m_{11}(\zeta_3)\ddot{\zeta}_{1r} + v_{11}\dot{\zeta}_{1r} + v_{13}\dot{\zeta}_3 + f_1 + g_1 + d_1 = B_1^{*T} S_1(Z_1) + \varepsilon_1(Z_1) \tag{7.69}$$

where the input vector $Z_1 = [\ddot{\zeta}_{1r}, \dot{\zeta}_{1r}, \dot{\zeta}_3, \zeta_3]^T \in \mathbb{R}^4$. Note that the input vector Z_1 is composed of real elements (i.e., $Z_1 \in \mathbb{R}^4$). So, if κ_1 fuzzy labels are assigned to each of these elements, the total number of fuzzy rules, L_1, in the FLC of ζ_1 is κ_1^4. For instance, if κ_1 was chosen to be three, the FLC of each of the robots has to fire 81 rules in order to compute and dispatch the control signals to the τ_1. Moreover, $\varepsilon_1(Z_1)$ is the approximation error satisfying $|\varepsilon_1(Z_1)| \le \bar{\varepsilon}_1$, where $\bar{\varepsilon}_1$ is an unknown positive constant; B_1^* are unknown ideal constant weights satisfying $\|B_1^*\| \le b_{1m}$, where b_{1m} is an unknown positive constant; and $S_1(Z_1)$ are the basis functions. By using $\hat{B}_1$ to approximate B_1^*, the error between the actual and the ideal FLSs can be expressed as

$$\hat{B}_1^T S_1(Z_1) - B_1^{*T} S_1(Z_1) = \tilde{B}_1^T S_1(Z_1) \tag{7.70}$$

where $\tilde{B}_1 = \hat{B}_1 - B_1^*$. As B_1^* is a constant vector, it is easy to obtain that

$$\dot{\tilde{B}}_1 = \dot{\hat{B}}_1 \tag{7.71}$$

Remark 7.19 If the fuzzy logic system $\hat{B}_1^T S_1(Z_1)$, are rich enough, that is if its input fuzzy labels uniformly and thoroughly span the whole input space and the number of tunable parameters is large enough to approximate the system dynamics, then from the universal approximation Theorem 7.17, the term can be approximated to zero with any degree of accuracy [124]. Since can be smaller than any machine precision, it can be considered to be practically zero

$$\varepsilon_1(Z_1) = 0 \tag{7.72}$$

Remark 7.20 Since fuzzy system approximation ability only holds on a compact set similar as NN, it is more rigorous to assume that the states are within a sufficiently large compact set beforehand, such that the NN or fuzzy system approximation will not be violated. Then, after the derivation, we can see that if we choose the fuzzy rule number to be large enough for fuzzy system approximation to cover the sufficiently large compact set, the states and weights in the closed-loop system will be remain in the bounded compact set and will eventually converge to another bounded compact set defined as similarly as in Lemma 1.2 of [41], and [47]. The analysis and adjustment of transient bounds of the weight errors and the tracking error and could be founded in Theorem 4.2 and Remark 4.13 of [47].

To analyze closed loop stability for the ζ_1-subsystem, consider the following Lyapunov function candidate

$$\mathbb{V}_1 = \frac{m_{11}(\zeta_3)}{2} r_1^2 + \frac{1}{2}\left(\int_0^t r_1\, ds\right) k_{1I} \int_0^t r_1\, ds + \frac{1}{2}\tilde{B}_1^T \Gamma_1^{-1} \tilde{B}_1 \tag{7.73}$$

where $k_{1I} > 0$, and Γ_1 is positive diagonal.

Its time derivative is given by

$$\begin{aligned}\dot{\mathbb{V}}_1 = {} & r_1\left(m_{11}(\zeta_3)\dot{r}_1 + \frac{1}{2}\frac{d}{dt}(m_{11}(\zeta_3))r_1 + k_{1I}\int_0^t r_1\, ds\right) \\ & + \tilde{B}_1^T \Gamma_1^{-1} \dot{\tilde{B}}_1\end{aligned} \tag{7.74}$$

Substituting (7.68) into (7.74), we have

$$\begin{aligned}\dot{\mathbb{V}}_1 = {} & r_1\left[\tau_1 - (m_{11}(\zeta_3)\ddot{\zeta}_{1r} + \upsilon_{11}\dot{\zeta}_{1r} + \upsilon_{13}\dot{\zeta}_3 + f_1 + g_1 + d_1) + k_{1I}\int_0^t r_1\, ds\right] \\ & + \tilde{B}_1^T \Gamma_1^{-1} \dot{\tilde{B}}_1\end{aligned} \tag{7.75}$$

Substituting (7.69) and (7.71) in (7.75), we have

$$\dot{\mathbb{V}}_1 = r_1\left[\tau_1 - B_1^{*T} S_1(Z_1) - \varepsilon_1(Z_1) + k_{1I}\int_0^t r_1\, ds\right] + \tilde{B}_1^T \Gamma_1^{-1} \dot{\hat{B}}_1 \tag{7.76}$$

Consider the FLSs control as

$$\tau_1 = -k_{1P} r_1 + \hat{B}_1^T S_1(Z_1) - k_{1I}\int_0^t r_1\, ds \tag{7.77}$$

where k_{1P} is a positive constant, and $k_{1P} > \frac{1}{2}$.

Substituting (7.77) in (7.76), we have

$$\dot{\mathbb{V}}_1 = r_1\big[-k_{1P}r_1 + \tilde{B}_1^T S_1(Z_1) - \varepsilon_1(Z_1)\big] + \tilde{B}_1^T \Gamma_1^{-1}\dot{\hat{B}}_1 \tag{7.78}$$

Consider the following FLSs weight adaptation law

$$\dot{\hat{B}}_1 = -\Gamma_1\big(S_1(Z_1)r_1 + \sigma_1|r_1|\hat{B}_1\big) \tag{7.79}$$

where $\sigma_1 > 0$.

Substituting (7.79) into (7.78), we have

$$\begin{aligned}\dot{\mathbb{V}}_1 &= r_1\big[-k_{1P}r_1 + \tilde{B}_1^T S_1(Z_1) - \varepsilon_1(Z_1)\big] - \tilde{B}_1^T\big(S_1(Z_1)r_1 + \sigma_1|r_1|\hat{B}_1\big)\\ &= -k_{1P}r_1^2 - r_1\varepsilon_1(Z_1) - \sigma_1|r_1|\tilde{B}_1^T\hat{B}_1\end{aligned} \tag{7.80}$$

Since $2\tilde{B}_1^T\hat{B}_1 \geq \|\tilde{B}_1\|^2 - \|B_1^*\|^2$, (7.80) becomes

$$\begin{aligned}\dot{\mathbb{V}}_1 &\leq -k_{1P}r_1^2 + |r_1|\bar{\varepsilon}_1 - \frac{1}{2}\sigma_1|r_1|\|\tilde{B}_1\|^2 + \frac{1}{2}\sigma_1|r_1|\|B_1^*\|^2\\ &\leq -k_{1P}r_1^2 + |r_1|\left(\bar{\varepsilon}_1 + \frac{1}{2}\sigma_1 b_{1m}^2\right)\\ &\leq -\left(k_{1P} - \frac{1}{2}\right)r_1^2 + \frac{1}{2}\alpha_1^2\end{aligned} \tag{7.81}$$

with

$$\alpha_1 = \bar{\varepsilon}_1 + \frac{1}{2}\sigma_1 b_{1m}^2 \tag{7.82}$$

From (7.81), we can obtain

$$\dot{\mathbb{V}}_1 \leq 0 \tag{7.83}$$

for $|r_1| \geq \sqrt{\frac{\frac{1}{2}\alpha_1^2}{K_{1P}-\frac{1}{2}}}$, and r_1 will converge to a compact set denoted by

$$\Omega_1 := \left\{r_1 : |r_1| \leq \sqrt{\frac{\frac{1}{2}\alpha_1^2}{k_{1P} - \frac{1}{2}}}\right\} \tag{7.84}$$

ζ_3-Subsystem Since $\dot{\zeta}_3 = \dot{\zeta}_{3r} + r_3$, $\ddot{\zeta}_3 = \ddot{\zeta}_{3r} + \dot{r}_3$, Eq. (5.19) becomes

$$\begin{aligned}&\frac{m_{22}m_{33} - m_{23}^2(\zeta_3)}{m_{23}(\zeta_3)}\dot{r}_3 - \upsilon_{23}r_3\\ &\quad = -\tau_2 - \left(\frac{m_{22}m_{33} - m_{23}^2(\zeta_3)}{m_{23}(\zeta_3)}\ddot{\zeta}_{3r} - \upsilon_{23}\dot{\zeta}_{3r}\right)\\ &\qquad - \frac{m_{22}}{m_{23}(\zeta_3)}(\upsilon_{31}\dot{\zeta}_1 + f_3 + g_3) + (f_2 + g_2 + d_2)\end{aligned} \tag{7.85}$$

Decompose $\upsilon_{23} = \hat{\upsilon}_{23} - \tilde{\upsilon}_{23}$ such that

$$\frac{d}{dt}\left(\frac{m_{22}m_{33} - m_{23}^2(\zeta_3)}{m_{23}(\zeta_3)}\right) - 2\tilde{\upsilon}_{23} = 0 \tag{7.86}$$

Let the unknown function in (7.85) as

$$\begin{aligned}\mu_3 = &\frac{m_{22}m_{33} - m_{23}^2(\zeta_3)}{m_{23}(\zeta_3)}\ddot{\zeta}_{3r} - \hat{\upsilon}_{23}r_3 - \upsilon_{23}\dot{\zeta}_{3r} \\ &+ \frac{m_{22}}{m_{23}(\zeta_3)}(\upsilon_{31}\dot{\zeta}_1 + f_3 + g_3) - (f_2 + g_2 + d_2)\end{aligned}$$

can be approximated by FLSs to arbitrary any accuracy as

$$\mu_3 = B_3^{*T} S_3(Z_3) + \varepsilon_3(Z_3) \tag{7.87}$$

where the input vector $Z_3 = [\ddot{\zeta}_{3r}, \dot{\zeta}_{3r}, r_3, \dot{\zeta}_1, \dot{\zeta}_3, \zeta_3]^T \in \mathbb{R}^6$ and $\varepsilon_3(Z_3)$ is the approximation error satisfying $|\varepsilon_3(Z_3)| \leq \bar{\varepsilon}_3$, where $\bar{\varepsilon}_2$ is an unknown positive constant; B_3^* are unknown ideal constant weights satisfying $\|B_3^*\| \leq b_{3m}$, where b_{3m} is an unknown positive constant; and $S_3(Z_3)$ are the basis functions.

Remark 7.21 Note that the input vector Z_3 is composed of real elements (i.e., $Z_3 \in \mathbb{R}^6$). So, if κ_3 fuzzy labels are assigned to each of these elements, the total number of fuzzy rules, L_3, in the FLC of ζ_3 is κ_3^6. For instance, if κ_3 was chosen to be three or two, the FLC has to fire 324 rules in order to compute and dispatch the control signals to the τ_2.

Remark 7.22 If the fuzzy logic system $\hat{B}_3^T S_1(Z_3)$, are rich enough, that is if its input fuzzy labels uniformly and thoroughly span the whole input space and the number of tunable parameters is large enough to approximate the system dynamics, then from the universal approximation Theorem 7.17, the term can be approximated to zero with any degree of accuracy [124]. Since can be smaller than any machine precision, it can be considered to be practically zero

$$\varepsilon_3(Z_3) = 0 \tag{7.88}$$

To analyze closed loop stability for the ζ_3-subsystem, consider the following Lyapunov function candidate

$$\begin{aligned}\mathbb{V}_3 = &\frac{1}{2}\frac{m_{22}m_{33} - m_{23}^2(\zeta_3)}{m_{23}(\zeta_3)}r_3^2 + \frac{1}{2}\left(\int_0^t r_3\,ds\right)k_{3I}\int_0^t r_3\,ds \\ &+ \frac{1}{2}\tilde{B}_3^T\Gamma_3^{-1}\tilde{B}_3\end{aligned} \tag{7.89}$$

where $k_{3I} > 0$ and Γ_3 is positive diagonal.

Its time derivative is given by

$$\begin{aligned}\dot{\mathbb{V}}_3 = &r_3\left(\frac{m_{22}m_{33} - m_{23}^2(\zeta_3)}{m_{23}(\zeta_3)}\dot{r}_3 + \frac{1}{2}\frac{d}{dt}\left(\frac{m_{22}m_{33} - m_{23}^2(\zeta_3)}{m_{23}(\zeta_3)}\right)r_3 + k_{3I}\int_0^t r_3\,ds\right) \\ &+ \tilde{B}_3^T\Gamma_3^{-1}\dot{\tilde{B}}_3\end{aligned} \tag{7.90}$$

Considering (7.86) and substituting (7.85) into (7.90), we have

$$\dot{\mathbb{V}}_3 = r_3\left[-\tau_2 + k_{3I}\int_0^t r_3\,ds - \mu_3\right] + \tilde{B}_3^T \Gamma_3^{-1}\dot{\tilde{B}}_3 \tag{7.91}$$

By using $\hat{B}_3$ to approximate B_3^*, the error between the actual and the ideal FLSs can be expressed as

$$\hat{B}_3^T S_3(Z_3) - B_3^{*T} S_3(Z_3) = \tilde{B}_3^T S_3(Z_3) \tag{7.92}$$

where $\tilde{B}_3 = \hat{B}_3 - B_3^*$. As B_3^* is a constant vector, it is easy to obtain that

$$\dot{\tilde{B}}_3 = \dot{\hat{B}}_3 \tag{7.93}$$

Substituting (7.87) and (7.93) into (7.91), we have

$$\dot{\mathbb{V}}_3 = r_3\left[-\tau_2 + k_{3I}\int_0^t r_3\,ds - B_3^{*T} S_3(Z_3) - \varepsilon_3(Z_3)\right] + \tilde{B}_3^T \Gamma_3^{-1}\dot{\hat{B}}_3 \tag{7.94}$$

Consider the FLs control as

$$\tau_2 = k_{3P} r_3 - \hat{B}_3^T S_3(Z_3) + k_{3I}\int_0^t r_3\,ds \tag{7.95}$$

where k_{3P} is a constant and $k_{3P} > \frac{1}{2}$.

Substituting (7.95) into (7.94), we have

$$\dot{\mathbb{V}}_3 = r_3\big[-k_{3P} r_3 + \tilde{B}_3^T S_3(Z_3) - \varepsilon_3(Z_3)\big] + \tilde{B}_3^T \Gamma_3^{-1}\dot{\hat{B}}_3 \tag{7.96}$$

Consider the following FLSs weight adaptation law

$$\dot{\hat{B}}_3 = -\Gamma_3\big(S_3(Z_3) r_3 + \sigma_3 |r_3| \hat{B}_3\big) \tag{7.97}$$

where $\sigma_3 > 0$.

Substituting (7.97) into (7.96), we have

$$\begin{aligned}
\dot{\mathbb{V}}_3 &= r_3\big[-k_{3P} r_3 + \tilde{B}_3^T S_3(Z_3) - \varepsilon_3(Z_3)\big] - \tilde{B}_3^T\big(S_3(Z_3) r_3 + \sigma_3 |r_3| \hat{B}_3\big)\\
&= -k_{3P} r_3^2 - r_3\varepsilon_3(Z_3) - \sigma_3|r_3|\tilde{B}_3^T\hat{B}_3\\
&\le -k_{3P} r_3^2 + |r_3|\bar{\varepsilon}_3 - \frac{1}{2}\sigma_3|r_3|\|\tilde{B}_3\|^2 + \frac{1}{2}\sigma_3|r_3|\big\|B_3^*\big\|^2\\
&\le -k_{3P} r_3^2 + |r_3|\left(\bar{\varepsilon}_3 + \frac{1}{2}\sigma_3 b_{3m}^2\right)\\
&\le -\left(k_{3P} - \frac{1}{2}\right) r_3^2 + \frac{1}{2}\alpha_3^2
\end{aligned} \tag{7.98}$$

with

$$\alpha_3 = \bar{\varepsilon}_3 + \frac{1}{2}\sigma_3 b_{3m}^2 \tag{7.99}$$

From (7.98), we can obtain

$$\dot{\mathbb{V}}_3 \leq 0 \tag{7.100}$$

for $|r_3| \geq \sqrt{\frac{\frac{1}{2}\alpha_3^2}{k_{3P}-\frac{1}{2}}}$, and r_3 will converge to a compact set denoted by

$$\Omega_3 := \left\{ r_3 : |r_3| \leq \sqrt{\frac{\frac{1}{2}\alpha_3^2}{k_{3P} - \frac{1}{2}}} \right\} \tag{7.101}$$

From Lemma 1.2 of [41], and [47], we know that $\hat{B}_3$ are bounded for bounded initial weights $\hat{B}_3(0)$, since the control signal τ_2 is a function of $\hat{B}_3$ and r_3, we know that it is bounded, moreover $\dot{\mathbb{V}}_3 \leq 0$, $\dot{r}_3, r_3$ are bounded, therefore, from Lemma 2.48 and Assumption 7.2, $\dot{\zeta}_3$ and ζ_3 are bounded. By $\dot{\mathbb{V}}_1 \leq 0$, we can also obtain that $\dot{\zeta}_1$ is bounded. From the dynamics of WIP, we know v_{23} is bounded. All the signals on the left-hand side of (7.85) are bounded, therefore, we can obtain f_2 is bounded, from Assumption 7.6, we can obtain $\dot{\zeta}_2$ is bounded.

ζ_2-Subsystem For system (5.17)–(5.19) under the control laws (5.25) and (5.32), the ζ_2-subsystem (5.18) can be rewritten as

$$\dot{\varphi} = f(\xi, \varphi, u) \tag{7.102}$$

where $\varphi = [\zeta_2, \dot{\zeta}_2]^T$, $\xi = [\zeta_1, \zeta_3, \dot{\zeta}_1, \dot{\zeta}_3]^T$, $u = [\tau_1, \tau_2]^T$. Then the zero dynamics of the system can be addressed as [42]

$$\dot{\varphi} = f\big(0, \varphi, u^*(0, \varphi)\big) \tag{7.103}$$

where $u^* = [\tau_1^*, \tau_2^*]^T$.

Lemma 7.23 *The ζ_2-subsystem* (5.18), *restricted in the set where $\xi = 0$, is stable.*

Remark 7.24 Since there is no objective of the control for ζ_2 subsystem, therefore, if the $\dot{\zeta}_2$ converges to a stable bounded velocity, ζ_2-subsystem is also stable.

Proof From (5.17), (5.18), (5.19), we choose the following function whose physical significance is the kinematic energy as

$$\mathbb{V}_2 = \frac{1}{2}\frac{m_{22}m_{33} - m_{23}^2(\zeta_3)}{m_{33}}\dot{\zeta}_2^2 \tag{7.104}$$

Remark 7.25 From Property 5.3, we know $\frac{m_{22}m_{33}-m_{23}^2(\zeta_3)}{m_{33}} > 0$.

Differentiating $\mathbb{V}_2$ and restricting it on $\xi = 0$ and $u^*(0, \varphi)$, we have

$$\dot{\mathbb{V}}_2 = \dot{\zeta}_2\frac{m_{22}m_{33} - m_{23}^2(\zeta_3)}{m_{33}}\ddot{\zeta}_2 + \frac{1}{2}\dot{\zeta}_2\frac{d}{dt}\left(\frac{m_{22}m_{33} - m_{23}^2(\zeta_3)}{m_{33}}\right)\dot{\zeta}_2 \tag{7.105}$$

Since m_{22} and m_{33} are constant, $\frac{d}{dt}(\frac{m_{22}m_{33}-m_{23}^2(\zeta_3)}{m_{33}}) = -\frac{2m_{23}(\zeta_3)}{m_{33}}\dot{m}_{23}\dot{\zeta}_3$. Moreover $\xi = 0 \Rightarrow r_3 = 0$, that is $\dot{\zeta}_3 = \dot{\zeta}_{3d} = 0$, we have $\frac{1}{2}\dot{\zeta}_2\frac{d}{dt}(\frac{m_{22}m_{33}-m_{23}^2(\zeta_3)}{m_{33}})\dot{\zeta}_2 = 0$.

Considering (5.18), we have

$$\dot{\mathbb{V}}_2 = \frac{1}{2}\dot{\zeta}_2\left(\frac{m_{23}(\zeta_3)}{m_{33}}(\upsilon_{31}\dot{\zeta}_1 + f_3 + g_3) + (\tau_2 - \upsilon_{23}\dot{\zeta}_3 - f_2 - g_2 - d_2)\right) \tag{7.106}$$

When $\xi = 0$, from (5.19), we have

$$\tau_2 - \upsilon_{23}\dot{\zeta}_3 - f_2 - g_2 - d_2 = -\frac{m_{22}}{m_{23}(\zeta_3)}(\upsilon_{31}\dot{\zeta}_1 + f_3 + g_3) \tag{7.107}$$

Integrating (7.107) into (7.106), we have

$$\dot{\mathbb{V}}_2 = \frac{1}{2}\dot{\zeta}_2\frac{m_{23}^2 - m_{22}m_{33}}{m_{33}m_{23}}(\upsilon_{31}\dot{\zeta}_1 + f_3 + g_3) \tag{7.108}$$

When $\xi = 0$ is achieved, g_3 is related with $\sin(\zeta_3)$, therefore, $g_3 = 0$ and considering Assumption 7.6, we have $f_3 = 0$, and from the dynamics of WIP, it can obtain $\upsilon_{31} = 0$, from all above, $\dot{\mathbb{V}}_2 = 0$ is achieved which means that if $\xi = 0$, the kinematic energy of $\mathbb{V}_2$ is unchanged. From the ζ_3 stability analysis, we can obtain $\dot{\zeta}_2$ is bounded. The internal dynamics system is bounded with respect to the output $\dot{\zeta}_2$ on the zero dynamics manifold, if $\xi = 0$. In addition, $\dot{\mathbb{V}}_2 = 0$, which means that the equilibrium point of the zero dynamics is stable. Therefore, the zero dynamics (7.103) is stable.

Assumption 7.8 From Lemma 7.23, the control input u is designed as a function of the states (ξ, φ) and the reference signal satisfies Assumption 7.2, and the function $f(\xi, \varphi, u)$ is Lipschitz in ξ, i.e., there exists Lipschitz constants L_ξ and L_f for $f(\xi, \varphi, u)$ such that

$$\left\| f(\xi, \varphi, u) - f(0, \varphi, u_\varphi) \right\| \le L_\xi \|\xi\| + L_f \tag{7.109}$$

where $u_\varphi = u^*(0, \varphi)$.

Under Assumption 7.8, by the converse theorem of Lyapunov [60], there exists a Lyapunov function $V_0(\varphi)$ which satisfies [42]

$$\sigma_2\|\varphi\|^2 \le V_0(\varphi) \le \sigma_1\|\varphi\|^2 \tag{7.110}$$

$$\frac{\partial V_0}{\partial \varphi} f(0, \varphi, u_\varphi) \le -\lambda_a\|\varphi\|^2 \tag{7.111}$$

$$\left\|\frac{\partial V_0}{\partial \varphi}\right\| \le \lambda_b\|\varphi\| \tag{7.112}$$

where σ_1, σ_2, λ_a and λ_b are positive constants.

From the previous stability analysis about the ζ_1-subsystem (5.17) and the ζ_3-subsystem (5.19), we know that $\zeta_1, \zeta_3, \dot{\zeta}_1$, and $\dot{\zeta}_3$ are bounded. Accordingly, ξ is bounded. We assume that

$$\|\xi\| \le \|\xi\|_{\max} \tag{7.113}$$

where $\|\xi\|_{\max}$ is a positive constant.

Lemma 7.26 *For the internal dynamics $\dot{\varphi} = f(\xi, \varphi, u)$ of the system, if Assumption 7.8 is satisfied, then there exist positive constants L_φ and T_0, such that*

$$\|\varphi(t)\| \le L_\varphi, \quad \forall t > T_0 \tag{7.114}$$

Proof According to Assumption 7.8, there exists a Lyapunov function $V_0(\varphi)$. Differentiating $V_0(\varphi)$ along (5.17), (5.18), (5.19) yields

$$\begin{aligned}\dot{V}_0(\varphi) &= \frac{\partial V_0}{\partial \varphi} f(\xi, \varphi, u)\\ &= \frac{\partial V_0}{\partial \varphi} f(0, \varphi, u_\varphi)\\ &\quad + \frac{\partial V_0}{\partial \varphi}\left[f(\xi, \varphi, u) - f(0, \varphi, u_\varphi)\right]\end{aligned} \tag{7.115}$$

Noting (7.109)–(7.112) and (7.115) can be written as

$$\dot{V}_0(\varphi) \le -\lambda_a\|\varphi\|^2 + \lambda_b\|\varphi\|\left(L_\xi\|\xi\| + L_f\right) \tag{7.116}$$

Noting (7.113), we have

$$\dot{V}_0(\varphi) \le -\lambda_a\|\varphi\|^2 + \lambda_b\|\varphi\|\left(L_\xi\|\xi\|_{\max} + L_f\right) \tag{7.117}$$

Therefore, $\dot{V}_0(\varphi) \le 0$, whenever

$$\|\varphi\| \ge \frac{\lambda_b}{\lambda_a}\left(L_\xi\|\xi\|_{\max} + L_f\right) \tag{7.118}$$

By letting $L_\varphi = \frac{\lambda_b}{\lambda_a}(L_\xi\|\xi\|_{\max} + L_f)$, we conclude that there exists a positive constant T_0, such that (7.114) holds. □

Theorem 7.27 *Consider the system* (5.17), (5.18), (5.19) *with Lemmas* 7.23 *and* 7.26, *under the action of control laws* (7.77) *and* (7.95) *and adaptation laws* (7.79) *and* (7.97). *For each compact set Ω_{10}, where $(\zeta_1(0), \dot{\zeta}_1(0), \hat{B}_1(0)) \in \Omega_{10}$, each compact set Ω_{30}, where $(\zeta_3(0), \dot{\zeta}_3(0), \hat{B}_3(0)) \in \Omega_{30}$, each compact set Ω_{30}, where $(\zeta_3(0), \dot{\zeta}_3(0)) \in \Omega_{30}$, the tracking errors r_1 and r_3 converge to the compact sets Ω_1 and Ω_3, respectively defined by* (7.84) *and* (7.101), *and all the signals in the closed loop system are bounded.*

Proof From the results (7.83) and (7.100), it is clear that the tracking errors r_1 and r_3 converge to the compact sets Ω_1 and Ω_3, respectively defined by (7.84) and (7.101). In addition, the signals $\tilde{B}_1$ are $\tilde{B}_3$ are bounded. From Lemma 2.48, we can know e_1, $\dot{e}_1$, e_3, $\dot{e}_3$ are also bounded. From the boundedness of ζ_{1d}, ζ_{3d} in Assumptions 7.2 and 7.7, we know that ζ_1, ζ_3 are bounded. Since $\dot{\zeta}_{1d}$, $\dot{\zeta}_{3d}$ are also bounded, it follows that $\dot{\zeta}_1$, $\dot{\zeta}_3$ are bounded. With B_1^*, B_3^* as constants, we know that $\hat{B}_1$, $\hat{B}_3$ are also bounded. From Lemmas 7.23 and 7.26, we know that the ζ_2-subsystem (5.19) is stable, and ζ_2, $\dot{\zeta}_2$ are bounded. This completes the proof. □

7.3.4 Simulation Studies

To verify the effectiveness of the proposed control algorithm, let us consider a wheeled inverted pendulum as shown in Fig. 3.1. In the simulation, we assume the parameters $m = 2.5$ kg, $I_m = 2.5\ \text{kg m}^2$, $I_w = 0.2\ \text{kg m}^2$, $M_w = 0.2$ kg, $l = 1$ m, $d = 1$ m, $r = 0.5$ m, $\mathcal{B} = \text{diag}(1.0)$, $\zeta(0) = [-0.2, 0, \pi/180]^T$, $\dot{\zeta}(0) = [0.0, 0.1, 0.0]^T$. The desired trajectories are chosen as $\theta_d = 0.2t$ rad, $\alpha_d = 0$ rad, initial velocity is 0.1 m/s. The external disturbances are set as $1.0\sin(t)$ and $1.0\cos(t)$.

The input vectors $Z_1 = [\ddot{\zeta}_{1r}, \dot{\zeta}_{1r}, \dot{\zeta}_3, \zeta_3]^T \in \mathbb{R}^4$, the each variable are divided into three fuzzy labels, each defined by a Gaussian membership function, $\mu_{A_i^j}(z_{1i}) = \exp[-\frac{(z_{1i}-c_{ij})^2}{\sigma_{ij}^2}]$ with centers c_{ij} $(i = 1, \dots, l_1)$ spaced in $[-1.0, 1.0] \times [-1.0, 1.0] \times [-1.0, 1.0] \times [-0.5, 0.5]$. Similarly, $Z_3 = [\ddot{\zeta}_{3r}, \dot{\zeta}_{3r}, r_3, \dot{\zeta}_1, \dot{\zeta}_3, \zeta_3]^T \in \mathbb{R}^6$, the variables are divided into three, three, two, two, three, three fuzzy labels, respectively, each defined by a Gaussian membership function, $\mu_{A_i^j}(z_{2i}) = \exp[-\frac{(z_{2i}-c_{ij})^2}{\sigma_{ij}^2}]$, with centers c_{ij} $(i = 1, \dots, l_2)$ spaced in $[-1.0, 1.0] \times [-1.0, 1.0] \times [-1.0, 1.0] \times [-1.0, -1.0] \times [-1.0, 1.0] \times [-1.0, 1.0]$. Fuzzy logic system $\hat{W}_1^T S_1(Z_1)$ contains 81 rules (i.e., $l_1 = 81$), $\hat{W}_3^T S_3(Z_3)$ contains 324 rules (i.e., $l_2 = 324$). The design parameters of the above controllers are: $K_{1P} = 10.0$, $\Lambda_1 = 10.0$, $K_{3P} = 10.0$, $K_{1I} = K_{3I} = 0.0$, $\Lambda_2 = 10.0$, $\Gamma_1 = \text{diag}[2000]$, $\Gamma_2 = \text{diag}[2500]$, $\hat{W}_1(0) = (0.1)$, $\hat{W}_2(0) = (0.1)$. The position tracking by the proposed control are shown in Figs. 7.7 and 7.8, and the input torques are shown in Fig. 7.9, the stable velocity is shown in Fig. 7.10. From these figures, even if the nominal parameters of the system, which have 30–50 % model uncertainty, and the initial disturbances boundedness from the environment are known, we can get good performance by the proposed control. From the simulation, we can see that θ is converged to the desired trajectory in Fig. 7.7, and α is converged to a bounded small set containing the origin in Fig. 7.8, and v is stable enough in Fig. 7.10, therefore, it can be obtained that the internal states are stable.

However, the performance of model-based approaches is sensitive to the accuracy of the model representation. Non-adaptive model-based control requiring functional and parametric certainty cannot handle functional uncertainty. In contrast, fuzzy logic control approaches are tolerant of modeling errors, and can be viewed as a key advantage over model-based control of wheeled inverted pendulums, for which accurate modeling of wheeled inverted pendulums dynamics is difficult, time-consuming and costly. Therefore, we will compare the performance of the wheeled inverted pendulums, in terms of tracking error, between the fuzzy logic control approaches and a model-based approach. In particular, we consider a non-adaptive model-based controller as $\tau_1 = -k_{1P}r_1 + m_{11}\ddot{\zeta}_{1r} + v_{11}\dot{\zeta}_{1r} + v_{13}\dot{\zeta}_3 + f_1 + g_1 + d_1$ and $\tau_2 = -k_{3P}r_3 - \mu_3$. Considering the unmodeled dynamics, we introduce the effects of 10 % parametric uncertainties in the dynamic model. In the simulation, we found that since the ζ_3 is very sensitive to the model, more than 10 % model uncertainty would cause the system volatilized under the model-based controller. The presence of parametric errors is a common problem for model-based controllers

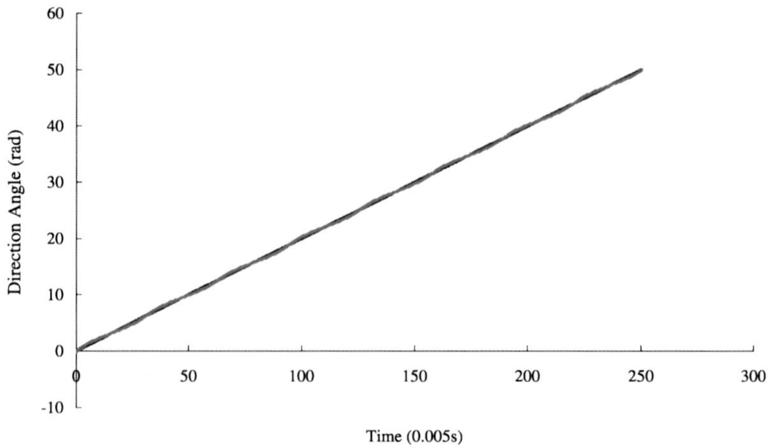

Fig. 7.7 Tracking the direction angle by the proposed control

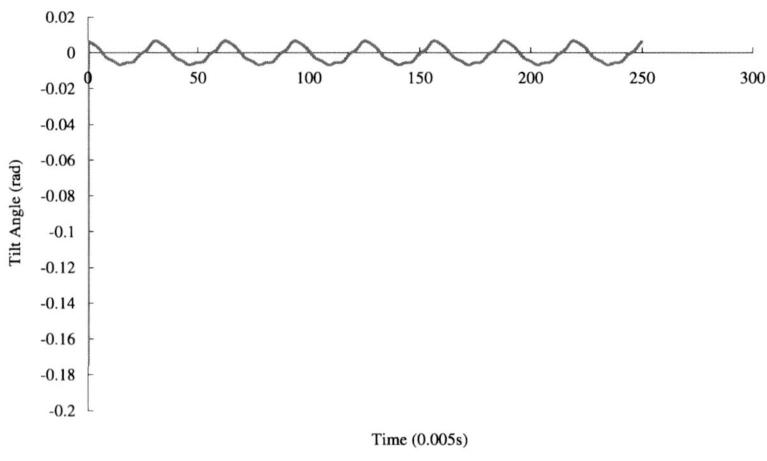

Fig. 7.8 Tracking the desired tilt angle by the proposed control

since the identification of dynamic parameters is error-prone. For instance, the controlled conditions in the test facility under which the parameters are identified are often very different from actual conditions, thus rendering the parameters inaccurate for real operating conditions. Fuzzy logic control presented in this chapter are not susceptible to this problem, since the unknown parameters are learn during the wheeled inverted pendulum operation in actual conditions. The position tracking using the model based control are shown in Figs. 7.11 and 7.12. The input torques are shown in Fig. 7.13. The final velocity is shown in Fig. 7.14, which fluctuates greatly. From these figures, the simulation results show the proposed control is with the better performance and is more realistic in practice, which validates the effectiveness of the control law in Theorem 7.27.

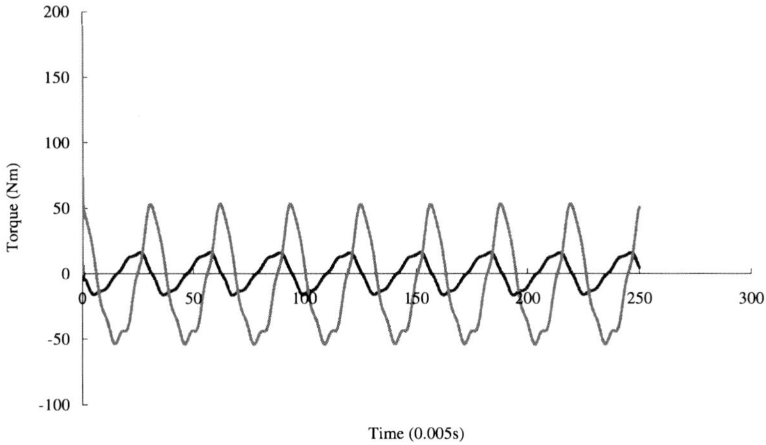

Fig. 7.9 Input torques by the proposed control

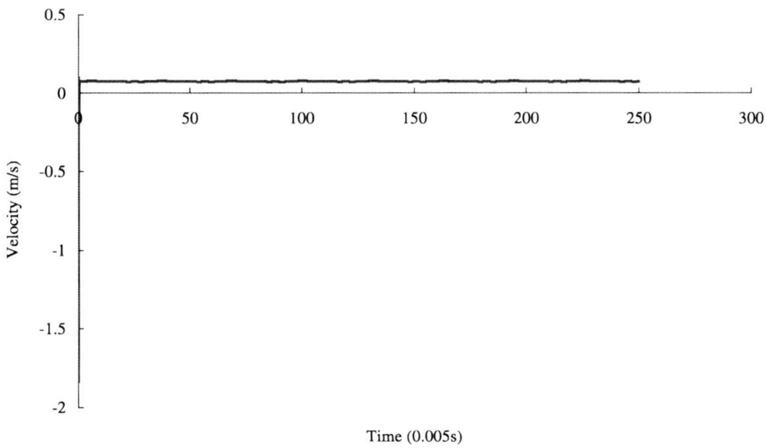

Fig. 7.10 The stable velocity by the proposed control

7.4 Neural Network Output Feedback Control

In this section, we consider NN based intelligent control of WIP systems. It is well known that NN systems have been credited in robotics controls and applications as powerful tools capable of approximating any continuous function of systems that are subjected to structured and unstructured uncertainties [44, 83, 87, 92]. The feasibility of applying neural networks to model unknown functions in dynamic systems has been demonstrated in our previous studies [87, 88]. In practical control applications, it is desirable to have systematic methods of ensuring stability, robustness, and performance of the overall system. Neural network implicit control approaches have been developed in [45, 47, 52, 132] for controlling nonaffine systems, and they guar-

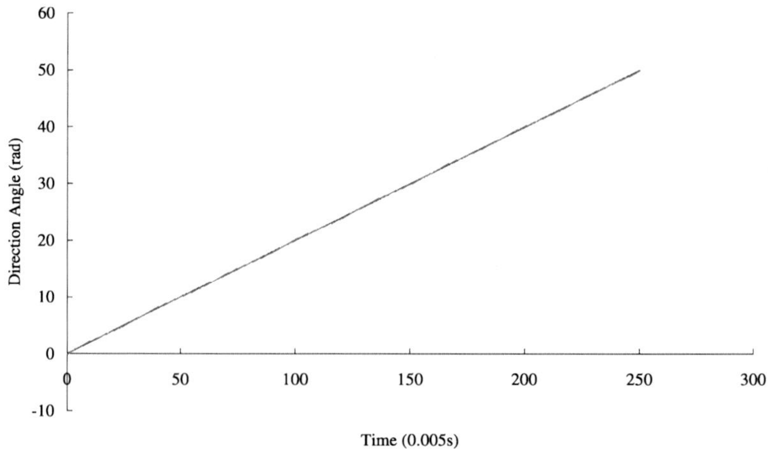

Fig. 7.11 Tracking the desired direction angle by the model based control

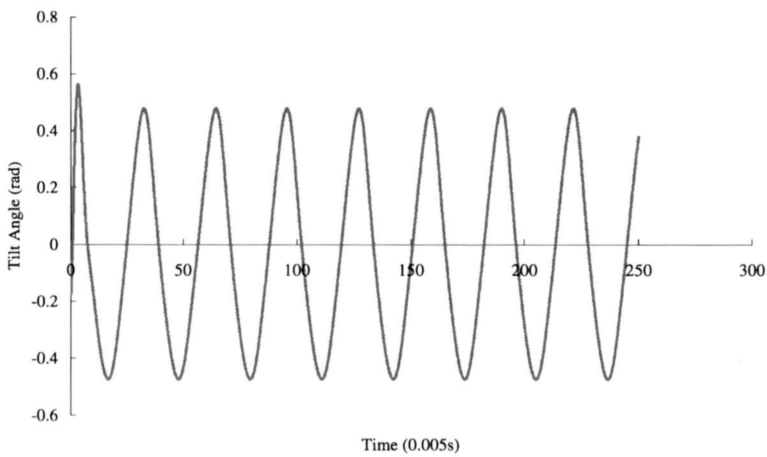

Fig. 7.12 Tracking the desired tilt angle by the model based control

antee the stability of the closed-loop system. However, the above mentioned papers concern little on the under-actuated systems such as wheeled inverted pendulum transportation systems with nonholonomic constraints, which is to be investigated in the chapter.

In this section, we consider the implicit control for dynamic balance and stable tracking of desired trajectories of WIP-car, in which both the dynamics and the dimension of the regulated system may be unknown. However, the relative degree of the regulated output is assumed to be known. After WIP-car is feedback linearizable, output feedback adaptive neural network incorporating a linear dynamic compensator is introduced to approximate the inversion dynamics so as to achieve stable dynamic balance and tracking desired trajectories.

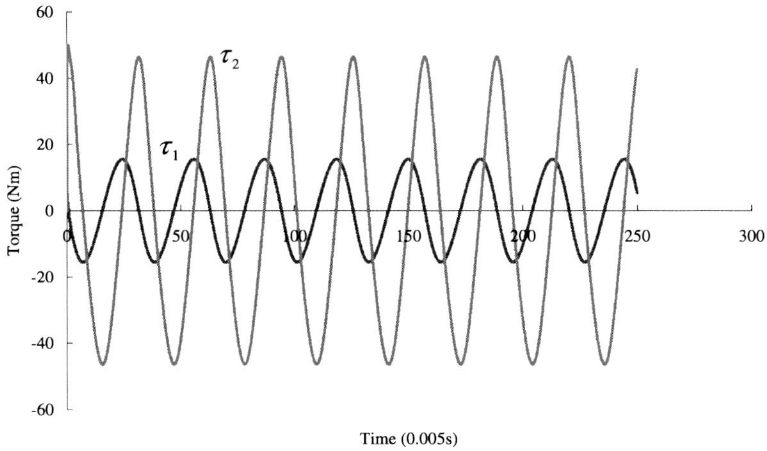

Fig. 7.13 Input torques by the model based control

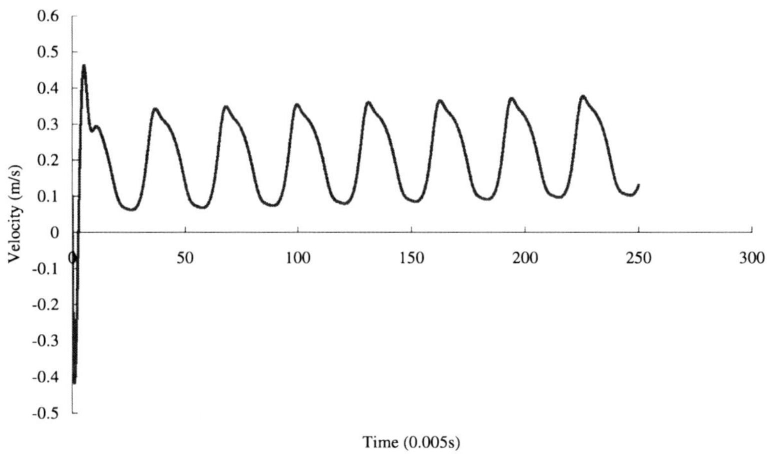

Fig. 7.14 The stable velocity by the model based control

7.4.1 Preliminaries

In the following, the notations used in this chapter are presented as well as some useful lemmas for stability analysis. As the control design relies on NN approximation, a brief review of a commonly used linear-in-parameter NN (LPNN) [47] will be given.

Lemma 7.28 [98] *Consider a C^r function $f : \mathbb{R}^{k+n} \to \mathbb{R}^n$ with $f(a,b) = \mathbf{0}_{[n]}$ and* $\mathrm{rank}(Df(a,b)) = n$ *where* $Df(a,b) = \frac{\partial f(x,y)}{\partial y}|_{(x,y)=(a,b)} \in \mathbb{R}^{n\times n}$. *Then, there exists a neighborhood A of a in $\mathbb{R}^k$ and a unique C^r function $g : A \to \mathbb{R}^n$ such that*

$g(a) = b$ and $f(x, g(x)) = \mathbf{0}_{[n]}$, $\forall x \in A$. If the condition is satisfied over $\mathbb{R}^k$, then the implicit function is a global map.

Lemma 7.29 *For any positive integers m, n and any real valued function $\chi(x, y) > 0$, the following inequality holds* [90]:

$$
\begin{aligned}
|x|^m |y|^n \le & \frac{m}{m+n} \chi(x, y) |x|^{m+n} \\
& + \frac{n}{m+n} \chi^{-m/n}(x, y) |y|^{m+n}
\end{aligned}
\tag{7.119}
$$

Lemma 7.30 *For $a, b \in \mathbb{R}$, and $a + b > 0$, the following inequality holds*:

$$
\frac{ab}{a+b} \le a \tag{7.120}
$$

Lemma 7.31 *Consider a first-order dynamical system*

$$
\dot{s} = -\lambda s + \phi(t) \tag{7.121}
$$

where $\phi(t)$ is a bounded and continuous nonnegative function and λ is a positive parameter. If a nonnegative initial value $s(0)$ is chosen, then the solution $s(t)$ is non-negative for all $t > 0$.

7.4.2 Neural Networks and Parametrization

Neural networks have been widely used in modeling and controlling of nonlinear systems because of their capabilities of nonlinear function approximation, learning, and fault tolerance. As a LPNN, radial basis function (RBF) NN is often used in practical control engineering due to its simple structure and nice approximation properties. In this chapter, the Gaussian RBF NNs will be employed to approximate a continuous function $h(\cdot) : \mathbb{R}^n \to \mathbb{R}$

$$
\hat{h}(Z) = \hat{W}^T \Phi(Z) \tag{7.122}
$$

where $Z \in \Omega_{NN} \subset \mathbb{R}^n$ is the input vector, $\hat{W} = [w_1, \dots, w_l]^T \in \mathbb{R}^l$ is the weight vector and the kernel vector is $\Phi(Z) = [s_1(Z), \dots, s_l(Z)]^T$ with active function $s_i(Z)$ being chosen as the commonly used Gaussian functions

$$
\phi_i(Z) = \exp\left[\frac{-\|Z - \mu_i\|^2}{\gamma_i^2}\right], \quad i = 1, 2, \dots, l \tag{7.123}
$$

where $\mu_i = [\mu_{i1}, \dots, \mu_{in}]^T$ is the center of the receptive field and γ_i is the width of the Gaussian function. By choosing enough nodes, neural network can approximate any continuous function over a compact region $\Omega_{NN} \subset \mathbb{R}^n$ with arbitrary accuracy

$$
h(Z) = W^{*T} \Phi(Z) + \delta, \quad \forall Z \in \Omega_{NN} \tag{7.124}
$$

The ideal weight vector W^* is an "artificial" quantity required for analytical purposes. It is defined as

$$W^* \triangleq \arg \min_{\hat{W} \in \mathbb{R}^n} \left\{ \sup_{Z \in \Omega_{NN}} \left| h(Z) - \hat{h}(Z) \right| \right\}$$

For an unknown continuous function vector $H(Z) \in \mathbb{R}^q$, it can be approximated by NN in the following manner

$$H(Z) = \left[W_M^{*T} \Phi(Z) \right] + E, \quad \forall Z \in \Omega_{NN} \tag{7.125}$$

where $W_M^* \in \mathbb{R}^{l \times q}$, $\Phi(Z) \in \mathbb{R}^l$ are the desired weights and basis function; and E is the collective NN reconstruction errors.

Since the functions approximated are assumed to be continuous, we have the following lemma.

Assumption 7.9 For $\forall Z \in \Omega_{NN}$, there exist ideal constant weights vector W_M^* such that $\| W_M^* \| \leq w_{\max}$, and $\| E \| \leq \delta_{\max}$ with bounds $w_{\max}, \delta_{\max} > 0$.

In Assumption 7.9, we can obtain an upper bound for the neural network parametrization, from which there is only one positive parameter to be adapted.

From Eq. (7.124) and Assumption 7.9, it shows

$$\begin{aligned} \left\| \left[W_M^{*T} \Phi(Z) \right] + E \right\| &\leq \left\| W^{*T} \right\| \left\| \Phi(Z) \right\| + \| E \| \\ &\leq \left\| \Phi(Z) \right\| w_{\max} + \delta_{\max} \\ &\leq \psi \beta(Z) \end{aligned} \tag{7.126}$$

where $\beta(Z) = \sqrt{\sum_{m=1}^{l} \phi_m^2(Z) + 1}$, and $\psi = \max\{\delta_{\max}, w_{\max}\}$.

Remark 7.32 The 2-norm is used to obtain $\beta(Z)$ so that it is differentiable, which is important for removing chattering phenomenon since it will be used in the following sliding mode control.

Remark 7.33 With neural network, the global Lipschitz assumption for unknown nonlinear functions can be avoided. Notice that the derivative on the NN basis function is bounded, thus the approximation by NN can naturally result in a semi-global Lipschitz condition. On the other side, due to the NN parametrization, the complexity and burden of computation can be reduced because the adaptive parameter becomes a scalar rather than a vector with many elements which can increase computation burden when large neural nodes are chosen.

7.5 Problem Formulation

It is observed that the dynamics of wheeled inverted pendulum (5.9) can be represented by the nonlinear non-affine MIMO form as followings:

$$\dot{x} = F(\zeta, \tau) \tag{7.127}$$

$$y = H(\zeta) \tag{7.128}$$

where $\zeta = [\zeta_1, \zeta_2, \zeta_3] \in \mathbb{R}^3$ is system states, $y \in \mathbb{R}^{n-l}$ with $n = 4$ and $l = 1$ denoting the system output, and it has derivatives up to third order. A matrix of y and its derivatives is constructed as $[y, y^{(1)}, y^{(2)}] \in \mathbb{R}^{(n-l)\times 3}$. The symbol $\tau \in \mathbb{R}^{n-l}$ denotes the system input, function $H : \mathbb{R}^3 \to \mathbb{R}^{n-l}$ is a partially unknown function, and function $F : \mathbb{R}^{3\times(n-l)} \to \mathbb{R}^{(n-l)}$, and $F(\zeta, \tau)$ is partially unknown vector field with respect to input τ.

System (7.127) is a general description of the dynamics of the nonlinear WIP-car, for which the control input is nonaffine. This is a realistic expression for the WIP-car with the inputs being position variables resulting in a non-affine form. We see that affine nonlinear systems and linear systems are special cases of (7.127), such that by designing a controller for (7.127), we actually include more general systems as well.

Define $\Theta_j(\zeta) = L_f^{j-1} H(\zeta)$ for $j = 1, \ldots, 3$, where $L_f H$ denotes the Lie derivative of the function $H(\zeta)$ with respect to the vector field $F(\zeta, \tau)$. As $n - l > 2$, we see that the system is input–output linearizable with strong relative degree such that there exist function Θ_4 independent of τ and that the mapping $\Theta(\zeta) = [\Theta_1(\zeta), \Theta_2(\zeta), \Theta_3(\zeta)]$ has a Jacobian matrix which is nonsingular for all $z \in \Omega_\zeta$. Thus, $\Theta(\zeta)$ is a diffeomorphism on Ω_ζ. Let $\phi = [\Theta_1(\zeta), \Theta_2(\zeta), \Theta_3(\zeta)]$ and $\vartheta = \Theta_4(\zeta)$, then system (7.127) can be expressed in the normal form:

$$\begin{aligned} \dot{\vartheta} &= \mathcal{Z}(\phi, \vartheta) \\ \dot{\phi}_j &= \phi_{j+1}, \quad j = 1, 2 \\ \dot{\phi}_3 &= \mathcal{N}(\phi, \vartheta, \tau) \\ y &= \phi_1 \end{aligned} \tag{7.129}$$

where $\mathcal{Z}(\phi, \vartheta) = [\phi, L_f\Theta_{\rho+1}(\zeta), \ldots, L_f\Theta_{n-l}(\zeta)]$, $\mathcal{N}(\phi, \vartheta, \tau) = L_f^3 H$ is C^1 for $(\phi, \vartheta, \tau) \in \mathbb{R}^{(n-l)\times(n-l+1)}$ and $\zeta = \Theta^{-1}(\phi, \vartheta)$, for $(\phi, \vartheta, \tau) \in V = \{(\phi, \vartheta, \tau) \mid (\phi, \vartheta) \in \Theta(\Omega_\zeta)\}$.

The control objective is to design an adaptive controller for system (7.129) such that the output y follows the desired trajectory y_d, namely the tracking error converges to neighborhood of zero, i.e., $\|y(t) - y_d(t)\| \to 0$, while all the signals in the closed-loop systems are bounded.

The reference trajectory $y_d(t)$ is given by the following reference model

$$\begin{aligned} \dot{\phi}_{di} &= \phi_{d(i+1)}, \quad 1 \le i \le \rho - 1 \\ \dot{\phi}_{d\rho} &= \mathcal{F}_d(\phi_d) \\ y_d &= \phi_{d1} \end{aligned} \tag{7.130}$$

where $\rho \ge 3$ is a constant index, $\phi_d = [\phi_{d1}, \phi_{d2}, \ldots, \phi_{d\rho}] \in \mathbb{R}^{3\times\rho}$ are the state matrix of the reference system, $y_d \in \mathbb{R}^3$ is the system output and $F_d : \mathbb{R}^{3\times\rho} \to \mathbb{R}^3$ is a known function.

Assumption 7.10 The reference trajectory $y_d(t)$ and its third derivatives remain bounded.

Assumption 7.11 The zero dynamics of system (7.129) is established by $\dot{\vartheta} = \mathcal{Z}(0, \vartheta)$ and they are exponentially stable. In addition, there exist Lipschitz constants p_1 and p_2 for $\mathcal{Z}(\phi, \vartheta)$ such that

$$\left\| \mathcal{Z}(\phi, \vartheta) - \mathcal{Z}(0, \vartheta) \right\| \le p_1 \|\phi\| + p_2, \quad \forall(\phi, \vartheta) \in \Theta(\Omega_\zeta)$$

Under Assumption 7.11, by the converse Lyapunov theorem, there exists a Lyapunov function $V_0(\vartheta)$ which satisfies the following inequalities:

$$\gamma_1 \|\vartheta\|^2 \le V_0(\vartheta) \le \gamma_2 \|\vartheta\|^2 \tag{7.131}$$

$$\frac{\partial V_0}{\partial \vartheta} \mathcal{Z}(0, \vartheta) \le -\lambda_a \|\vartheta\|^2 \tag{7.132}$$

$$\left\| \frac{\partial V_0}{\partial \vartheta} \right\| \le \lambda_b \|\vartheta\| \tag{7.133}$$

where γ_1, γ_2, λ_a, λ_b are positive constants.

Define the desired trajectory as $y_d = \zeta_d$, and the tracking error as $e = y - y_d = \zeta - \zeta_d$. Let us introduce ϕ_d and Υ as

$$\phi_d = \left[y_d, y_d^{(1)}, y_d^{(2)}\right], \quad \phi_d \in \mathbb{R}^{(n-l)\times 3} \tag{7.134}$$

$$\Upsilon = \phi - \phi_d \tag{7.135}$$

and define the filtered tracking error as

$$\mathbf{r} = \Upsilon[\Lambda, 1]^T \in \mathbb{R}^3 \tag{7.136}$$

where the vector $\Lambda = [\lambda_1, \lambda_2]$ so that $s^2 + \lambda_2 s + \lambda_1$ is Hurwitz and as a result, $\Upsilon \to 0$ as $\mathbf{r} \to 0$. Then, the time derivative of the filtered tracking error can be written as

$$\dot{\mathbf{r}} = \mathcal{N}(\phi, \vartheta, \tau) - y_d^{(3)}(t) + \Upsilon[0\ \Lambda]^T \tag{7.137}$$

where $\mathbf{r} = [\mathbf{r}_1, \mathbf{r}_2, \mathbf{r}_3]^T$.

Add and subtract $\hat{\mathcal{N}}(\phi, \vartheta, \tau)$ on the right-hand side of Eq. (7.137), we obtain

$$\begin{aligned}\dot{\mathbf{r}} &= \mathcal{N}(\phi, \vartheta, \tau) - \hat{\mathcal{N}}(\phi, \vartheta, \tau) + \hat{\mathcal{N}}(\phi, \vartheta, \tau) \\ &\quad - y_d^{(3)}(t) + \Upsilon[0\ \Lambda]^T \\ &= \Delta + \nu - y_d^{(3)}(t) + \Upsilon[0\ \Lambda]^T\end{aligned} \tag{7.138}$$

where Δ is modeling error:

$$\Delta = \mathcal{N}(\phi, \vartheta, \tau) - \hat{\mathcal{N}}(\phi, \vartheta, \tau)$$

and ν is the pseudo control, which is defined as $\nu = \hat{\mathcal{N}}(\phi, \vartheta, \tau)$. According to [64], the pseudo control is chosen as

$$\nu = -K_p \mathbf{r} + y_d^{(3)} - \Upsilon[0\ \Lambda]^T + \nu_{dc} - \nu_{ad} \tag{7.139}$$

where K_p is diagonal positive, $y_d^{(3)}$ is the third derivative of the reference output, ν_{dc} is the output of a linear dynamic compensator, and ν_{ad} is the adaptive control signal designed to cancel Δ. Substituting (7.139) into the output dynamics (7.138), we have

$$\dot{\mathbf{r}} = -K_p\mathbf{r} + \nu_{dc} - \nu_{ad} + \Delta \tag{7.140}$$

The following linear dynamic compensator is introduced

$$\begin{aligned} \dot{\eta} &= A_c\eta + B_c\mathbf{r}(t) \\ \upsilon_{dc} &= C_c\eta + D_c\mathbf{r}(t) \end{aligned} \tag{7.141}$$

where $\eta \in \mathbb{R}^3$. Define

$$X = \begin{bmatrix} \mathbf{r} \\ \eta \end{bmatrix} \tag{7.142}$$

and then we can obtain the augmented tracking error dynamics as follows:

$$\dot{X} = \mathcal{A}X + \mathcal{B}[\upsilon_{ad} - \Delta] \tag{7.143}$$

$$\mathcal{A} = \begin{bmatrix} -K_p + D_c & C_c \\ B_c & A_c \end{bmatrix}, \qquad \mathcal{B} = \begin{bmatrix} -I \\ 0 \end{bmatrix}$$

where I is a identity matrix and A_c, B_c, C_c and D_c in (7.143) should be designed such that $\mathcal{A}$ is Hurwitz.

7.5.1 Output Feedback Control

The observer is adopted as follows:

$$\dot{\hat{X}} = \mathcal{A}\hat{X} + K(X - \hat{X}) \tag{7.144}$$

where K is a gain matrix and should be chosen such that $\mathcal{A} - K$ is asymptotically stable.

Define observing error as $\tilde{X} \triangleq X - \hat{X}$, the error dynamics is obtained by subtracting (7.144) from (7.143)

$$\dot{\tilde{X}} = A_K\tilde{X} + \mathcal{B}(\upsilon_{ad} - \Delta) \tag{7.145}$$

where $A_K = \mathcal{A} - K$.

Remark 7.34 The matrices $\mathcal{A}$ and A_K are Hurwitz and satisfy

$$\mathcal{A}^T P_1 + P_1\mathcal{A} = -Q_1 \tag{7.146}$$

$$A_K^T P_2 + P_2 A_K = -Q_2 \tag{7.147}$$

where $P_1 = P_1^T > 0$, $P_2 = P_2^T > 0$, $Q_1 = Q_1^T > 0$ and $Q_2 = Q_2^T > 0$.

Since the smooth function vector Δ is unknown, it can be approximated by NN over the compact set Ω_c as follows

$$\Delta = \left[W_M^{*T}\Phi\left(X, y_d^{(n)}\right)\right] + E \tag{7.148}$$

where W_M^* is the ideal NN weight vector, and E is the approximating error. Both W_M^* and E satisfy Assumption 7.9.

With (7.148) and the technique of NN parametrization, it has

$$\begin{aligned}\|\Delta\| &\le \Big\| W_M^{*T}\Phi\left(\hat{X}, y_d^{(n)}\right) \\ &\quad + W_M^{*T}\left(\Phi\left(X, y_d^{(n)}\right) - \Phi\left(\hat{X}, y_d^{(n)}\right)\right) + E\Big\| \\ &\le \psi^*\beta\left(\hat{X}, y_d^{(n)}\right)\end{aligned} \tag{7.149}$$

in which

$$\psi^* = \max\{w_{\max}, \delta_{\max} + 2w_{\max}\Phi_{\max}\}$$

$$\beta\left(\hat{X}, y_d^{(n)}\right) = \sqrt{\sum_{m=1}^{l}\phi_m^2\left(\hat{X}, y_d^{(n)}\right) + 1}$$

where neural nodes, $\Phi_{\max}$ is an unknown constant such that $\|\Phi(\cdot)\| \le \Phi_{\max}$ and $|\phi_m(\cdot)| < 1$.

Define

$$\tilde{\psi} \triangleq \hat{\psi} - \psi^* \tag{7.150}$$

in which $\hat{\psi}$ is the estimation of an unknown constant scalar ψ^*.

We employ the following adaptive control law for v_{ad}:

$$\begin{aligned} v_{ad} &= -\mathcal{K}\left(\hat{X}^T P_1 \mathcal{B}\right)^T - \hat{\psi}\frac{(\hat{X}^T P_1 \mathcal{B})^T \beta^2(\hat{X}, y_d^{(n)})}{\|\hat{X}^T P_1 \mathcal{B}\|\beta(\hat{X}, y_d^{(n)}) + \varepsilon} \\ \dot{\hat{\psi}} &= -\lambda_{\hat{\psi}}\hat{\psi} + \frac{\|\hat{X}^T P_1 \mathcal{B}\|^2\beta^2(\hat{X}, y_d^{(n)})}{\|\hat{X}^T P_1 \mathcal{B}\|\beta(\hat{X}, y_d^{(n)}) + \varepsilon}\end{aligned} \tag{7.151}$$

where $\mathcal{K}$ is diagonal definite and satisfying $\lambda_{\min}(\mathcal{K}) \ge 1$, ε and $\lambda_{\hat{\psi}}$ are positive design parameters.

7.5.2 Stability Analysis

In this section, the main result is summarized in Theorem 7.35 in which the stability analysis for the closed-loop system under the proposed control will be provided, and the analysis on how to specify the parameters for the observer and controller will also be included in the following remarks.

Theorem 7.35 *Consider the adaptation and control law* (7.139), (7.141) *and* (7.151) *over the compact set* $\Omega_c \subset \Omega_{NN}$ *in which the NN approximation is valid. For initial condition* $\hat{e}(0), e(0)$ *and* $\hat{\psi}(0)$ *starting in any compact set* $\Omega_0 \subset \Omega_c$, *if condition* (7.154) *is satisfied, then all closed-loop signals are semi-globally uniformly bounded over the following compact sets. In addition,* (i) *for* $\forall t \geq 0$, *the transient boundedness of the closed-loop system is depicted as*

$$\begin{aligned}\Omega_{X(t)} &= \left\{ X(t) \in \mathbb{R} \mid \|X(t)\| \leq \sqrt{\frac{2[V(0)\gamma + \rho]}{\gamma \lambda_{\min}(P_1)}} \right\} \\ \Omega_{\tilde{X}(t)} &= \left\{ \tilde{X}(t) \in \mathbb{R} \mid \|\tilde{X}(t)\| \leq \sqrt{\frac{2[V(0)\gamma + \rho]}{\gamma \lambda_{\min}(P_2)}} \right\} \\ \Omega_{\tilde{\psi}(t)} &= \left\{ \tilde{\psi}(t) \in \mathbb{R} \mid \|\tilde{\psi}(t)\| \leq \sqrt{\frac{2[V(0)\gamma + \rho]}{\gamma}} \right\}\end{aligned} \tag{7.152}$$

(ii) *for* $t \to \infty$, *the steady-state stability of the closed-loop system is depicted as*

$$\begin{aligned}\Omega_{X_\infty} &= \left\{ X(t) \in \mathbb{R} \mid \|X(t)\| \leq \sqrt{\frac{2\rho}{\gamma \lambda_{\min}(P_1)}} \right\} \\ \Omega_{\tilde{X}_\infty} &= \left\{ \tilde{X}(t) \in \mathbb{R} \mid \|\tilde{X}(t)\| \leq \sqrt{\frac{2\rho}{\gamma \lambda_{\min}(P_2)}} \right\} \\ \Omega_{\tilde{\psi}_\infty} &= \left\{ \tilde{\psi}(t) \in \mathbb{R} \mid \|\tilde{\psi}(t)\| \leq \sqrt{\frac{2\rho}{\gamma}} \right\}\end{aligned} \tag{7.153}$$

where c_2 *is a positive design parameter,*

$$\begin{aligned}c_1 &= 2\beta(\cdot)\|(P_1 + P_2)\mathcal{B}\| \\ c_3 &= \|\mathcal{K}\|^2\|(P_1 + P_2)\mathcal{B}\|^2 \\ \gamma &= \min\left\{ \frac{\lambda_{\min}(Q_1)}{2\lambda_{\max}(P_1)}, \frac{\lambda_{\min}(Q_2) - 2c_1c_2 - 2c_3}{\lambda_{\max}(P_2)}, \lambda_{\hat{\psi}} - \frac{c_1}{c_2} \right\} \\ \rho &= \frac{\lambda_{\hat{\psi}}}{2}\psi^{*2} + \psi^*\varepsilon \\ \lambda_{\min}(Q_2) &> 2c_1c_2 + 2c_3 \\ \lambda_{\hat{\psi}} &> \frac{c_1}{c_2}\end{aligned} \tag{7.154}$$

Proof Consider the following Lyapunov candidate:

$$V = \frac{1}{2}X^T P_1 X + \frac{1}{2}\tilde{X}^T P_2 \tilde{X} + \frac{1}{2}\tilde{\psi}^2 \tag{7.155}$$

with (7.143), (7.146), (7.147), (7.144) and (7.151), it has

$$\dot{V} = -\frac{1}{2}\|X\|^2_{Q_1} - \frac{1}{2}\|\tilde{X}\|^2_{Q_2} + X^T P_1 \mathcal{B}(v_{ad} - \Delta) + \tilde{X}^T P_2 \mathcal{B}(v_{ad} - \Delta) + \tilde{\psi}\dot{\tilde{\psi}} \tag{7.156}$$

with $X = \tilde{X} + \hat{X}$, we can rewrite (7.156) as

$$\dot{V} = -\frac{1}{2}\|X\|^2_{Q_1} - \frac{1}{2}\|\tilde{X}\|^2_{Q_2} + \tilde{X}^T (P_1 + P_2)\mathcal{B}(v_{ad} - \Delta) + \hat{X}^T P_1 \mathcal{B}(v_{ad} - \Delta) + \tilde{\psi}\dot{\tilde{\psi}} \tag{7.157}$$

From (7.150), we have $\psi^* = \hat{\psi} - \tilde{\psi}$, and from (7.149) and (7.150), the control law can be parameterized as

$$\|v_{ad}\| \le \|\mathcal{K}\| \big\| \big(\hat{X}^T P_1 \mathcal{B}\big)^T \big\| + \beta\big(\hat{X}, y_d^{(n)}\big)|\hat{\psi}| \tag{7.158}$$

and thus, we have

$$\begin{aligned}
&\big\|\tilde{X}^T (P_1 + P_2)\mathcal{B}(v_{ad} - \Delta)\big\| \\
&\quad\le \beta(\cdot)\big\|\tilde{X}^T (P_1 + P_2)\mathcal{B}\big\| \big(|\hat{\psi}| + |\psi^*|\big) \\
&\qquad + \big\|\tilde{X}^T (P_1 + P_2)\mathcal{B}\big\| \big\|\mathcal{K}\big(\hat{X}^T P_1 \mathcal{B}\big)^T\big\| \\
&\quad\le \big\|\tilde{X}^T (P_1 + P_2)\mathcal{B}\big\| \big(|\hat{\psi}| + \big(|\hat{\psi}| + |\tilde{\psi}|\big)\big) \\
&\qquad + \big\|\tilde{X}^T (P_1 + P_2)\mathcal{B}\big\| \big\|\mathcal{K}\big(\hat{X}^T P_1 \mathcal{B}\big)^T\big\| \\
&\quad\le \beta(\cdot)\big\|\tilde{X}^T (P_1 + P_2)\mathcal{B}\big\| \big(2|\hat{\psi}| + |\tilde{\psi}|\big) \\
&\qquad + \big\|\tilde{X}^T (P_1 + P_2)\mathcal{B}\big\| \big\|\mathcal{K}\big(\hat{X}^T P_1 \mathcal{B}\big)^T\big\| \\
&\quad\le \beta(\cdot)\big\|(P_1 + P_2)\mathcal{B}\big\| \|\tilde{X}\| 2\big(|\hat{\psi}| + |\tilde{\psi}|\big) \\
&\qquad + \big\|\tilde{X}^T (P_1 + P_2)\mathcal{B}\big\| \big\|\mathcal{K}\big(\hat{X}^T P_1 \mathcal{B}\big)^T\big\| \\
&\quad\le c_1\|\tilde{X}\|\big(|\hat{\psi}| + |\tilde{\psi}|\big) \\
&\qquad + \|\mathcal{K}\| \big\|\tilde{X}^T (P_1 + P_2)\mathcal{B}\big\| \big\|\big(\hat{X}^T P_1 \mathcal{B}\big)^T\big\|
\end{aligned} \tag{7.159}$$

where $c_1 = 2\beta(\cdot)\|(P_1 + P_2)\mathcal{B}\|$. From Lemma 7.29, we have

$$\begin{aligned}
&\|\mathcal{K}\| \big\|\tilde{X}^T (P_1 + P_2)\mathcal{B}\big\| \big\|\big(\hat{X}^T P_1 \mathcal{B}\big)^T\big\| \\
&\quad\le \|\mathcal{K}\|^2 \big\|(P_1 + P_2)\mathcal{B}\big\|^2 \|\tilde{X}\|^2 + \big\|\big(\hat{X}^T P_1 \mathcal{B}\big)^T\big\|^2 \\
&\quad= c_3\|\tilde{X}\|^2 + \big\|\big(\hat{X}^T P_1 \mathcal{B}\big)^T\big\|^2
\end{aligned} \tag{7.160}$$

with $c_3 = \|\mathcal{K}\|^2 \|(P_1 + P_2)\mathcal{B}\|^2$.

Considering (7.160), we can rewrite (7.159) as

$$\begin{aligned}
&\big\|\tilde{X}^T (P_1 + P_2)\mathcal{B}(v_{ad} - \Delta)\big\| \\
&\quad\le c_1\|\tilde{X}\|\big(|\hat{\psi}| + |\tilde{\psi}|\big) + c_3\|\tilde{X}\|^2 + \big\|\big(\hat{X}^T P_1 \mathcal{B}\big)^T\big\|^2 \\
&\quad\le \frac{c_1 c_2}{2}\|\tilde{X}\|^2 + \frac{c_1}{2c_2}\|\hat{\psi}\|^2 + \frac{c_1 c_2}{2}\|\tilde{X}\|^2 + \frac{c_1}{2c_2}\|\tilde{\psi}\|^2
\end{aligned}$$

$$
\begin{aligned}
&+ c_3\|\tilde{X}\|^2 + \left\|\left(\hat{X}^T P_1 \mathcal{B}\right)^T\right\|^2 \\
&\le (c_1c_2 + c_3)\|\tilde{X}\|^2 + \frac{c_1}{2c_2}\left(|\hat{\psi}|^2 + |\tilde{\psi}|^2\right) + \left\|\left(\hat{X}^T P_1 \mathcal{B}\right)^T\right\|^2
\end{aligned} \tag{7.161}
$$

where c_2 is a design parameter.

For the item $\hat{X}^T P_1 \mathcal{B}(v_{ad} - \Delta)$, it shows

$$
\begin{aligned}
&\hat{X}^T P_1 \mathcal{B}(v_{ad} - \Delta) \\
&\quad = -\hat{X}^T P_1 \mathcal{B}\mathcal{K}\left(\hat{X}^T P_1 \mathcal{B}\right)^T - \hat{\psi}\frac{\|\hat{X}^T P_1 \mathcal{B}\|^2 \beta^2}{\|\hat{X}^T P_1 \mathcal{B}\|\beta + \varepsilon} - \hat{X}^T P_1 \mathcal{B}\Delta \\
&\quad \le -\hat{X}^T P_1 \mathcal{B}\mathcal{K}\left(\hat{X}^T P_1 \mathcal{B}\right)^T - \hat{\psi}\frac{\|\hat{X}^T P_1 \mathcal{B}\|^2 \beta^2}{\|\hat{X}^T P_1 \mathcal{B}\|\beta + \varepsilon} + \left\|\hat{X}^T P_1 \mathcal{B}\Delta\right\| \\
&\quad \le -\hat{X}^T P_1 \mathcal{B}\mathcal{K}\left(\hat{X}^T P_1 \mathcal{B}\right)^T - \hat{\psi}\frac{\|\hat{X}^T P_1 B\|\beta^2}{\|\hat{X}^T P_1 B\|\beta + \varepsilon} + \psi^*\beta\left\|\hat{X}^T P_1 \mathcal{B}\right\| \\
&\quad \le -\hat{X}^T P_1 \mathcal{B}\mathcal{K}\left(\hat{X}^T P_1 \mathcal{B}\right)^T + \psi^*\frac{\|\hat{X}^T P_1 \mathcal{B}\|^2 \beta^2}{\|\hat{X}^T P_1 \mathcal{B}\|\beta + \varepsilon} - \hat{\psi}\frac{\|\hat{X}^T P_1 \mathcal{B}\|^2 \beta^2}{\|\hat{X}^T P_1 B\|\beta + \varepsilon} \\
&\qquad + \psi^*\beta\left\|\hat{X}^T P_1 \mathcal{B}\right\| - \psi^*\frac{\|\hat{X}^T P_1 B\|^2 \beta^2}{\|\hat{X}^T P_1 B\|\beta + \varepsilon}
\end{aligned} \tag{7.162}
$$

where $\beta(\cdot)$ is denoted as β.

In (7.150), $\hat{\psi}$ is positive with positive initial value according to Lemma 7.31 and Eqs. (7.148) and (7.151) are used. With Lemma 7.30, it has

$$
\begin{aligned}
&\psi^*\beta\left\|\hat{X}^T P_1 \mathcal{B}\right\| - \psi^*\frac{\|\hat{X}^T P_1 \mathcal{B}\|^2 \beta^2}{\|\hat{X}^T P_1 \mathcal{B}\|\beta + \varepsilon} \\
&\quad = \psi^*\frac{\|\hat{X}^T P_1 \mathcal{B}\|\beta\varepsilon}{\|\hat{X}^T P_1 \mathcal{B}\|\beta + \varepsilon} \le \psi^*\varepsilon
\end{aligned} \tag{7.163}
$$

In view of (7.163), (7.162) becomes

$$
\begin{aligned}
\hat{X}^T P_1 \mathcal{B}(v_{ad} - \Delta) \le{}& -\hat{X}^T P_1 \mathcal{B}\mathcal{K}\left(\hat{X}^T P_1 \mathcal{B}\right)^T \\
& - \tilde{\psi}\frac{\|\hat{X}^T P_1 \mathcal{B}\|^2 \beta^2}{\|\hat{X}^T P_1 \mathcal{B}\|\beta + \varepsilon} + \psi^*\varepsilon
\end{aligned} \tag{7.164}
$$

With (7.151), (7.159) and (7.164), (7.157) is written as

$$
\begin{aligned}
\dot{V} \le{}& -\frac{1}{2}\|X\|_{Q1}^2 - \frac{1}{2}\|\tilde{X}\|_{Q2}^2 + (c_1c_2 + c_3)\|\tilde{X}\|^2 \\
& + \frac{c_1}{2c_2}\left(|\hat{\psi}|^2 + |\tilde{\psi}|^2\right) + \left\|\left(\hat{X}^T P_1 \mathcal{B}\right)^T\right\|^2 \\
& - \hat{X}^T P_1 \mathcal{B}\mathcal{K}\left(\hat{X}^T P_1 \mathcal{B}\right)^T - \tilde{\psi}\frac{\|\hat{X}^T P_1 \mathcal{B}\|^2 \beta^2}{\|\hat{X}^T P_1 \mathcal{B}\|\beta + \varepsilon} \\
& + \psi^*\varepsilon - \hat{\psi}\tilde{\psi} + \tilde{\psi}\frac{\|\hat{X}^T P_1 \mathcal{B}\|^2 \beta^2}{\|\hat{X}^T P_1 \mathcal{B}\|\beta + \varepsilon}
\end{aligned}
$$

$$
\begin{aligned}
&\le -\frac{1}{2}\|X\|^2_{Q1} - \frac{1}{2}\|\tilde{X}\|^2_{Q2} + (c_1c_2 + c_3)\|\tilde{X}\|^2 \\
&\quad + \frac{c_1}{2c_2}\big(|\hat{\psi}|^2 + |\tilde{\psi}|^2\big) - \big(\lambda_{\min}(\mathcal{K}) - 1\big)\big\|\big(\hat{X}^T P_1 \mathcal{B}\big)^T\big\|^2 \\
&\quad - \tilde{\psi}\frac{\|\hat{X}^T P_1 \mathcal{B}\|^2\beta^2}{\|\hat{X}^T P_1 \mathcal{B}\|\beta + \varepsilon} + \psi^*\varepsilon - \hat{\psi}\tilde{\psi} \\
&\quad + \tilde{\psi}\frac{\|\hat{X}^T P_1 \mathcal{B}\|^2\beta^2}{\|\hat{X}^T P_1 \mathcal{B}\|\beta + \varepsilon}
\end{aligned} \tag{7.165}
$$

By completing the squares

$$
\tilde{\psi}\hat{\psi} = \frac{1}{2}\tilde{\psi}^2 + \frac{1}{2}\hat{\psi}^2 - \frac{1}{2}\psi^{*2} \tag{7.166}
$$

Equation (7.157) becomes

$$
\begin{aligned}
\dot{V} &\le -\frac{\lambda_{\min}(Q_1)}{2}\|X\|^2 + \frac{c_1}{2c_2}\big(|\hat{\psi}|^2 + |\tilde{\psi}|^2\big) \\
&\quad - \frac{1}{2}\big[\lambda_{\min}(Q_2) - 2c_1c_2 - 2c_3\big]\|\tilde{X}\|^2 - \frac{1}{2}\lambda_{\hat{\psi}}\tilde{\psi}^2 \\
&\quad - \frac{1}{2}\lambda_{\hat{\psi}}\hat{\psi}^2 + \frac{1}{2}\lambda_{\hat{\psi}}\psi^{*2} \\
&\quad + \psi^*\varepsilon \\
&\le -\frac{\lambda_{\min}(Q_1)}{2\lambda_{\max}(P_1)}\|X\|^2_{P_1} - \frac{1}{2}\left[\lambda_{\hat{\psi}} - \frac{c_1}{c_2}\right]\tilde{\psi}^2 \\
&\quad - \frac{[\lambda_{\min}(Q_2) - 2c_1c_2 - 2c_3]\|\tilde{X}\|^2_{P_2}}{2\lambda_{\max}(P_2)} \\
&\quad - \frac{1}{2}\left[\lambda_{\hat{\psi}} - \frac{c_1}{c_2}\right]\hat{\psi}^2 + \frac{1}{2}\lambda_{\hat{\psi}}\psi^{*2} + \psi^*\varepsilon \\
&\le -\gamma V + \rho
\end{aligned} \tag{7.167}
$$

where γ and ρ are given by (7.154).

Multiplying both sides of (7.167) by $e^{-\gamma t}$ and integrating over $[0, t]$, it can be shown that

$$
\big\|X(t)\big\|^2_{P_1} \le 2V(t) \le 2V(0)e^{-\gamma t} + 2\frac{\rho}{\gamma}\big(1 - e^{-\gamma t}\big) \tag{7.168}
$$

which can be written into

$$
\begin{aligned}
\big\|X(t)\big\| &\le \sqrt{\frac{2[V(0)\gamma + \rho]}{\gamma\lambda_{\min}(P_1)}}, \quad \forall t \ge 0 \\
\big\|X(t)\big\| &\le \sqrt{\frac{2\rho}{\gamma\lambda_{\min}(P_1)}}, \quad t \to \infty
\end{aligned} \tag{7.169}
$$

Similarly, we can obtain the rest result of (7.152) and (7.153) as follows

$$\begin{aligned}
\|\tilde{X}(t)\| &\leq \sqrt{\frac{2[V(0)\gamma+\rho]}{\gamma\lambda_{\min}(P_2)}}, \quad \forall t\geq 0\\
\|\tilde{X}(t)\| &\leq \sqrt{\frac{2\rho}{\gamma\lambda_{\min}(P_2)}}, \quad t\to\infty\\
\|\tilde{\psi}(t)\| &\leq \sqrt{\frac{2[V(0)\gamma+\rho]}{\gamma}}, \quad \forall t\geq 0\\
\|\tilde{\psi}(t)\| &\leq \sqrt{\frac{2\rho}{\gamma}}, \quad t\to\infty
\end{aligned} \tag{7.170}$$

This completes the proof. □

7.5.3 Simulation Studies

To demonstrate the effectiveness of the proposed control algorithm, let us consider the autonomous vehicle based on wheeled inverted pendulum as shown in Fig. 3.1.

In the simulation, we choose the parameters $I_w = 0.5\ \text{kg}\,\text{m}^2$, $M_w = 0.2$ kg, $I_m = 2.5\ \text{kg}\,\text{m}^2$, $M = 50.0$ kg, $m = 5.0$ kg, $l = 1.0$ m, $d = 0.5$ m, $r = 0.5$ m, $\zeta(0) = [-0.2, 0, \pi/18]^T$, $\dot{\zeta}(0) = [0.0, 0.1, 0.0]^T$. The disturbances from environments on the system are introduced as $1.0\sin(t)$, $1.0\cos(t)$ in the simulation model. The desired trajectories are chosen as $\theta_d = 0.2t$ rad, $\alpha_d = 0$ rad, initial velocity is 0.1 m/s. The system state is observed through the noisy linear measurement channel, zero-mean Gaussian noises are added to the state information. All noises are assumed to be mutually independent. The noises have variances corresponding to a 5 % noise to signal radio. We designed the following dynamic compensator $\dot{\eta} = -5\eta + \mathbf{r}(t)$, $v_{dc} = -35\eta + \mathbf{r}(t)$ with $\eta(0) = 0$. The observer gain is adopted as $\mathcal{K} = \text{diag}[1, 2]$.

The neural-adaptive network control without any knowledge of system dynamics under the random noise inputting to the controllers. The input vectors is $Z_1 = [\ddot{\zeta}_{1r}, \dot{\zeta}_{1r}, \dot{\zeta}_3, \zeta_3]^T \in \mathbb{R}^4$. Neural networks $\hat{W}_1^T S_1(Z_1)$ contains 32 nodes, with centers μ_l ($l = 1, \dots, l_1$) evenly spaced in $[-1.0, 1.0] \times [-1.0, 1.0] \times [-1.0, 1.0] \times [-1.0, 1.0]$. Neural networks $\hat{W}_3^T S_3(Z_3)$ contains 729 nodes, with centers μ_l ($l = 1, \dots, l_2$) evenly spaced in $Z_3 = [\ddot{\zeta}_{3r}, \dot{\zeta}_{3r}, r_3, \dot{\zeta}_1, \dot{\zeta}_3, \zeta_3]^T \in \mathbb{R}^6$, $[-1.0, 1.0] \times [-0.1, 0.1] \times [-1.0, 1.0] \times [-3.0, 3.0] \times [-1.0, 1.0] \times [-1.0, 1.0]$. The design parameters of the above controllers are: $Q_1 = \text{diag}[1, 2]$, and $Q_2 = \text{diag}[6]$, $K_p = \text{diag}[200.0, 6000]$, $\lambda_{\hat{\psi}} = 0.5$, $\varepsilon = \frac{1}{(1+t)^2}$, $\Lambda = \text{diag}[5.0, 10.0]$, $\hat{W}(0) = [0.1, 0.1; 0.1, 0.1]$.

The direction angles tracking by three control approaches are shown in Fig. 7.15, and the input torques are shown respectively in Fig. 7.16, the tilt angles for the

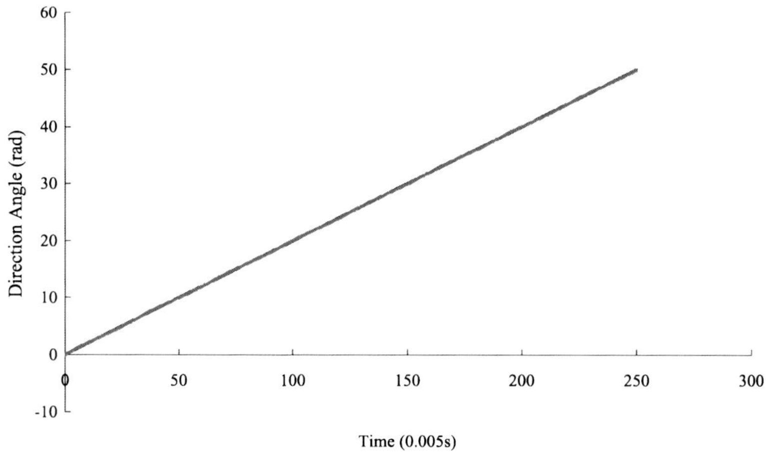

Fig. 7.15 Tracking the direction angle by the neural-adaptive network

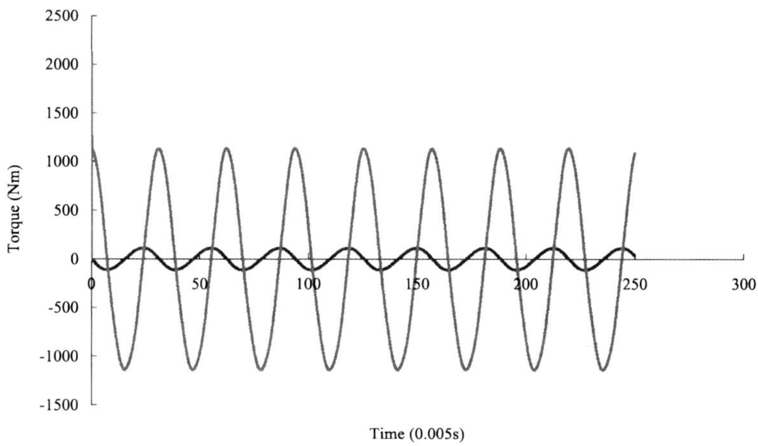

Fig. 7.16 Input torques by neural-adaptive network control

dynamic balance and the stable velocities under three control approaches are shown in Figs. 7.17 and 7.18, respectively. From these figures, even if without the prior knowledge of the system, we can obtain good performance by the proposed neural network control.

7.5.4 Conclusions

In this chapter, firstly, an adaptive neural network implicit control design has been carried out for dynamic balance and stable tracking of WIP-car, in the presence of

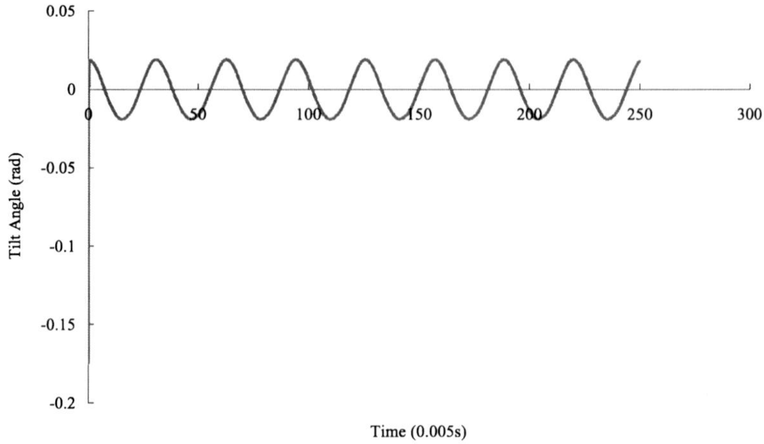

Fig. 7.17 Tracking the desired tilt angle by neural-adaptive network

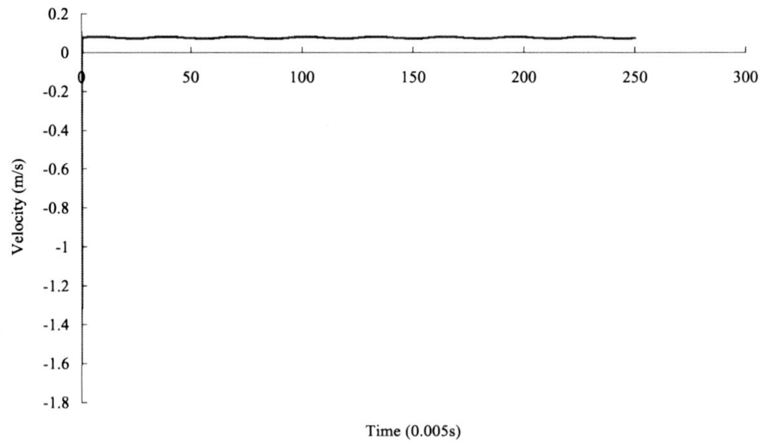

Fig. 7.18 The stable velocity by the neural-adaptive network control

unmodeled dynamics, or parametric/functional uncertainties. The control has been rigorously shown to guarantee semi-globally uniformly boundedness stability, and the steady state compact sets to which the closed-loop error signals converge are derived.

Second, LS-SVM based control design is carried out for dynamic balance and stable tracking of desired trajectories of mobile wheeled inverted pendulum, in the presence of unmodeled dynamics, or parametric/functional uncertainties. The control is mathematically shown to guarantee semi-globally uniformly bounded stability, and the steady state compact sets to which the closed loop error signals converge are derived. The size of compact sets can be made small through appropriate choice of control design parameters. Simulation results demonstrate that the system is able to track reference signals satisfactorily, with all closed loop signals uniformly bounded.

Third, adaptive fuzzy logic control design is carried out for dynamic balance and stable tracking of desired trajectories of wheeled inverted pendulum, in the presence of unmodeled dynamics, or parametric/functional uncertainties. The controllers are mathematically shown to guarantee semi-globally uniformly bounded stability, and the steady state compact sets to which the closed loop error signals converge are derived. The size of compact sets can be made small through appropriate choice of control design parameters. Simulation results demonstrate that the system is able to track reference signals satisfactorily, with all closed loop signals uniformly bounded.

Chapter 8
Optimized Model Reference Adaptive Control

8.1 Introduction

Similarly as in the previous chapters, we break down the nonlinear WIP system into a tilt and yaw angular motion dynamics subsystem and a forward velocity subsystem. Aiming at shaping the controlled angular motion dynamics to be of minimized motion tracking errors and angle accelerations, we employ the linear quadratic regulation (LQR) optimization technique to obtain an optimal reference model. The property of the optimized reference model as well as the minimized yaw and tilt angle accelerations help to enhance rider's comfort. In addition, due to the under-actuated mechanism of WIP, the vehicle forward velocity dynamics cannot be independently controlled separating from the pendulum tilt angle dynamics. In the previous chapters, the vehicle forward velocity dynamics is regarded as zero dynamics and zero dynamics theories have been used to analyze the stability. As the vehicle forward velocity dynamics is directly affected by the pendulum tilt angle dynamics, in this chapter, we develop a new framework to manipulate the forward velocity by using the tilt angle reference trajectory as a virtual "controller". A neural network (NN) based reference trajectory generator has been designed for the pendulum tilt angle, such that the vehicle forward velocity can be indirectly manipulated to follow the desired forward velocity.

As we know, a critical problem in the control design of WIP systems is the uncertain dynamics in addition to the unstable balance. In the literature, many works have been done on the adaptive model reference control of robots. In [115], a model reference adaptive control design is proposed for impedance control of a robot manipulator and this work was motivated by the model reference adaptive control design for motion control as in [112]. In [94], another two adaptive model reference control methods are developed. Although the system stability and the tracking performance can be guaranteed by using the methods in [94, 115], it is noted that they require the precise knowledge of the manipulator structure, which is often difficult to obtain in practice. To solve this problem, several methods that require little dynamics information are proposed in [28] and [128], while in this chapter, a new model reference control using little dynamics information is developed based on the sliding mode

Z. Li et al., *Advanced Control of Wheeled Inverted Pendulum Systems*,
DOI 10.1007/978-1-4471-2963-9_8, © Springer-Verlag London 2013

approach, such that the dynamics of the fully actuated subsystem can exactly follow the reference model within finite time horizon. A novel model matching error is introduced, which is used in both controller and parameter update law, and plays a key role in the convergence and stability analysis.

The main contributions of this chapter lie in: (i) a reference model for the yaw and tilt angles subsystems of the WIP system is derived using the LQR optimization approach which guarantees motion tracking and achieves the minimized angular accelerations for better riding comfort; (ii) variable structure control technique has been employed to design the adaptive reference control in order to make the controlled dynamics to match the reference model dynamics in finite time; and (iii) instead of leaving the unactuated forward velocity dynamics uncontrolled, the reference trajectory for the tilt angle is designed to indirectly affect the forward velocity such that the desired velocity can be achieved. High order neural network (HONN) has been employed to construct a reference trajectory generator of the tilt angle.

The rest of this chapter is organized as follows. In Sect. 8.2, notations, preliminary knowledge of NN approximation and LQR optimization are presented. In Sect. 8.3, two subsystems of the under-actuated WIP systems are derived with the first subsystem fully actuated while the second unactuated. A reference trajectory generator using HONN for the tilt angle is designed in Sect. 8.5 such that the forward velocity is indirectly manipulated to follow its desired trajectory. In Sect. 8.6, simulation studies are carried out to verify the effectiveness of the proposed method. Concluding remarks are given in Sect. 8.7.

8.2 Preliminaries

8.2.1 Finite Time Linear Quadratic Regulator

Given a linear system with completely stabilizable pair $[A, B]$ [12]

$$\dot{x} = Ax + Bu, \quad x(t_0) = x_0, \; x \in R^n, \; u \in R^n \tag{8.1}$$

the optimal control $u^*(t)$, $t > 0$, that minimizes the following performance index

$$\begin{aligned} J &= \int_{t_0}^{t_f} \big((x - x_d)^T Q(x - x_d) + u^T Ru\big)\, dt \\ R &= R^T > 0, \quad Q = Q^T > 0 \end{aligned} \tag{8.2}$$

is given by

$$u^* = -R^{-1}B^T(Px + s) \tag{8.3}$$

where P the solution of the following Riccati equation

$$-\dot{P} = PA + A^T P - PBR^{-1}B^T P + Q, \quad P(t_f) = 0_{[n,n]} \tag{8.4}$$

and s is the solution of

$$-\dot{s} = \big(A - BR^{-1}B^T P\big)^T s + Qx_d, \quad s(t_f) = 0_{[n]} \tag{8.5}$$

8.2.2 HONN Approximation

There are many well developed approaches used to approximate an unknown function. Neural network is one of the mostly frequently employed approximation methods due to the fact that NN is shown to be capable of universally approximating any unknown continuous function to arbitrary precision [16, 17, 24, 27]. Similar to biological neural networks, NN consists of massive simple processing units which correspond to biological neurons. With the highly parallel structure, NN is of powerful computing ability and intelligence of learning and adaptation with respect to fresh and unknown data. Higher order neural network (HONN) is a kind of linearly parameterized neural network (LPNN) [47], and it has been shown to have strong storage capacity, approximation and learning capability. HONN satisfies the conditions of the Stone–Weierstrass theorem and can therefore approximate any continuous function over a compact set [78, 102]. It is pointed in [54] that by utilizing *a priori* information, HONN is very efficient in solving problems because the order or structure of HONN can be tailored to the order or structure of a given problem. The structure of HONN is expressed as follows:

$$\phi(W, z) = W^T S(z) W, \quad S(z) \in R^l$$

$$S(z) = \left[s_1(z), s_2(z), \dots, s_l(z)\right]^T \tag{8.6}$$

$$s_i(z) = \prod_{j \in I_i} \left[s(z_j)\right]^{d_j(i)}, \quad i = 1, 2, \dots, l \tag{8.7}$$

where $z \in \Omega_z \subset R^m$ is the input to HONN, l the NN nodes number, $\{I_1, I_2, \dots, I_l\}$ a collection of l not-ordered subsets of $\{1, 2, \dots, m\}$, e.g., $I_1 = \{1, 3, m\}$, $I_2 = \{2, 4, m\}$, $d_j(i)$'s nonnegative integers, W an adjustable synaptic weight vector, and $s(z_j)$ a monotonically increasing and differentiable sigmoidal function. In this chapter, it is chosen as a hyperbolic tangent function, i.e., $s(z_j) = \frac{e^{z_j} - e^{-z_j}}{e^{z_j} + e^{-z_j}}$.

For a smooth function $\varphi(z)$ over a compact set $\Omega_z \subset R^m$, given a small constant real number $\mu^* > 0$, if l is sufficiently large, there exist a set of ideal bounded weights W^* such that

$$\max\left|\varphi(z) - \phi\left(W^*, z\right)\right| < \mu(z), \quad \left|\mu(z)\right| < \mu^* \tag{8.8}$$

From the universal approximation results for neural networks [58], it is known that the constant μ^* can be made arbitrarily small by increasing the NN nodes number l.

Lemma 8.1 [105] *Consider the basis functions of HONN* (8.6) *with z being the input vector. The following properties of HONN will be used in the proof of closed-loop system stability*

$$\lambda_{\max}\left[S(z)S^T(z)\right] < 1, \quad S^T(z)S(z) < l \tag{8.9}$$

where $\lambda_{\max}(M)$ denotes the max eigenvalue of M.

8.3 Dynamics of Wheeled Inverted Pendulums

Follow the techniques developed in previous chapters, we see that the dynamics of wheeled inverted pendulum can be expressed as

$$D(\zeta)\ddot{\zeta} + C(\zeta, \dot{\zeta})\dot{\zeta} + G(\zeta) + f_d = \tau \tag{8.10}$$

where

$$D(\zeta) = \begin{bmatrix} \Phi^T D_v \Phi & \Phi^T D_{v\alpha} \\ D_{\alpha v}\Phi & D_\alpha \end{bmatrix} = \begin{bmatrix} d_{11}(\zeta_3) & 0 & 0 \\ 0 & d_{22} & d_{23}(\zeta_3) \\ 0 & d_{32}(\zeta_3) & d_{33} \end{bmatrix}$$

$$C(\zeta, \dot{\zeta}) = \begin{bmatrix} \Phi^T D_v \dot{\Phi} + \Phi^T C_v \Phi & \Phi^T C_{v\alpha} \\ D_{\alpha v}\dot{\Phi} + C_{\alpha v}\Phi & C_\alpha \end{bmatrix} = \begin{bmatrix} c_{11}(\zeta_3, \dot{\zeta}_3) & 0 & c_{13}(\dot{\zeta}_1, \zeta_3) \\ 0 & 0 & c_{23}(\zeta_3, \dot{\zeta}_3) \\ c_{31}(\dot{\zeta}_1, \zeta_3) & 0 & 0 \end{bmatrix}$$

$$f_d = \begin{bmatrix} \Phi^T d_v \\ d_\alpha \end{bmatrix} = \begin{bmatrix} \bar{f}_1 \\ \bar{f}_2 \\ \bar{f}_3 \end{bmatrix}, \qquad G(\zeta) = \begin{bmatrix} \Phi^T G_v \\ G_\alpha \end{bmatrix} = \begin{bmatrix} 0 \\ 0 \\ g_3(\zeta_3) \end{bmatrix} \tag{8.11}$$

$$\tau = \begin{bmatrix} \Phi^T B_v \tau_v \\ 0 \end{bmatrix} = \begin{bmatrix} \tau_1 \\ \tau_2 \\ 0 \end{bmatrix}$$

with $d_{11}(\zeta_3) = d^2 m/2 + I_M d^2/2R^2 + I_\omega + ML^2 \sin^2\alpha$, $d_{22} = 2m + 2I_M/R^2 + M$, $d_{33} = ML^2 + I_M$ and $d_{23} = d_{32} = ML\cos\alpha$; $c_{11}(\zeta_3, \dot{\zeta}_3) = \frac{1}{2}ML^2\dot{\alpha}\sin^2 2\alpha$, $c_{13}(\dot{\zeta}_1, \zeta_3) = \frac{1}{2}\omega ML^2 \sin 2\alpha$, $c_{23}(\zeta_3, \dot{\zeta}_3) = -ML\dot{\alpha}\sin\alpha$ and $c_{31}(\dot{\zeta}_1, \zeta_3) = -\frac{1}{2}\omega \times ML^2 \sin 2\alpha$; $g_3(\zeta_3) = -MgL\sin\alpha$.

Remark 8.2 It should be mentioned that due to the unknown system parameters in the above dynamics formulation, the dynamics matrices are actually unknown for control design.

The following two properties are well known for the Lagrange–Euler formulation of robotic dynamics:

Property 8.3 The matrix $D(\zeta)$ is symmetric and positive definite.

Property 8.4 The matrix $2C(\zeta, \dot{\zeta}) - \dot{D}(\zeta)$ is a skew-symmetric matrix.

Expanding Eq. (8.10), we obtain three equations in which the first one can be regarded as ζ_1-subsystem described as follows:

$$\Sigma_{\zeta_1}: \quad d_{11}(\zeta_3)\ddot{\zeta}_1 + c_{11}\dot{\zeta}_1 + c_{13}\dot{\zeta}_3 + \bar{f}_1 = \tau_1 \tag{8.12}$$

The second and the third equations are:

$$\begin{aligned} d_{22}\ddot{\zeta}_2 + d_{23}(\zeta_3)\ddot{\zeta}_3 + c_{23}\dot{\zeta}_3 + \bar{f}_2 &= \tau_2 \\ d_{32}(\zeta_3)\ddot{\zeta}_2 + d_{33}\ddot{\zeta}_3 + c_{31}\dot{\zeta}_1 + \bar{f}_3 + g_3 &= 0 \end{aligned} \tag{8.13}$$

from which we could clearly see the under-actuated configuration, i.e., it is not possible to control ζ_2 and ζ_3 independently.

By substituting the third equation into the second one, we obtain the ζ_2-subsystem and ζ_3-subsystem as follows:

$$\begin{aligned}\Sigma_{\zeta_2}:\quad & \frac{d_{22}d_{33} - d_{23}^2(\zeta_3)}{d_{33}}\ddot{\zeta}_2 - \frac{d_{23}(\zeta_3)}{d_{33}}\left(c_{31}\dot{\zeta}_1 + g_3(\zeta_3) + \bar{f}_3\right) \\ & + c_{23}\dot{\zeta}_3 + \bar{f}_2(\dot{\zeta}_2) = \tau_2 \qquad (8.14)\\ \Sigma_{\zeta_3}:\quad & \frac{d_{22}d_{33} - d_{23}^2(\zeta_3)}{d_{23}(\zeta_3)}\ddot{\zeta}_3 + \frac{d_{22}}{d_{23}(\zeta_3)}\left(c_{31}\dot{\zeta}_1 + g_3(\zeta_3) + \bar{f}_3\right) \\ & - c_{23}\dot{\zeta}_3 - \bar{f}_2(\dot{\zeta}_2) = -\tau_2 \qquad (8.15)\end{aligned}$$

Remark 8.5 Note that an implicit assumption is made to derive (8.15) that the pendulum tilt angle $\zeta_3 = \alpha \in (-\frac{\pi}{2}, \frac{\pi}{2})$ such that $d_{23}(\zeta_3) \neq 0$. This assumption is very reasonable as the dynamics when the tilt angle is beyond $-\frac{\pi}{2}$ or $\frac{\pi}{2}$ is out of the research interest.

8.4 Control of ζ_1 and ζ_3-Subsystems

8.4.1 Subsystem Dynamics

For convenience, let us combine the dynamics of subsystems Σ_{ζ_1} and Σ_{ζ_3} together as follows:

$$\mathcal{M}\ddot{\xi} + \mathcal{C}\dot{\xi} + h = \tau \qquad (8.16)$$

where

$$\begin{aligned}\mathcal{M} &= \begin{bmatrix} d_{11}(\xi_2) & 0 \\ 0 & \frac{d_{22}d_{33}-d_{23}^2(\xi_2)}{d_{23}(\xi_2)} \end{bmatrix}\\ \mathcal{C} &= \begin{bmatrix} c_{11} & c_{13} \\ c_{31} & \frac{1}{2}\frac{d}{dt}\frac{d_{22}d_{33}-d_{23}^2(\xi_2)}{d_{23}(\xi_2)} \end{bmatrix}\\ h &= \begin{bmatrix} \bar{f}_1 \\ \frac{d_{22}}{d_{23}(\xi_2)}(g_3(\xi_2) + \bar{f}_3) - \bar{f}_2 \end{bmatrix}\\ &\quad + \begin{bmatrix} 0 & 0 \\ \frac{d_{22}-d_{23}(\xi_2)}{d_{23}(\xi_2)}c_{31} & -c_{23} - \frac{1}{2}\frac{d}{dt}\frac{d_{22}d_{33}-d_{23}^2(\xi_2)}{d_{23}(\xi_2)} \end{bmatrix}\begin{bmatrix} \dot{\xi}_1 \\ \dot{\xi}_2 \end{bmatrix}\\ \tau &= \begin{bmatrix} \tau_1 & -\tau_2 \end{bmatrix}^T, \qquad \xi = \begin{bmatrix} \zeta_1 & \zeta_3 \end{bmatrix}^T\end{aligned}$$

Recall that the inertia matrix $D(\zeta)$ is positive definite and symmetric, so we know that the terms $d_{11}(\xi_2)$ and $d_{22}d_{33} - d_{23}^2(\xi_2)$ are positive. And according to Remark 8.5 we know that $\frac{d_{22}d_{33}-d_{23}^2(\xi_2)}{d_{23}(\xi_2)}$ are also positive. In addition, due to the fact that $\dot{D} - 2C$ is skew-symmetric, we have the following properties:

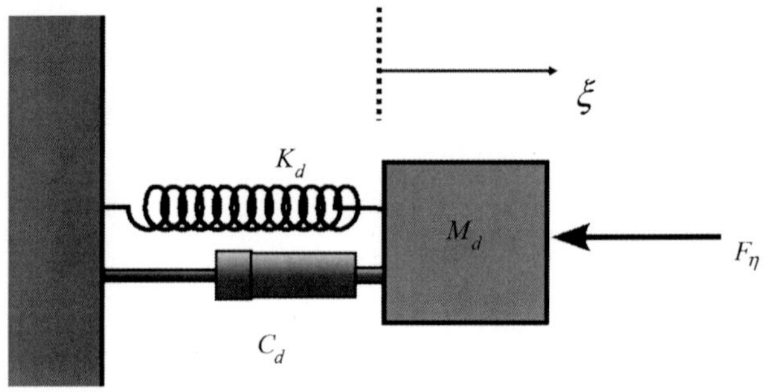

Fig. 8.1 Mass–spring–damper impedance model

Property 8.6 The inertia matrix $\mathcal{M}$ is positive definite and symmetric.

Property 8.7 The matrix $\dot{\mathcal{M}} - 2\mathcal{C}$ is skew-symmetric such that $x^T(\dot{\mathcal{M}} - 2\mathcal{C})x = 0$, $\forall x \in R^2$.

In addition, it is well known in the robotics literature that the following property hold.

Property 8.8 [110] There exist unknown positive scalars $\theta_M, \theta_C, \theta_{h1}, \theta_{h2}$ such that $\|\mathcal{M}(\xi)\| \le \theta_M$, $\|\mathcal{C}(\xi, \dot{\xi})\| \le \theta_C \|\dot{\xi}\|$ and $\|h(\xi, \dot{\xi})\| \le \theta_{h1} \|\xi\| \|\dot{\xi}\| + \theta_{h2}$.

8.4.2 Optimal Reference Model

To design the model reference control of subsystem dynamics (8.16), we need to construct a reference model for the ζ_1 and ζ_3 subsystem. We consider a second order system as follows:

$$M_d\ddot{\xi} + C_d\dot{\xi} + K_d\xi = -F_\eta(\xi_d, \dot{\xi}_d) \tag{8.17}$$

where M_d, C_d, K_d are the desired inertia, damping and stiffness matrices, respectively, and F_η can be regarded as an artificial force. This reference model can be illustrated by a mass–spring–damper system shown in Fig. 8.1. As the control objective is to make the closed-loop dynamics of the controlled subsystem (8.16) match the dynamics of the reference model (8.17), we should suitably choose the parameters of the reference model such that it not only guarantees the motion tracking but also take into consideration of the rider's comfort, e.g., it must be overdamped to guarantee no overshooting and no oscillation exist. Actually the virtual impedance model provides kind of cushion effect for better riding experience and when there is no artificial force F_η (e.g., the rider let go the wheel and pedal), the yaw and tilt angles ζ_1 and ζ_3 will tend to rest on the zero position.

In order to choose the optimal values of the reference model parameters, we introduce the following performance index:

$$I_P = \int_{t_0}^{t_f} \left(e^T Qe + \ddot{\xi}^T M_d \ddot{\xi}\right) dt \tag{8.18}$$

which minimizes both the motion tracking error

$$e = \xi - \xi_d \tag{8.19}$$

and the yaw and tilt angular accelerations. We aim to enhance rider's comfort by reducing any unnecessary angular accelerations, while at the same time to ensure motion tracking performance. Now let us consider how to minimize the performance index I_P by suitably designing C_d, K_d and F_η. In order to apply the LQR optimization technique, we rewrite the reference model (8.17) as

$$\dot{\bar{\xi}} = A\bar{\xi} + Bu \tag{8.20}$$

with

$$\bar{\xi} = [\xi, \dot{\xi}], \qquad \bar{\xi}_d = [\xi_d, \dot{\xi}_d] \tag{8.21}$$

$$A = \begin{bmatrix} 0_{[2,2]} & I_{[2,2]} \\ 0_{[2,2]} & 0_{[2,2]} \end{bmatrix}, \quad B = \left[0_{[2,2]}, I_{[2,2]}\right]^T, \quad Q = \begin{bmatrix} q_1 & 0 \\ 0 & q_2 \end{bmatrix} \tag{8.22}$$

$$u = -M_d^{-1}[K_d, C_d]\bar{\xi} - M_d^{-1} F_\eta(\xi_d, \dot{\xi}_d) \tag{8.23}$$

Noting that $u = \ddot{\xi}$ and introducing $\bar{Q}$ defined as

$$\bar{Q} = \begin{bmatrix} Q & 0_{[2,2]} \\ 0_{[2,2]} & 0_{[2,2]} \end{bmatrix} \tag{8.24}$$

we can then rewrite the performance index (8.18) as

$$P_{\bar{I}} = \int_{t_0}^{t_f} \left((\bar{\xi} - \bar{\xi}_d)^T \bar{Q}(\bar{\xi} - \bar{\xi}_d) + u^T M_d u\right) dt \tag{8.25}$$

If we regard u as the control input to system (8.20), then the minimization of (8.25) subject to dynamics constraint (8.20) becomes a typical LQR control design problem. According to the LQR optimal control technique reviewed in Sect. 8.2.1, the solution of u that minimizes (8.25) is

$$u = -M_d^{-1} B^T P\bar{\xi} - M_d^{-1} B^T s \tag{8.26}$$

where P is the solution of the following differential equation

$$-\dot{P} = PA + A^T P - PBM_d^{-1} B^T P + \bar{Q}, \quad P(t_f) = 0_{[4,4]} \tag{8.27}$$

and s is the solution of the following differential equation

$$-\dot{s} = \left(A - BM_d^{-1} B^T P\right)^T g + \bar{Q}\bar{\xi}_d, \quad s(t_f) = 0_{[4]} \tag{8.28}$$

Comparing Eqs. (8.23) and (8.26), we can see that the matrices K_d and C_d can be calculated in the following manner:

$$[K_d, C_d] = B^T P, \qquad F_\eta = B^T s \tag{8.29}$$

Remark 8.9 To obtain a reference model with minimizing (8.18), we first need to choose values of q_1 and q_2 for the weighting matrix Q as well as the desired mass M_d, then we can calculate K_d and C_d according to (8.27), (8.28) and (8.29).

Remark 8.10 It should be emphasized that the performance index (8.25) is a finite time integration such that when the minimization of (8.25) is achieved using (8.26), we always have $e = 0_{[2]}$, $\forall t > t_f$.

8.4.3 Model Matching Error

By using $e = \xi - \xi_d$, the reference model (8.17) can be rewritten as

$$M_d\ddot{e} + C_d\dot{e} + K_d e = -\eta \tag{8.30}$$

with

$$\eta = F_\eta + M_d\ddot{\xi}_d + C_d\dot{\xi}_d + K_d\xi_d \tag{8.31}$$

We can see that the objective of the control design becomes to look for a proper control input torque τ in (8.16) such that the dynamics (8.16) match the desired reference model dynamics (8.30). In order to measure the difference between the subsystem dynamics (8.16) and the reference model dynamics (8.30), we introduce the following matching error defined as

$$w = M_d\ddot{e} + C_d\dot{e} + K_d e + \eta \tag{8.32}$$

such that when the following condition is achieved

$$w(t) = 0_{[2]}, \quad t > t_f \tag{8.33}$$

the subsystem dynamics (8.16) would exactly match the desired reference model dynamics (8.30).

For the convenience of the following analysis, we define an augmented matching error as

$$\bar{w} = K_\eta w = \ddot{e} + C_m\dot{e} + K_m e + K_\eta\eta \tag{8.34}$$

where $C_m = M_d^{-1}C_d$, $K_m = M_d^{-1}K_d$ and $K_\eta = M_d^{-1}$.

Remark 8.11 The virtual mass matrix M_d is always chosen to be positive definite such that it is invertible and $\bar{w}$ in (8.34) is well defined.

By choosing two positive definite matrices Λ and Γ such that $\Lambda + \Gamma = C_m$ and $\Gamma\Lambda + \dot{\Lambda} = K_m$, we could further rewrite the augmented matching error $\bar{w}$ as:

$$\bar{w} = \ddot{e} + (\Lambda + \Gamma)\dot{e} + (\Gamma\Lambda + \dot{\Lambda})e + \dot{\eta}_l + \Gamma\eta_l \tag{8.35}$$

where η_l satisfies

$$\dot{\eta}_l + \Gamma \eta_l = K_\eta \eta \tag{8.36}$$

By defining a filtered matching error

$$z = \dot{e} + \Lambda e + \eta_l \tag{8.37}$$

we see that the augmented matching error $\bar{w}$ can be written as

$$\bar{w} = \dot{z} + \Gamma z \tag{8.38}$$

which implies that z could be obtained by passing $\bar{w}$ through a filter. From (8.38) and (8.34), we see that $z = 0_{[2]}$ and subsequently $\dot{z} = 0_{[2]}$ will lead to $w = 0_{[2]}$, i.e., matching error diminished. Then, the yaw and tilt angle subsystem dynamics (8.16) would exactly match the reference model (8.17). According to Remark 8.10, after a finite time t_f, we will have $\xi = \xi_d$. In next section, we will design an adaptive controller which guarantee there exist a finite time $t_z \ll t_f$ such that $z = 0_{[2]}, t > t_z$.

8.4.4 Adaptive Control Design

8.4.5 Controller Structure

In this section, we are ready to discuss the details of the adaptive control design. We propose the control input of the subsystem (8.16) as

$$\tau = \tau_{ct} + \tau_{fb} \tag{8.39}$$

where τ_{ct} and τ_{fb} are the computed torque control input and the feedback torque control input, respectively. The feedback torque control input is given by

$$\tau_{fb} = -K \operatorname{sgn}(z) \tag{8.40}$$

where K is a diagonal positive definite matrix with $k_{\min}$ denoting the minimal element on the diagonal, and $\operatorname{sgn}(\cdot)$ is the sign function.

The computed torque control input is designed as

$$\tau_{ct} = -Y(\ddot{\xi}_r, \dot{\xi}_r, z)\hat{\Theta} \tag{8.41}$$

where the trajectories ξ_r and $\dot{\xi}_r$ are defined as follows:

$$\begin{aligned} \dot{\xi}_r &= \dot{\xi}_d - \Lambda e - \eta_l \\ \ddot{\xi}_r &= \ddot{\xi}_d - \Lambda \dot{e} - \dot{\eta}_l \end{aligned} \tag{8.42}$$

and $\hat{\Theta}$ is the estimate of $\Theta = [\theta_D, \theta_C, \theta_{h1}, \theta_{h2}]^T$ (refer to Property 8.8) and

$$Y(\ddot{\xi}_r, \dot{\xi}_r, \dot{\xi}, \xi, z) = \begin{bmatrix} \|\ddot{\xi}_r\| \operatorname{sgn}(z_1) & \|\dot{\xi}\| \|\dot{\xi}_r\| \operatorname{sgn}(z_1) & \|\dot{\xi}\| \|\xi\| \operatorname{sgn}(z_1) & \operatorname{sgn}(z_1) \\ \|\ddot{\xi}_r\| \operatorname{sgn}(z_2) & \|\dot{\xi}\| \|\dot{\xi}_r\| \operatorname{sgn}(z_2) & \|\dot{\xi}\| \|\xi\| \operatorname{sgn}(z_2) & \operatorname{sgn}(z_2) \end{bmatrix} \tag{8.43}$$

In the following, for the convenience, we use Y instead of $Y(\ddot{\xi}_r, \dot{\xi}_r, \dot{\xi}, \xi, z)$ where it does not result in any confusion.

Remark 8.12 The potential drawback caused by the sgn(·) function used in (8.40) and (8.43), such as chattering problems, can be avoided by replacing sgn(·) with hyperbolic tangent function tanh(·) or saturation function sat(·) [74]. For the convenience of our analysis, we keep using sgn(·) in the following theoretic analysis. It is trivial but tedious to use tanh(·) or sat(·) for theoretic analysis.

To obtain $\hat{\Theta}$, we develop the following parameter estimator:

$$\dot{\hat{\Theta}} = \Gamma_{\Theta}^{-1} Y^T z \tag{8.44}$$

where Γ_{Θ} is a diagonal positive definite matrix.

8.4.6 Control Performance Analysis

Lemma 8.13 *Consider the closed-loop control system consisting of the yaw and tilt angle subsystem* (8.16) *and the controller* (8.39). *The following results hold*: (i) *all the signals in the closed-loop are uniformly bounded, and* (ii) *the filtered matching error z will converge to zero, i.e.*, $\lim_{t\to 0} z = 0_{[2]}$.

Proof To prove the above lemma, let us consider the following Lyapunov-like composite energy function:

$$V_1(t) = U_1(t) + U_2(t) \tag{8.45}$$

where

$$U_1(t) = \frac{1}{2} z^T \mathcal{M}(\xi) z, \qquad U_2(t) = \frac{1}{2} \tilde{\Theta}^T(t) \Gamma_{\Theta}^T \tilde{\Theta}(t) \tag{8.46}$$

with $\tilde{\Theta}(t) = \Theta(t) - \hat{\Theta}(t)$.

According to Property 8.8 and the definitions of Y in (8.43), we have

$$\begin{aligned}
&-z^T \big(\mathcal{M}(\xi)\ddot{\xi}_r + \mathcal{C}(\xi,\dot{\xi})\dot{\xi}_r + h(\xi,\dot{\xi})\big) \\
&\quad \le \|z\| \big(\|\mathcal{M}(\xi)\ddot{\xi}_r\| + \|\mathcal{C}(\xi,\dot{\xi})\dot{\xi}_r\| + \|h(\xi,\dot{\xi})\|\big) \\
&\quad \le \|z\| \big(\|\mathcal{M}(\xi)\| \|\ddot{\xi}_r\| + \|\mathcal{C}(\xi,\dot{\xi})\| \|\dot{\xi}_r\| + \|h(\xi,\dot{\xi})\|\big) \\
&\quad \le \|z\| \big(\theta_M \|\ddot{\xi}_r\| + \theta_C \|\dot{\xi}\| \|\dot{\xi}_r\| + \theta_{h1} \|\xi\| \|\dot{\xi}\| + \theta_{h2}\big) \\
&\quad = z^T \operatorname{sgn}(z) \big(\theta_M \|\ddot{\xi}_r\| + \theta_C \|\dot{\xi}\| \|\dot{\xi}_r\| + \theta_{h1} \|\xi\| \|\dot{\xi}\| + \theta_{h2}\big) \\
&\quad = z^T Y \Theta
\end{aligned} \tag{8.47}$$

Then, by considering the closed-loop dynamics (8.37) and the controller (8.39), we obtain

$$\begin{aligned}
\dot{U}_1 &= z^T \mathcal{M}(\xi)\dot{z} + \frac{1}{2} z^T \dot{\mathcal{M}}(\xi) z \\
&= z^T \mathcal{M}(\xi)\dot{z} + z^T \mathcal{C}(\xi,\dot{\xi}) z
\end{aligned}$$

$$
\begin{aligned}
&= z^T\big(-\mathcal{M}(\xi)\ddot{\xi}_r - \mathcal{C}(\xi,\dot{\xi})\dot{\xi}_r - h(\xi,\dot{\xi}) + u\big)\\
&\le z^T\big(Y\Theta - Y\hat{\Theta} - K\,\mathrm{sgn}(z)\big)\\
&= z^T\big(Y\tilde{\Theta} - K\,\mathrm{sgn}(z)\big)
\end{aligned} \tag{8.48}
$$

where we have used Property 8.7 in the second equality.

On the other hand, we have

$$
\dot{U}_2 = -\dot{\hat{\Theta}}^T \Gamma_\Theta^T \tilde{\Theta} = -z^T Y\tilde{\Theta} \tag{8.49}
$$

According to (8.45), (8.48) and (8.49), we obtain the boundedness of $V_1(t)$ according to the following derivation:

$$
\dot{V}_1 = \dot{U}_1 + \dot{U}_2 \le -z^T K\,\mathrm{sgn}(z) \le 0 \tag{8.50}
$$

This implies that both U_1 and U_2 are bounded and consequently z and $\tilde{\Theta}$ are bounded. Integrating both sides of (8.50) and noting the fact

$$
-z^T K\,\mathrm{sgn}(z) = -K|z| \le -\|K\|\|z\| \le -k_{\min}\|z\| \le 0 \tag{8.51}
$$

we see that

$$
\int_0^\infty k_{\min}\|z\|\,dt \le V_1(0) \tag{8.52}
$$

from which one can immediately obtain $\lim_{t\to 0}\|z\| = 0$, which is equivalent to $\lim_{t\to 0} z = 0_{[2]}$. This completes the proof. □

Theorem 8.14 *There exist a finite time t_z such that $z = 0_{[2]}$, $t > t_z \ll t_f$.*

Proof This proof uses a contradiction argument. Assume $\|z\| > 0, t > 0$.

There exist two constants $\bar{m}_{\min}$ and $\bar{m}_{\max}$ such that $\bar{m}_{\min} \le \|\mathcal{M}\| \le \bar{m}_{\max}$ and consequently the following inequalities can be obtained from (8.46)

$$
\frac{1}{2}\bar{m}_{\min}\|z\|^2 \le U_1 \le \frac{1}{2}\bar{m}_{\max}\|z\|^2 \tag{8.53}
$$

At the same time, we also have that

$$
\frac{d}{dt}\frac{U_1}{\|z\|} = \frac{\dot{U}_1}{\|z\|} - \frac{U_1}{\|z\|^2}\frac{d\|z\|}{dt} \tag{8.54}
$$

Integrating both sides of the above equation, we arrive at

$$
\int_0^t \frac{\dot{U}_1}{\|z\|}\,dt = \frac{U_1}{\|z\|}\bigg|_0^t + \int_0^t \frac{U_1}{\|z\|^2}\,d\|z\| \tag{8.55}
$$

Combining the equation above with (8.53), we have

$$
\bar{m}_{\min}\|z\| - \bar{m}_{\min}\|z(0)\| \le \int_0^t \frac{\dot{U}_1}{\|z\|}\,dt \le \bar{m}_{\max}\|z\| - \bar{m}_{\max}\|z(0)\| \tag{8.56}
$$

Recalling the boundedness of $\|z\|$ and the fact $\lim_{t\to 0}\|z\| = 0$, we obtain the boundedness of $\int_0^\infty \frac{\dot{U}_1}{\|z\|}\,dt$, which implies

$$\lim_{t\to\infty} \frac{\dot{U}_1}{\|z\|} = 0 \tag{8.57}$$

On the other hand, according to (8.48) and (8.52), we can prove that

$$\dot{U}_1 \leq -\big(k_{\min} - \|Y\|\|\tilde{\Theta}\|\big)\|z\| \tag{8.58}$$

And if $k_{\min}$ is chosen to be sufficiently large, then there exist a constant k' such that $k' = k_{\min} - \|Y\|\|\tilde{\Theta}\| > 0$. If $\|z\| > 0$, $\forall t > 0$ then we would have $\frac{\dot{U}_1}{\|z\|} \leq -k' < 0$. This obviously conflicts with (8.57), such that we see there must exist a finite t_z such that $z = 0_{[2]}$, $t > t_z$. Further analysis shows that the larger the $k_{\min}$ is chosen the smaller t_z will be. Therefore, we could always be able to properly choose t_f and $k_{\min}$ to guarantee that $t_z \ll t_f$. This completes the proof. □

8.5 Reference Trajectory Generator for ζ_2 Subsystem

From the experience of WIP vehicle riders, it can be concluded that human riders are usually able to adjust the tilt angle suitable to maintain the forward velocity at a desired value. However, until now, to our best knowledge, there has been very little study on the automatic control of forward velocity. In the next chapter, we will set up a framework of adaptive generator of implicit control trajectory (AGICT) to design the tilt angle reference trajectory for manipulating the forward velocity to track the desired trajectory. Please refer to Sect. 9.5 in next chapter for detail.

The overall system scheme combining both adaptive controller and NN based reference trajectory generator is illustrated in Fig. 8.2.

8.6 Simulation Studies

The simulation study is carried out to verify the efficiency of both controller and trajectory generator. In the simulation study, the parameters of the WIP system are specified as follows: $M = 15.0$ kg, $I_w = 1.0$ kg m^2, $M_w = 0.2$ kg, $I_M = 2.0$ kg m^2, $I_p = 1.0$ kg m^2, $m = 2.0$ kg, $L = 1$ m, $d = 1.0$ m, and $R = 0.5$ m. The simulation is carried out in 5 seconds and the horizon t_f for LQR performance index is chosen as 3 seconds. The disturbances from environments on the system are introduced as $d_\alpha = 0.5\sin(t)$ and $d_v = [0.3\cos(2t), 0.3\sin(t/2), 0.3\sin(t)]$ in the simulation model. The desired trajectory for yaw angle is $\zeta_{1d} = -0.05t$ (rad) and the initial yaw angle is set as $\zeta_1 = -3$ (rad); the desired forward velocity is set as $\zeta_{2d} = 0$ while the initial velocity is 0.084 m/s. The design parameters of the LQR optimization performance index is $Q = \text{diag}[5, 10]$ and $M_d = \text{diag}[15, 15]$. The initial value of $\hat{\Theta}$ is $\hat{\Theta} = [0.27, 0.28, 0.16, 0.25]$ and the tuning gain matrix $\Gamma_\Theta = \text{diag}[2, 2]$. The control gain matrix $K = \text{diag}[40, 50]$. The HONN is constructed with $l = 65$ neurons and for the initial weight estimate $\hat{W} \in R^l$, each element is selected as an

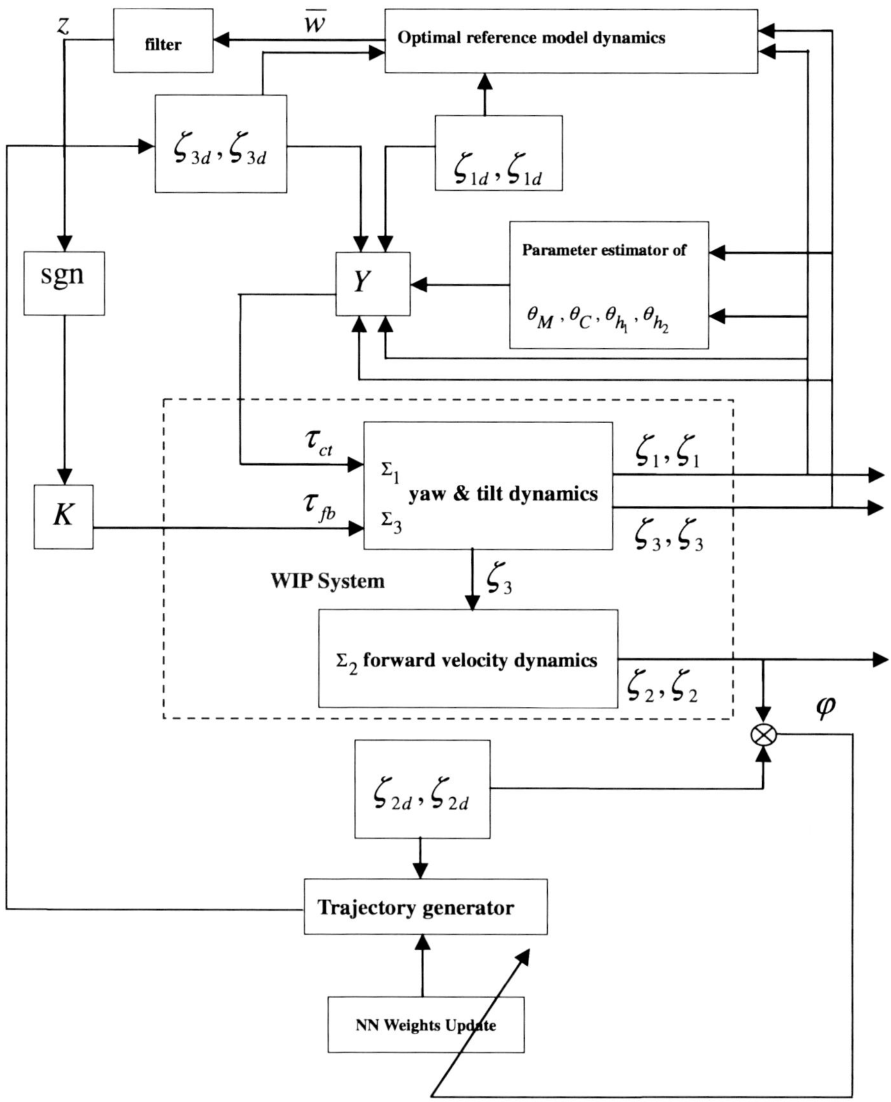

Fig. 8.2 Scheme of overall system: controller and trajectory generator

random variable with magnitude 0.05. The tuning gain matrix as well as forgetting factor of NN weights estimator are chosen as $\Gamma_W = \text{diag}[0.5, 1]$ and $\sigma = 0.05$.

To demonstrate the efficiency of the proposed controller, we compare with the model based controller under the same operation condition and using same NN reference trajectory generator for ξ_{3d}. The model based controller presented as $\tau_1 = D_{11}\ddot{\zeta}_{1d} + C_{11}\dot{\zeta}_1 + C_{13}\dot{\zeta}_3 - k_{11}(\dot{\zeta}_1 - \dot{\zeta}_{1d}) - k_{12}(\zeta_1 - \zeta_{1d})$ and $\tau_2 = -\frac{D_{22}D_{33}-D_{23}^2(\zeta_3)}{D_{23}(\zeta_3)}\ddot{\zeta}_{3d} - k_{31}(\dot{\zeta}_3 - \dot{\zeta}_{3d}) - k_{32}(\zeta_3 - \zeta_{3d}) - C_{23}\dot{\zeta}_3 + \frac{D_{22}}{D_{23}(\zeta_3)}(C_{31}\dot{\zeta}_1 + f_3 +$

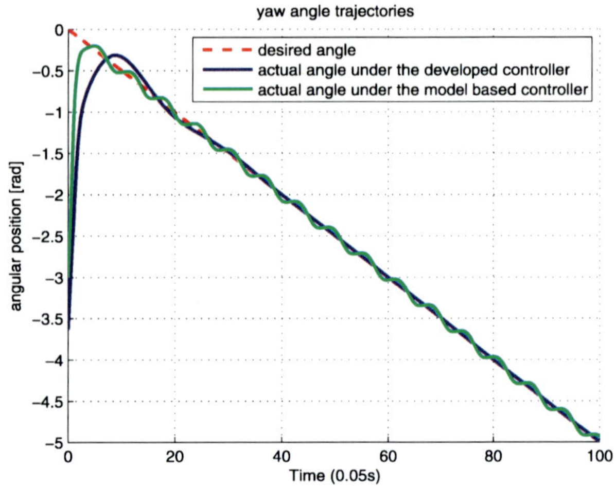

Fig. 8.3 Trajectories of yaw angle

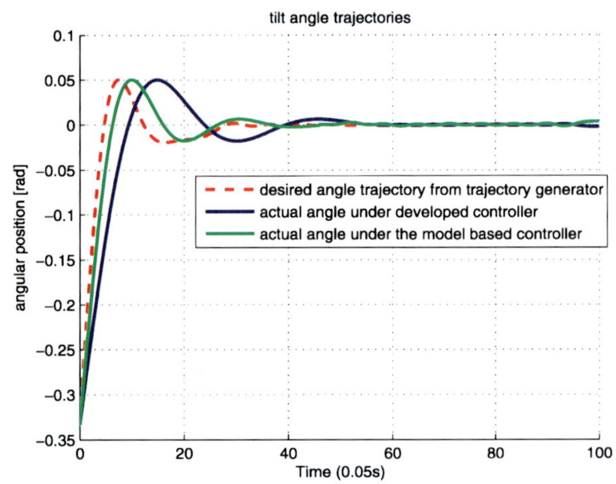

Fig. 8.4 Trajectories of tilt angle

$g_3 + d_3) - (f_2 + g_2 + d_2))$ with $k_{11} = k_{12} = k_{31} = k_{32} = 12.0$, and it is assumed that there is 10 % model uncertainty.

The tracking performance of the desired yaw angle is illustrated in Fig. 8.3, from which we see that though the proposed controller response less quicker but the controlled yaw trajectory exactly track the desired trajectory and there is no steady state error, while the model based controller could only made the controlled yaw trajectory oscillating around the desired trajectory. The slightly slower response is due to the learning process of the adaptive controller, e.g., the estimated parameters need enough time to be adapted to suitable values. The tilt angle trajectories are shown in Fig. 8.4, which shows that the reference trajectory of the tilt angle generated by the HONN will eventually be around zero in order to maintain the forward velocity at zero. Similar as the response of the yaw angle, the proposed controller response a

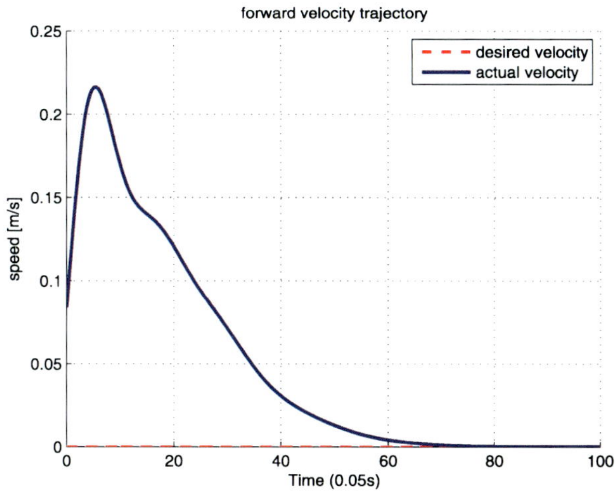

Fig. 8.5 Trajectories of forward velocity

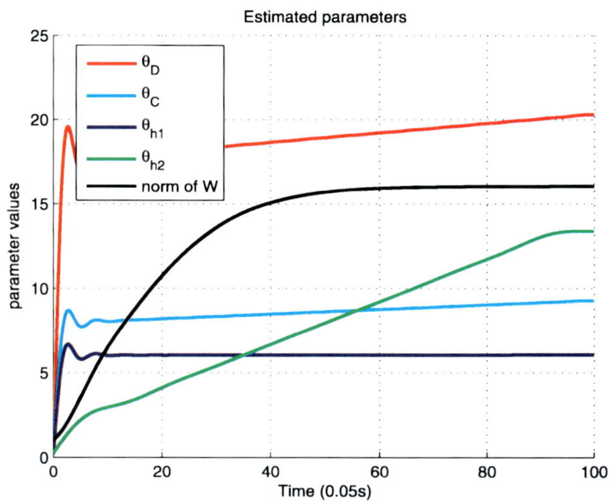

Fig. 8.6 Estimates of parameters

little bit slower but the exact tracking is guaranteed. Figure 8.5 illustrates that under the effect of the reference tilt angle, the forward velocity will eventually maintain at zero as we desire. The convergence of the forward velocity is slower than the convergence of tilt and yaw angles, because in addition to the adaptive controller parameter estimator adaptation, there is also the HONN weight adaptation. The boundedness of both controller parameter adaptation and HONN weight adaptation are shown in Fig. 8.6. The control torques inputs τ_1 and τ_2 are shown in Fig. 8.7 and the ground floor trajectory of the center of WIP is shown in Fig. 8.8.

As clearly shown by the simulation results, in the presence of totally unknown system parameters and external disturbance, the proposed adaptive sliding mode controller is able to guarantee exact tracking of the tilt and yaw subsystem, while the HONN generated the tilt angle reference trajectory is able to maintain the for-

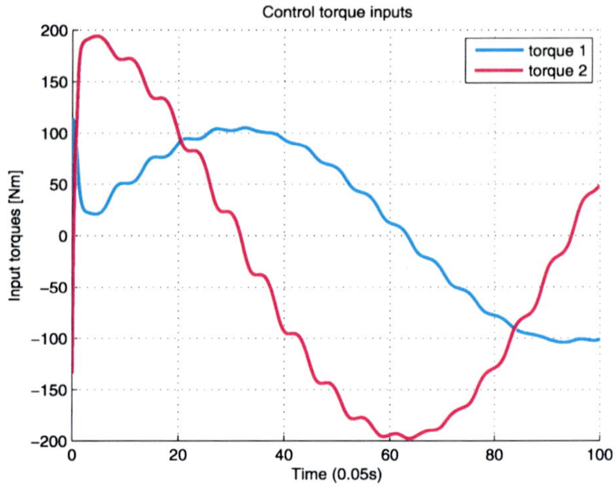

Fig. 8.7 Control input torques

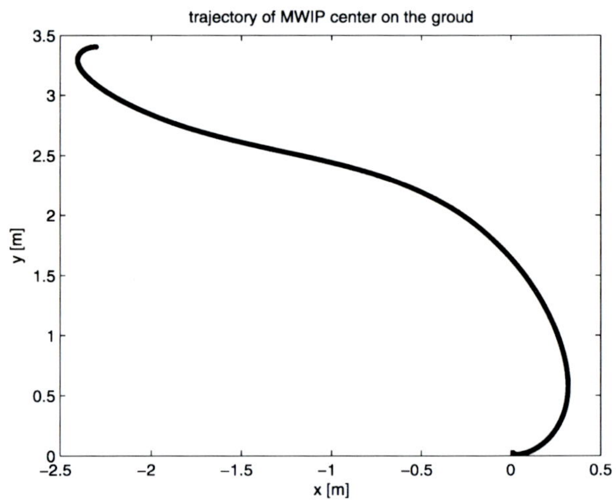

Fig. 8.8 Trajectory of center of WIP on the ground

ward velocity at the desired level. Therefore, the proposed controller is efficient in the presence of unknown nonlinear dynamic systems and environments. While on the other hand, the model-based approach is sensitive to the accuracy of the dynamic model as shown in Fig. 8.3 and Fig. 8.4, and more than 10 % model parameter uncertainty will tend to make the closed-loop system unstable. In contrast, our proposed adaptive controller is able to tolerate complete unknown parameter uncertainty. This is one of the key advantages over model-based controller. As the controlled conditions in the test facility under which the parameters are identified are often very different from actual conditions, thus rendering the parameters inaccuracy for real operating conditions. In addition, the controlled closed-loop dynamics of the yaw and tilt subsystem are shaped through the LQR technique such that the optimization in term of the performance index is achieved. It is also worth to mention that in the "learning" mechanism of both the adaptive controller parameter and the HONN weights adaptation, once the estimates converge they do not need to be "relearned" as long as there are significant changes of system dynamics and external conditions.

8.7 Conclusion

In this chapter, adaptive control has been designed on WIP systems for dynamic balance and motion tracking of desired trajectories. The dynamics of the subsystem consisting of the pendulum tilt angle and the mobile platform yaw angle has been shaped to follow a reference model, which is derived by using the LQR optimization technique to minimize both the motion tracking error and the transient acceleration to enhance driver's comfort. The proposed control method considers the presence of various uncertainties including both parametric and functional uncertainties. Simulation results have demonstrated the efficiency of the proposed method.

Chapter 9
Neural Network Based Model Reference Control

9.1 Introduction

In this chapter, we further study the model reference control design for the angular motion subsystem, we propose to employ Radial Basis Function Neural Network (RBFNN) for control design. In this chapter, adaptive NN model reference control is developed together with variable structure control for the fully actuated angular motion subsystem of WIP such that the closed-loop subsystem dynamics can exactly follow an optimized reference model within a finite time horizon. In addition, instead of leaving the unactuated forward velocity dynamics uncontrolled, the reference trajectory for the tilt angle is generated using NN to indirectly affect the forward velocity such that the desired velocity can be tracked. The rest of this chapter is organized as follows. In Sect. 9.2, preliminary knowledge of NN approximation as well as some useful theorems and lemmas are presented. In Sect. 9.3, an angular motion subsystem and a forward velocity subsystem of the under-actuated WIP systems are derived with the first subsystem fully actuated while the second unactuated. Optimized model reference control of angular motion subsystem is developed in Sect. 9.4. A reference trajectory generator using RBFNN for the tilt angle is designed in Sect. 9.5 such that the forward velocity is indirectly manipulated to follow its desired trajectory. In Sect. 9.6, simulation studies are carried out to verify the effectiveness of the proposed method. Concluding remarks are given in Sect. 9.7.

9.2 Preliminaries

9.2.1 Radial Basis Function Neural Network

Consider the following RBFNN used to approximate a function $h(z) : R^m \to R$,

$$\phi(W, z) = W^T S(z) \tag{9.1}$$

Z. Li et al., *Advanced Control of Wheeled Inverted Pendulum Systems*,
DOI 10.1007/978-1-4471-2963-9_9, © Springer-Verlag London 2013

where the input vector $z \in \Omega_z \subset \mathbf{R}^m$ is of NN input dimension m, weight vector $W = [w_1, w_2, \ldots, w_l]^T \in R^l$, the NN node number $l > 1$, and basis function $S(z) = [s_1(z), \ldots, s_l(z)]^T$, with $s_i(z)$ chosen as Gaussian functions as follows:

$$s_i(z) = \exp\left[\frac{-(z-\mu_i)^T(z-\mu_i)}{\eta_i^2}\right], \quad i = 1, 2, \ldots, l \tag{9.2}$$

where $\mu_i = [\mu_{i1}, \mu_{i2}, \ldots, \mu_{iq}]^T$ is the center of the receptive field and η_i is the width of the Gaussian function.

It has been proven that the RBFNN (9.1) can approximate any continuous function over a compact set $\Omega_z \subset \mathbf{R}^q$ to arbitrary accuracy as [84]

$$\phi(z) = W^{*T} S(z) + \epsilon_z, \quad \forall z \in \Omega_z \tag{9.3}$$

where W^* is ideal constant weights, and ϵ_z is the approximation error.

The ideal weight vector W^* is an "artificial" quantity required for analytical purposes. W^* is defined as the value of W that minimizes $|\epsilon_z|$ for all $z \in \Omega_z$ in a compact region, i.e.,

$$W^* \stackrel{\text{def}}{=} \arg \min_{W \in R^l} \left\{ \sup \left| h(z) - W^T S(z) \right| \right\}, \quad z \in \Omega_z$$

Lemma 9.1 [84] *For the Gaussian RBFNN, if $\hat{z} = z - \epsilon\bar{\psi}$ where $\bar{\psi}$ is a bounded vector and constant $\epsilon > 0$, then*

$$S(\hat{z}) = S(z) + \epsilon S_t \tag{9.4}$$

where S_t is a bounded function vector.

9.2.2 Block Matrix Operation

For convenience of approximation of a nonlinear matrix function with each element a unknown scalar function, we use the following block matrix operators as introduced in [44]. We define block matrices $\{W\}$ and $\{S\}$ as follows:

$$\begin{aligned}
\{W\} &\triangleq \begin{bmatrix} W_{11} & W_{11} & \cdots & W_{1m} \\ W_{21} & W_{11} & \cdots & W_{1m} \\ \vdots & \vdots & \ddots & \vdots \\ W_{n1} & W_{11} & \cdots & W_{nm} \end{bmatrix} \in R^{nl \times m} \\
\{S\} &\triangleq \begin{bmatrix} S_{11} & S_{11} & \cdots & S_{1m} \\ S_{21} & S_{11} & \cdots & S_{1m} \\ \vdots & \vdots & \ddots & \vdots \\ S_{n1} & S_{11} & \cdots & S_{nm} \end{bmatrix} \in R^{nl \times m}
\end{aligned} \tag{9.5}$$

with each block $W_{ij} \in \mathbf{R}^l$ a column vector of NN weight and $S_{ij} \in \mathbf{R}^l$ a column vector of NN basis function. A "transpose" operation $\langle T \rangle$ of the block matrix $\{W\}$ is defined in such a way

$$\{W\}^{\langle T \rangle} \triangleq \begin{bmatrix} W_{11}^T & W_{11}^T & \cdots & W_{1m}^T \\ W_{21}^T & W_{22}^T & \cdots & W_{2m}^T \\ \vdots & \vdots & \ddots & \vdots \\ W_{n1}^T & W_{n2}^T & \cdots & W_{nm}^T \end{bmatrix} \in R^{n \times ml} \tag{9.6}$$

that each block of the column vector is transposed to a row vector while the relative location of each block in the matrix $\{W\}$ is not changed. Furthermore, we define block matrix multiplication between $\{W\}^{\langle T \rangle}$ and S as

$$\{W\}^{\langle T \rangle}\langle \cdot \rangle S \triangleq \begin{bmatrix} W_{11}^T S_{11} & W_{12}^T S_{12} & \cdots & W_{1m}^T S_{1m} \\ W_{21}^T S_{21} & W_{22}^T S_{22} & \cdots & W_{2m}^T S_{2m} \\ \vdots & \vdots & \ddots & \vdots \\ W_{n1}^T S_{n1} & W_{n2}^T S_{n2} & \cdots & W_{nm}^T S_{nm} \end{bmatrix} \in R^{n \times m} \tag{9.7}$$

In this manner, we could easily use $\{W\}^{\langle T \rangle}\langle \cdot \rangle S$ to emulate a unknown matrix of dimension $n \times m$.

9.3 Dynamics of Wheeled Inverted Pendulums

Following the same techniques in Chap. 8, we see that the dynamics of wheeled inverted pendulum can be expressed as the following three dynamic equations:

$$d_{11}(\zeta_3)\ddot{\zeta}_1 + c_{11}\dot{\zeta}_1 + c_{13}\dot{\zeta}_3 + \tau_{v1} = \tau_1 \tag{9.8}$$

$$d_{22}\ddot{\zeta}_2 + d_{23}(\zeta_3)\ddot{\zeta}_3 + c_{23}\dot{\zeta}_3 + \tau_{v2} = \tau_2 \tag{9.9}$$

$$d_{32}(\zeta_3)\ddot{\zeta}_2 + d_{33}\ddot{\zeta}_3 + c_{31}\dot{\zeta}_1 + \tau_{v3} + g_3 = 0 \tag{9.10}$$

Assume that the external disturbance τ_{vi}, $i = 1, 2, 3$, depends on velocities, i.e., there exist unknown functions dependent on state ζ such that

$$\tau_{vi} = \bar{v}_i(\zeta)^T \dot{\zeta} = v_{i1}(\zeta)\dot{\zeta}_1 + v_{i2}(\zeta)\dot{\zeta}_2 + v_{i3}(\zeta)\dot{\zeta}_3 \tag{9.11}$$

from which we could clearly see the under-actuated configuration, i.e., using only τ_1 and τ_2 we could not manipulate three variables ζ_1, ζ_2 and ζ_3 independently. Looking back to (9.10) and substituting the third equation into the second one, we can obtain the following two equations:

$$\begin{aligned} &\frac{d_{22}d_{33} - d_{23}^2}{d_{23}(\zeta_3)}\ddot{\zeta}_3 - c_{23}\dot{\zeta}_3 + \frac{d_{22}}{d_{23}}(c_{31}\dot{\zeta}_1 + \tau_{v3}) - \tau_{v2} + \frac{d_{22}}{d_{23}}g_3 \\ &= -\tau_2 \end{aligned} \tag{9.12}$$

$$\begin{aligned} &\frac{d_{22}d_{33} - d_{23}^2}{d_{33}}\ddot{\zeta}_2 + c_{23}\dot{\zeta}_3 - \frac{d_{23}}{d_{33}}\left(c_{31}\dot{\zeta}_1 + g_3'\zeta_3 + \tau_{v3}\right) + \tau_{v2} - \frac{d_{22}}{d_{23}}g_3 \\ &= \tau_2 \end{aligned} \tag{9.13}$$

Combination of (9.8) and (9.12) yields the following dynamics of tilt and yaw angular motion dynamics subsystem:

$$\Sigma_a: \quad \mathcal{M}\ddot{\zeta}_a + \mathcal{C}_a\dot{\zeta}_a + \mathcal{C}_b\dot{\zeta}_b = \tau_a - \mathcal{G} \tag{9.14}$$

where $\tau_a = [\tau_1 \ -\tau_2]^T$, $\zeta_a = [\zeta_1 \ \zeta_3]^T$, $\zeta_b = \zeta_2$ and according to the definition of ν_{ij} in (9.11), we have

$$\mathcal{M} = \begin{bmatrix} d_{11}(\zeta_2) & 0 \\ 0 & \frac{d_{22}d_{33}-d_{23}^2(\zeta_2)}{d_{23}(\zeta_2)} \end{bmatrix}, \qquad \mathcal{G} = \begin{bmatrix} 0 \\ \frac{d_{22}}{d_{23}} g_3 \end{bmatrix}$$

$$\mathcal{C}_a = \begin{bmatrix} c_{11} + \nu_{11} & \nu_{12} \\ \frac{d_{22}c_{31}(1+\nu_{31})}{d_{23}} - \nu_{21} & \frac{d_{22}\nu_{32}}{d_{23}} - \nu_{22} \end{bmatrix}, \qquad \mathcal{C}_b = \begin{bmatrix} c_{13} + \nu_{13} \\ -c_{23} - \nu_{23} + \frac{d_{22}\nu_{32}}{\nu_{33}} \end{bmatrix}$$

Let $\bar{\zeta}_b = [\zeta_b, \dot{\zeta}_b]^T = [\bar{\zeta}_{b1}, \bar{\zeta}_{b2}]^T$, $\bar{\zeta}_a = [\zeta_a^T, \dot{\zeta}_a^T]^T$, and $\pi = \ddot{\zeta}_{3d}$, then from (9.10) we can obtain dynamics of forward velocity subsystem as follows:

$$\Sigma_b: \quad \begin{aligned} \dot{\bar{\zeta}}_{b1} &= \bar{\zeta}_{b2} \\ \dot{\bar{\zeta}}_{b2} &= h_b(\bar{\zeta}_b, \bar{\zeta}_a, \pi) \end{aligned} \tag{9.15}$$

with

$$h_b(\bar{\zeta}_b, \bar{\zeta}_a, \pi) = b(\zeta_3)\pi + \frac{1}{ML\cos\zeta_3}\left(\frac{1}{2}\dot{\zeta}_1 ML^2 \sin 2\zeta_3 + MgL\sin\zeta_3 - \tau_{\nu 3}\right) \tag{9.16}$$

$$b(\zeta_3) = -\frac{1}{ML\cos\zeta_3}(ML^2 + I_r)\ddot{\zeta}_3 \tag{9.17}$$

9.4 Control of Angular Motion Subsystems

9.4.1 Optimized Reference Model

The angular motion subsystem Σ_a (9.14) can be rewritten as

$$\dot{\bar{\zeta}}_a = A_a\bar{\zeta}_a + A_b\bar{\zeta}_b + B\mathcal{M}^{-1}(\tau_a - \mathcal{G}) \tag{9.18}$$

where

$$A_a \triangleq \begin{bmatrix} 0_{[2,2]} & I_{[2,2]} \\ 0_{[2,2]} & -\mathcal{M}^{-1}\mathcal{C}_a \end{bmatrix} \in R^{4\times 4} \tag{9.19}$$

$$A_b \triangleq \begin{bmatrix} 0_{[2,2]} \\ -\mathcal{M}^{-1}\mathcal{C}_b \end{bmatrix} \in R^{4\times 2} \tag{9.20}$$

$$B \triangleq \left[0_{[2,2]}, I_{[2,2]}\right]^T \tag{9.21}$$

For convenience, the argument of A and $\mathcal{M}$ is omitted. Next, we will construct a reference model with optimized dynamics and turn the control objective of subsystem (9.14) to be matching the reference model. Accordingly, the chosen reference model must matchable, i.e., there exist suitable control input τ_a to make angular motion dynamics (9.14) exactly same as the reference model. Consider choosing reference model in the following form where the subscript "r" stands for "reference":

$$\dot{\bar{\zeta}}_{ar} = A_r \bar{\zeta}_{ar} + B M_r^{-1} f_r \tag{9.22}$$

$$A_r \triangleq \begin{bmatrix} 0_{[2,2]} & I_{[2,2]} \\ -M_r^{-1} K_r & -M_r^{-1} C_r \end{bmatrix} \in R^{4\times 4} \tag{9.23}$$

where $f_r \triangleq f_r(\bar{\zeta}_d)$ is a virtual force caused by the desired trajectory $\bar{\zeta}_d = [\zeta_d, \dot{\zeta}_d]^T$ with subscript "d" standing for "desired", to be explained later, $\bar{\zeta}_{ar} = [\zeta_{ar}, \dot{\zeta}_{ar}]^T \in R^4$ is the desired angular position and velocity response and M_r the predefined positive definite desired apparent mass/inertia matrix and K_r and C_r to be defined later. The reference model (9.22) is actually derived from the following impedance model

$$M_r \ddot{\zeta}_{ar} + C_r \dot{\zeta}_{ar} + K_r \zeta_{ar} = f_r(\bar{\zeta}_d) \tag{9.24}$$

where C_r and K_r are the desired damping and stiffness matrices, respectively. As mentioned above M_r is the desired mass/inertia predefined and f_r is an artificial force that drive the reference states ζ_{ar} tracking the desired trajectory ζ_d. This reference model can be illustrated by a mass–spring–damper system shown in Fig. 8.1. As the control objective is to make the closed-loop dynamics of the controlled subsystem (9.18) match the dynamics of the reference model (9.24), we should suitably choose the parameters of the reference model such that it not only guarantees the motion tracking but also minimize the acceleration to make rider comfortable. Actually the impedance model provides kind of cushion effect for better riding experience and when there is no artificial force f_r (e.g., the rider let go the wheel and pedal), the yaw and tilt angles ζ_1 and ζ_3 will tend to rest on the zero position.

In order to choose the optimal values of the reference model parameters C_r and K_r, we introduce the following performance index in finite time horizon $[t_0, t_f]$:

$$P_I = \int_{t_0}^{t_f} \left((\zeta_{ar} - \zeta_{ad})^T Q (\zeta_{ar} - \zeta_{ad}) + \ddot{\zeta}_{ar}^T M_r \ddot{\zeta}_{ar} \right) dt \tag{9.25}$$

with a positive definite weight matrix

$$Q = \begin{bmatrix} q_1 & 0 \\ 0 & q_2 \end{bmatrix} \tag{9.26}$$

which minimizes both the motion tracking error and the yaw and tilt angular accelerations. We aim to enhance rider's comfort by reducing any unnecessary angular accelerations, while at the same time to ensure motion tracking performance. Now let us consider how to minimize the performance index P_I by suitably designing C_r, K_r and f_r. It is easy to derive from (9.24) that

$$\ddot{\zeta}_{ar} = -M_r^{-1}[K_r, \ C_d]\bar{\zeta}_{ar} - M_r^{-1} f_r(\zeta_d, \dot{\zeta}_d) \tag{9.27}$$

such that by denoting $u = \ddot{\zeta}_{ar}$ we could reformulate reference model (9.24) as follows

$$\dot{\bar{\zeta}}_{ar} = A_r \bar{\zeta}_{ar} + Bu \tag{9.28}$$

with

$$A_r = \begin{bmatrix} 0_{[2,2]} & I_{[2,2]} \\ 0_{[2,2]} & 0_{[2,2]} \end{bmatrix} \tag{9.29}$$

such that we could follow the standard LQR technique to minimize the performance index (9.25), which can be rewritten as follows

$$P_{\bar{I}} = \int_{t_0}^{t_f} \left((\bar{\zeta}_{ar} - \bar{\zeta}_d)^T \bar{\zeta}(\bar{\zeta}_{ar} - \bar{\zeta}_d) + u^T M_r u\right) dt \tag{9.30}$$

with

$$\bar{Q} = \begin{bmatrix} Q & 0_{[2,2]} \\ 0_{[2,2]} & 0_{[2,2]} \end{bmatrix} \tag{9.31}$$

According to the LQR optimal control technique reviewed in last chapter, the solution of u that minimizes (9.30) is

$$u = -M_r^{-1} B^T P \bar{\zeta}_{ar} - M_r^{-1} B^T s \tag{9.32}$$

where P is the solution of the following differential equation

$$-\dot{P} = PA_r + A_r^T P - PBM_r^{-1} B^T P + \bar{Q}, \quad P(t_f) = 0_{[4,4]} \tag{9.33}$$

and s is the solution of the following differential equation

$$-\dot{s} = \left(A_r - BM_r^{-1} B^T P\right)^T s + \bar{Q}\bar{\zeta}_d, \quad s(t_f) = 0_{[4]} \tag{9.34}$$

Comparing Eqs. (9.27) and (9.32), we can see that the matrices K_d and C_d can be calculated in the following manner:

$$[K_r, C_r] = B^T P, \qquad f_r = -B^T s \tag{9.35}$$

9.4.2 Adaptive NN Model Reference Control

In this section, we design control input τ_a for angular motion subsystem (9.18) such that its dynamics matches with the optimized reference model given by (9.22) and (9.35). As the reference model (9.22) satisfies the matching condition, it is trivial to show that there must exist three gain matrix

$$K_a(\bar{\zeta}) \in R^{2\times 4}, \qquad K_b(\bar{\zeta}) \in R^{2\times 2}, \qquad L(\bar{\zeta}) \in R^{2\times 2} \tag{9.36}$$

such that if

$$\tau_a = K_a(\bar{\zeta})\bar{\zeta}_a + K_b(\bar{\zeta})\bar{\zeta}_b + L(\bar{\zeta})f_r + \mathcal{G}(\bar{\zeta}) \tag{9.37}$$

then the closed-loop system is the same as the reference model (9.22).

By substituting the control law (9.37) into the system equation (9.18), the closed-loop system is given by

$$\begin{aligned}\dot{\bar{\zeta}}_a &= A_a\bar{\zeta}_a + A_b\bar{\zeta}_b + B\mathcal{M}^{-1}\big[K_a(\bar{\zeta})\bar{\zeta}_a + K_b(\bar{\zeta})\bar{\zeta}_b + L(\bar{\zeta})f_r + \mathcal{G}(\bar{\zeta})\big]\\ &= \big[A_a + B\mathcal{M}^{-1}K_a(\bar{\zeta})\big]\bar{\zeta}_a + \big[A_b + B\mathcal{M}^{-1}K_b(\bar{\zeta})\big]\bar{\zeta}_b\\ &\quad + B\mathcal{M}^{-1}L(\bar{\zeta})[f_r + \mathcal{G}]\end{aligned} \tag{9.38}$$

Comparing it to match the reference model (9.22), we obtain

$$\begin{aligned}A_a + B\mathcal{M}^{-1}K_a(\bar{\zeta}) &= A_r\\ A_b + B\mathcal{M}^{-1}K_a(\bar{\zeta}) &= 0_{[2,2]}\\ \mathcal{M}^{-1}L(\bar{\zeta}) &= M_r^{-1}\end{aligned} \tag{9.39}$$

By the definitions of A, B, A_r, we have

$$K_a(\bar{\zeta}) = \mathcal{M}\big([A_{m1}\ A_{m2}] - [A_{a1}\ A_{a2}]\big) \tag{9.40}$$

$$K_b(\bar{\zeta}) = -\mathcal{M}[A_{b1}\ A_{b2}] \tag{9.41}$$

$$L(\bar{\zeta}) = \mathcal{M}M_r^{-1} \tag{9.42}$$

As the matrices $A_a(\bar{\zeta})$, $A_b(\bar{\zeta})$, and $\mathcal{M}$ are unknown for control design we could calculate the desired gains $K(\bar{\zeta})$ and $L(\bar{\zeta})$ accordingly. Thus, NN will be employed to emulate the desired $K_a(\bar{\zeta})$, $K_b(\bar{\zeta})$ and $L(\bar{\zeta})$ used in (9.37) as below

$$K_a(\bar{\zeta}) = K_a^*(\bar{\zeta}) + \varepsilon_{K_a} \tag{9.43}$$

$$K_b(\bar{\zeta}) = K_b^*(\bar{\zeta}) + \varepsilon_{K_b} \tag{9.44}$$

$$L(\bar{\zeta}) = L^*(\bar{\zeta}) + \varepsilon_L \tag{9.45}$$

$$\mathcal{G}(\bar{\zeta}) = \mathcal{G}^*(\bar{\zeta}) + \varepsilon_{\mathcal{G}} \tag{9.46}$$

with

$$\begin{aligned}K_a^*(\bar{\zeta}) &= W_{K_a}^{*\langle T\rangle}\langle\cdot\rangle S_{K_a}(\bar{\zeta}) \in \mathbf{R}^{2\times 4}\\ &= \begin{bmatrix} W_{K_a1,1}^{*T}S_{K_a1,1}(\bar{\zeta}) & \cdots & W_{K_a1,4}^{*T}S_{K_a1,4}(\bar{\zeta})\\ W_{K_a2,1}^{*T}S_{K_a2,1}(\bar{\zeta}) & \cdots & W_{K_a2,4}^{*T}S_{K_a2,4}(\bar{\zeta})\end{bmatrix}\end{aligned} \tag{9.47}$$

$$\begin{aligned}K_b^*(\bar{\zeta}) &= W_{K_b}^{*\langle T\rangle}\langle\cdot\rangle S_{K_b}(\bar{\zeta}) \in \mathbf{R}^{2\times 2}\\ &= \begin{bmatrix} W_{K_b1,1}^{*T}S_{K_b1,1}(\bar{\zeta}) & W_{K_b1,4}^{*T}S_{K_b1,4}(\bar{\zeta})\\ W_{K_b2,1}^{*T}S_{K_b2,1}(\bar{\zeta}) & W_{K_b2,4}^{*T}S_{K_b2,4}(\bar{\zeta})\end{bmatrix}\end{aligned} \tag{9.48}$$

$$L^*(\bar{\zeta}) = W_L^{*\langle T\rangle}\langle\cdot\rangle S_L(\bar{\zeta}) \in \mathbf{R}^{2\times 4}$$
$$= \begin{bmatrix} W_{L1,1}^{*T} S_{L1,1}(\bar{\zeta}) & \cdots & W_{L1,4}^{*T} S_{L1,2}(\bar{\zeta}) \\ W_{L2,1}^{*T} S_{L2,1}(\bar{\zeta}) & \cdots & W_{L2,4}^{*T} S_{L2,2}(\bar{\zeta}) \end{bmatrix} \quad (9.49)$$

$$\mathcal{G}^*(\bar{\zeta}) = \left[W_{\mathcal{G}}^{\langle T\rangle}\langle\cdot\rangle S_{\mathcal{G}}(\bar{\zeta})\right] = \begin{bmatrix} W_{\mathcal{G}1}^{*T} S_{\mathcal{C}1}(\bar{\zeta}) \\ \vdots \\ W_{\mathcal{G}2}^{T} S_{\mathcal{G}2}(\bar{\zeta}) \end{bmatrix} \quad (9.50)$$

Obviously, $W_{K_a}^*$ and $W_{K_a}^{*\langle T\rangle}$ are defined as

$$W_{K_a}^* = \begin{bmatrix} W_{K_a 1}^* \\ W_{K_a 2}^* \end{bmatrix}, \quad W_{K_a i}^* = \left[W_{K_a i,1}^* \cdots W_{K_a i,4}^*\right], \quad i = 1, 2$$
$$W_{K_a}^{*\langle T\rangle} = \begin{bmatrix} W_{K_a 1}^{*\langle T\rangle} \\ W_{K_a 2}^{*\langle T\rangle} \end{bmatrix}, \quad W_{K_a i}^{*\langle T\rangle} = \left[W_{K_a i,1}^{*T} \cdots W_{K_a i,4}^{*T}\right], \quad i = 1, 2$$

and $W_{K_b}^*$, $W_{K_b}^{*\langle T\rangle}$, W_L^*, $W_L^{*\langle T\rangle}$, $W_{\mathcal{G}}$, $W_{\mathcal{G}}^{\langle T\rangle}$ are correspondingly appropriate dimensional matrices and for convenience, they are omitted here. $W_{K_a i,j}^*, W_{K_b i,k}^*$, $W_{Li,s}^*, W_{\mathcal{G}i}^* \in \mathbf{R}^{l\times 1}$ are the NN ideal weights for $K_{ai,j}^*(\bar{\zeta})$, $K_{bi,s}^*(\bar{\zeta})$, $L_{i,k}^*(\bar{\zeta})$, $G_i^*(\bar{\zeta})$, respectively ($i = 1, 2$; $j = 1, 2, 3, 4$; $k = 1, 2$; $s = 1, 2$), l is the number of the neurons. $S_{K_a}(\bar{\zeta})$, $S_{K_b}(\bar{\zeta})$, $S_L(\bar{\zeta})$, $S_{\mathcal{G}}(\bar{\zeta})$ are the outputs of the bounded basis functions, and ε_{K_a}, ε_{K_b}, ε_L, $\varepsilon_{\mathcal{G}}$ are the NN approximation errors. For a fixed number of nodes, we know that $\|\varepsilon_{K_a}\|$, $\|\varepsilon_{K_b}\|$, $\|\varepsilon_L\|$, $\|\varepsilon_{\mathcal{G}}\|$ are bounded, W_{K_a}, W_{K_b}, W_L, $W_{\mathcal{G}}$ are unknown constant parameters.

Consider the following NN based control law

$$\tau_a = \hat{K}_a(\bar{\zeta})\bar{\zeta}_a + \hat{K}_b(\bar{\zeta})\bar{\zeta}_b + \hat{L}(\bar{\zeta}) f_r + \tau_r \quad (9.51)$$

Consider the following NN based control law

$$\begin{aligned} \tau_a &= \hat{K}_a(\bar{\zeta})\bar{\zeta}_a + \hat{K}_b(\bar{\zeta})\bar{\zeta}_b + \hat{L}(\bar{\zeta}) f_r + \hat{\mathcal{G}}(\bar{\zeta}) + \tau_r \\ &= \left[\hat{W}_{K_a}^{\langle T\rangle}\langle\cdot\rangle S_{K_a}(\bar{\zeta})\right]\bar{\zeta}_a + \left[\hat{W}_{K_b}^{\langle T\rangle}\langle\cdot\rangle S_L(\bar{\zeta})\right]\bar{\zeta}_b \\ &\quad + \left[W_L^{\langle T\rangle}\langle\cdot\rangle S_L(\bar{\zeta})\right] f_r + \left[W_{\mathcal{G}}^{\langle T\rangle}\langle\cdot\rangle S_{\mathcal{G}}(\bar{\zeta})\right] + \tau_r \end{aligned}$$

Define

$$\begin{aligned} e_{\bar{\zeta}_a} &= \bar{\zeta}_a - \bar{\zeta}_{ar} \\ \tilde{W}_{K_a} &= \hat{W}_{K_a} - W_{K_a}^* \\ \tilde{W}_{K_b} &= \hat{W}_{K_b} - W_{K_b}^* \\ \tilde{W}_L &= \hat{W}_L - W_L^* \\ \tilde{W}_{\mathcal{G}} &= \hat{W}_{\mathcal{G}} - W_{\mathcal{G}} \end{aligned} \quad (9.52)$$

Substituting the control law (9.51) into the subsystem dynamics (9.18), and using (9.52), we have

$$
\begin{aligned}
\dot{e}_a &= \dot{\bar{\zeta}}_a - \dot{\bar{\zeta}}_{ar} \\
&= \big(A_a + B\mathcal{M}^{-1}\big[\hat{W}_{K_a}^{\langle T\rangle}\langle\cdot\rangle S_{K_a}(\bar{\zeta})\big]\big)\bar{\zeta}_a + \big(A_b + B\mathcal{M}^{-1}\big[\hat{W}_{K_b}^{\langle T\rangle}\langle\cdot\rangle S_{K_b}(\bar{\zeta})\big]\big)\bar{\zeta}_b \\
&\quad + B\mathcal{M}^{-1}\big[\hat{W}_{L}^{\langle T\rangle}\langle\cdot\rangle S_{L}(\bar{\zeta})\big]f_r + B\mathcal{M}^{-1}\big[\hat{W}_{\mathcal{G}}^{\langle T\rangle}\langle\cdot\rangle S_{\mathcal{G}}(\bar{\zeta})\big] \\
&\quad - A_r\bar{\zeta}_r - BM_d^{-1}f_r + B\mathcal{M}^{-1}\tau_r \\
&= \big(A_a + B\mathcal{M}^{-1}\big[W_{K_a}^{\langle T\rangle}\langle\cdot\rangle S_{K_a}(\bar{\zeta})\big]\big)\bar{\zeta}_a + \big(A_b + B\mathcal{M}^{-1}\big[W_{L}^{\langle T\rangle}\langle\cdot\rangle S_{L}(\bar{\zeta})\big]\big)\bar{\zeta}_b \\
&\quad + B\mathcal{M}^{-1}\big[W_{L}^{\langle T\rangle}\langle\cdot\rangle S_{L}(\bar{\zeta})\big]f_r + B\mathcal{M}^{-1}\big(\big[W_{\mathcal{G}}^{\langle T\rangle}\langle\cdot\rangle S_{\mathcal{G}}(\bar{\zeta})\big]\big) \\
&\quad - A_r\bar{\zeta}_r - BM_d^{-1}f_r + B\mathcal{M}^{-1}\big[\tilde{W}_{K_a}^{\langle T\rangle}\langle\cdot\rangle S_{K_a}(\bar{\zeta})\big]\bar{\zeta}_a \\
&\quad + B\mathcal{M}^{-1}\big[\tilde{W}_{L}^{\langle T\rangle}\langle\cdot\rangle S_{L}(\bar{\zeta})\big]\bar{\zeta}_b + B\mathcal{M}^{-1}\big[\tilde{W}_{L}^{\langle T\rangle}\langle\cdot\rangle S_{L}(\bar{\zeta})\big]f_r \\
&\quad + B\mathcal{M}^{-1}\big[\tilde{W}_{G}^{\langle T\rangle}\langle\cdot\rangle S_{G}(\bar{\zeta})\big] + B\mathcal{M}^{-1}\tau_r
\end{aligned}
\tag{9.53}
$$

By applying the NN approximations (9.40)–(9.42), we obtain the following error equation

$$
\begin{aligned}
\dot{e}_a &= \big(A_a + B\mathcal{M}^{-1}K(\bar{\zeta})\big)\bar{\zeta}_a + \big(A_b + B\mathcal{M}^{-1}L(\bar{\zeta})\big)\bar{\zeta}_b \\
&\quad + B\mathcal{M}^{-1}T(\bar{\zeta})f_r + B\mathcal{M}^{-1}G(\bar{\zeta}) - A_r\bar{\zeta}_r - BM_d^{-1}f_r \\
&\quad + B\mathcal{M}^{-1}\big[\tilde{W}_{K_a}^{\langle T\rangle}\langle\cdot\rangle S_{K_a}(\bar{\zeta})\big]\bar{\zeta}_a \\
&\quad + B\mathcal{M}^{-1}\big[\tilde{W}_{K_b}^{\langle T\rangle}\langle\cdot\rangle S_{K_b}(\bar{\zeta})\big]\bar{\zeta}_b \\
&\quad + B\mathcal{M}^{-1}\big[\tilde{W}_{L}^{\langle T\rangle}\langle\cdot\rangle S_{L}(\bar{\zeta})\big]f_r + B\mathcal{M}^{-1}\big[\tilde{W}_{\mathcal{G}}^{\langle T\rangle}\langle\cdot\rangle S_{\mathcal{G}}(\bar{\zeta})\big] \\
&\quad - B\mathcal{M}^{-1}(\varepsilon_{K_a}\bar{\zeta}_a + \varepsilon_{K_b}\bar{\zeta}_b + \varepsilon_L f_r + \varepsilon_{\mathcal{G}}) + B\mathcal{M}^{-1}\tau_r
\end{aligned}
\tag{9.54}
$$

Substituting (9.39) into (9.54) leads to

$$
\begin{aligned}
\dot{e}_a &= A_r e_a + B\mathcal{M}^{-1}(\tau_r - \varepsilon_{K_a}\bar{\zeta}_a - \varepsilon_{K_b}\bar{\zeta}_b - \varepsilon_L f_r - \varepsilon_{\mathcal{G}}) \\
&\quad + B\mathcal{M}^{-1}\big[\tilde{W}_{K_a}^{\langle T\rangle}\langle\cdot\rangle S_{K_a}(\bar{\zeta})\big]\bar{\zeta}_a + B\mathcal{M}^{-1}\big[\tilde{W}_{K_b}^{\langle T\rangle}\langle\cdot\rangle S_{K_b}(\bar{\zeta})\big]\bar{\zeta}_b \\
&\quad + B\mathcal{M}^{-1}\big[\tilde{W}_{L}^{\langle T\rangle}\langle\cdot\rangle S_{L}(\bar{\zeta})\big]f_r + B\mathcal{M}^{-1}\big[\tilde{W}_{\mathcal{G}}^{\langle T\rangle}\langle\cdot\rangle S_{\mathcal{G}}(\bar{\zeta})\big]
\end{aligned}
\tag{9.55}
$$

For stable A_r of the reference model, let P_r be the symmetric positive definite solution of the Lyapunov equation

$$
P_m A_r + A_r^T P_r = -Q_r \tag{9.56}
$$

where Q_r is symmetric positive definite. The following theorem states the stability of the adaptive NN sliding-mode control.

Theorem 9.2 *For the system* (9.18), *consider the NN based control laws* (9.51). *If the updating laws of the weights of the adaptive NNs are given by*

$$
\begin{aligned}
\big(\dot{\hat{W}}_{K_ai}^{\langle T\rangle}\big)^T &= -\Gamma_{K_ai}\langle\cdot\rangle S_{K_ai}(\bar{\zeta})\bar{\zeta}e_{\bar{\zeta}_a}^T P_r(B)_i \\
\big(\dot{\hat{W}}_{K_bi}^{\langle T\rangle}\big)^T &= -\Gamma_{K_bi}\langle\cdot\rangle S_{K_bi}(\bar{\zeta})\bar{\zeta}e_{\bar{\zeta}_b}^T P_r(B)_i
\end{aligned}
\tag{9.57}
$$

$$\begin{aligned}
\left(\dot{\hat{W}}_{Li}^{\langle T\rangle}\right)^T &= -\Gamma_{Li}\langle\cdot\rangle S_{Li}(\bar{\zeta}) f_r e_{\bar{\zeta}_a}^T P_r(B)_i \\
\dot{\hat{W}}_{\mathcal{G}i} &= -\Gamma_{\mathcal{G}i} S_{\mathcal{G}i}(\bar{\zeta})\bar{\zeta}_b e_a^T P_r(B)_i
\end{aligned}$$

and

$$\begin{aligned}
\tau_r &= -k_r \operatorname{sgn}\left(B^T P_m e_a\right) - k_2 e_a \\
k_r &= k_1 + k_{r1} + k_{r2}
\end{aligned} \tag{9.58}$$

where $(B)_i$ stands for the ith column of B, k_1, k_2 are the positive constants, k_{r1}, k_{r2} and k_{r3} are enough large constants, $\Gamma_{K_ai} \in R^{4l\times 4l}$, $\Gamma_{K_bi} \in R^{2l\times 2l}$ and $\Gamma_{Li} \in R^{(2\cdot l)\times(2\cdot l)}$, $\Gamma_{\mathcal{G}i} \in \mathbf{R}^{l\times 1}$ are the symmetric positive definite matrices, then the adaptive NN controller ensures that all states of the system are bounded, and the matching error e_a will converge to zero, i.e.,

$$\lim_{t\to\infty} \|e_a\| = 0 \tag{9.59}$$

Remark 9.3 In τ_r, we introduced last term $-k_2 e_a$, which is only used in the practical controller design to improve the controller's smoothness and doesn't affect our stability proof.

Proof Choose the following Lyapunov function

$$V_1 = U_1 + U_2 \tag{9.60}$$

with

$$\begin{aligned}
U_1 &= \frac{1}{\bar{m}} e_a^T P_m e_a \\
U_2 &= \sum_{i=1}^{2} \tilde{W}_{K_ai}^{\langle T\rangle} \Gamma_{K_ai}^{-1} \left(\tilde{W}_{K_ai}^{\langle T\rangle}\right)^T + \sum_{i=1}^{2} \tilde{W}_{K_bi}^{\langle T\rangle} \Gamma_{K_bi}^{-1} \left(\tilde{W}_{K_bi}^{\langle T\rangle}\right)^T \\
&\quad + \sum_{i=1}^{2} \tilde{W}_{Li}^{\langle T\rangle} \Gamma_{Li}^{-1} \left(\tilde{W}_{Li}^{\langle T\rangle}\right)^T + \sum_{i=1}^{2} \tilde{W}_{\mathcal{G}i}^T \Gamma_{\mathcal{G}i}^{-1} \tilde{W}_{\mathcal{G}i}
\end{aligned} \tag{9.61}$$

Differentiating the Lyapunov candidate along (9.55) and noting $\underline{m} \le |\mathcal{M}^{-1}| \le \bar{m}$ yields the following expression

$$\begin{aligned}
\dot{V}_1 &= -\frac{1}{\bar{m}} e_a^T Q_m e_a + 2e_a^T P_m B \frac{\mathcal{M}^{-1}}{\bar{m}} (\tau_r - \varepsilon_{K_a}\bar{\zeta}_a - \varepsilon_{K_b}\bar{\zeta}_b - \varepsilon_L f_r - \varepsilon_{\mathcal{G}}) \\
&\quad + 2e_a^T P_m B \frac{\mathcal{M}^{-1}}{\bar{m}} \left[\tilde{W}_{K_a}^{\langle T\rangle}\langle\cdot\rangle S_K(\bar{\zeta})\right]\bar{\zeta}_a + 2e_a^T P_m B \frac{\mathcal{M}^{-1}}{\bar{m}} \left[\tilde{W}_L^{\langle T\rangle}\langle\cdot\rangle S_L(\bar{\zeta})\right] f_r \\
&\quad + 2e_a^T P_m B \frac{\mathcal{M}^{-1}}{\bar{m}} \left[\tilde{W}_{K_b}^{\langle T\rangle}\langle\cdot\rangle S_{K_b}(\bar{\zeta})\right]\bar{\zeta}_b + 2e_a^T P_m B \frac{\mathcal{M}^{-1}}{\bar{m}} \left[\tilde{W}_G^{\langle T\rangle}\langle\cdot\rangle S_{\mathcal{G}}(\bar{\zeta})\right] \\
&\quad + 2\sum_{i=1}^{n-1} \tilde{W}_{Ki}^{\langle T\rangle} \Gamma_{Ki}^{-1} \left(\dot{\tilde{W}}_{Ki}^{\langle T\rangle}\right)^T + 2\sum_{i=1}^{n-1} \tilde{W}_{Ti}^{\langle T\rangle} \Gamma_{Ti}^{-1} \left(\dot{\tilde{W}}_{Ti}^{\langle T\rangle}\right)^T
\end{aligned}$$

$$
\begin{aligned}
&+2\sum_{i=1}^{n-1}\tilde{W}_{Li}^{\langle T\rangle}\Gamma_{Li}^{-1}\big(\dot{\tilde{W}}_{Li}^{\langle T\rangle}\big)^T+2\sum_{i=1}^{n-1}\tilde{W}_{\mathcal{G}i}^T\Gamma_{\mathcal{G}i}^{-1}\dot{\tilde{W}}_{\mathcal{G}i}\\
&\le-\frac{1}{\bar{m}}e_a^TQ_me_a+2e_a^TP_mB(\tau_r-\varepsilon_{K_a}\bar{\zeta}_a-\varepsilon_{K_b}\bar{\zeta}_b-\varepsilon_Tf_r-\varepsilon_{\mathcal{G}})\\
&+2\sum_{i=1}^{n-1}\big[\tilde{W}_{K_ai}^{\langle T\rangle}\langle\cdot\rangle S_{K_ai}(\bar{\zeta})\big]\bar{\zeta}_ae_a^TP_r(B)_i+2\sum_{i=1}^{n-1}\big[\tilde{W}_{Li}^{\langle T\rangle}\langle\cdot\rangle S_{Li}(\bar{\zeta})\big]f_re_a^TP_r(B)_i\\
&+2\sum_{i=1}^{n-1}\big[\tilde{W}_{K_bi}^{\langle T\rangle}\langle\cdot\rangle S_{K_bi}(\bar{\zeta})\big]\bar{\zeta}_be_a^TP_r(B)_i+2\sum_{i=1}^{n-1}\tilde{W}_{\mathcal{G}i}^TS_{\mathcal{G}i}(\bar{\zeta})e_a^TP_r(B)_i\\
&+2\sum_{i=1}^{n-1}\tilde{W}_{K_ai}^{\langle T\rangle}\Gamma_{K_ai}^{-1}\big(\dot{\tilde{W}}_{K_ai}^{\langle T\rangle}\big)^T+2\sum_{i=1}^{n-1}\tilde{W}_{Li}^{\langle T\rangle}\Gamma_{Li}^{-1}\big(\dot{\tilde{W}}_{Li}^{\langle T\rangle}\big)^T\\
&+2\sum_{i=1}^{n-1}\tilde{W}_{K_bi}^{\langle T\rangle}\Gamma_{K_bi}^{-1}\big(\dot{\tilde{W}}_{K_bi}^{\langle T\rangle}\big)^T+2\sum_{i=1}^{n-1}\tilde{W}_{\mathcal{G}i}^T\Gamma_{Ti}^{-1}\dot{\tilde{W}}_{\mathcal{G}i}
\end{aligned}
\tag{9.62}
$$

Remark 9.4 Here we applied $\mathcal{M}^{-1}$'s 1-norm because we need to compute $(B\mathcal{M}^{-1})$ column-wise at the beginning.

Substituting the adaptive laws (9.57) to (9.62), we have

$$
\dot{V}_1\le-\frac{1}{\bar{m}}e_a^TQ_me_a+2e_a^TP_mB(\tau_r-\varepsilon_{K_a}\bar{\zeta}_a-\varepsilon_{K_b}\bar{\zeta}_b-\varepsilon_Lf_r-\varepsilon_{\mathcal{G}})
\tag{9.63}
$$

Further substituting τ_r from (9.58), leads to

$$
\begin{aligned}
\dot{V}_1&=-\frac{1}{\bar{m}}e_a^TQ_me_a-2(k_1+k_{r2})e_a^TP_mB\,\mathrm{sgn}\big(B^TP_me_a\big)-2k_2e_a^TP_mBe_a\\
&\le-k_0\|e_a\|<0,\quad \|e_a\|\ne0
\end{aligned}
\tag{9.64}
$$

with $k_0=2k_1\|P_mB\|>0$. Expression (9.64) implies both U_1 and U_2 are bounded and consequently e_a, $\tilde{W}_{K_a}$, $\tilde{W}_L$ are bounded. And then e_a asymptotically converges to zero. This completes the proof. □

Theorem 9.5 *There exist a finite time t_e such that $\|e_a\|=0$, $t>t_e\ll t_f$.*

Proof This proof uses a contradiction argument. Assume

$$
\|e_a\|>0,\quad t>0
$$

There exist two constants $\underline{P}_r$ and $\bar{P}_r$ such that $\underline{P}_r\le\|P_r\|\le\bar{P}_r$ and consequently the following inequalities can be obtained from (9.61)

$$
\frac{1}{2}\underline{P}_r\|e_a\|^2\le U_1\le\frac{1}{2}\bar{P}_r\|e_a\|^2
\tag{9.65}
$$

At the same time, we also have that

$$\frac{d}{dt}\frac{U_1}{\|e_a\|} = \frac{\dot{U}_1}{\|e_a\|} - \frac{U_1}{\|e_a\|^2}\frac{d\|e_a\|}{dt} \tag{9.66}$$

Integrating both sides of the above equation, we arrive at

$$\int_0^t \frac{\dot{U}_1}{\|e_a\|} dt = \frac{U_1}{\|e_a\|}\bigg|_0^t + \int_0^t \frac{U_1}{\|e_a\|^2} d\|e_a\| \tag{9.67}$$

Combining the equation above with (9.65), we have

$$\underline{P}_r\|e_a\| - \underline{P}_r\|e_a(0)\| \le \int_0^t \frac{\dot{U}_1}{\|e_a\|} dt \le \bar{P}_r\|e_a\| - \bar{P}_r\|e_a(0)\| \tag{9.68}$$

Recalling the boundedness of $\|e_{\bar{\zeta}_a}\|$ and the fact $\lim_{t\to 0}\|e_{\bar{\zeta}_a}\| = 0$, we obtain the boundedness of $\int_0^\infty \frac{\dot{U}_1}{\|e_a\|} dt$, which implies

$$\lim_{t\to\infty} \frac{\dot{U}_1}{\|e_a\|} = 0 \tag{9.69}$$

On the other hand, we have

$$\begin{aligned}\dot{U}_1 \le & -e_a^T Q_m e_a + 2e_a^T P_m B\bar{m}(\tau_r - \varepsilon_{K_a}\bar{\zeta} - \varepsilon_L r_r) \\ & + 2e_a^T P_m B\bar{m}\big[\tilde{W}_{K_a}^{\langle T\rangle}\langle\cdot\rangle S_{K_a}(\bar{\zeta})\big]\bar{\zeta} + 2e_a^T P_m B\bar{m}\big[\tilde{W}_L^{\langle T\rangle}\langle\cdot\rangle S_L(\bar{\zeta})\big]r_r\end{aligned} \tag{9.70}$$

And if $k_{\min}$ is sufficiently large, then there exist a constant k' such that $k' = k_{\min} - \|Y\|\|\tilde{\Theta}\| > 0$. If $\|e_a\| > 0$, $\forall t > 0$ then we would have $\frac{\dot{U}_1}{\|e_a\|} \le -k' < 0$. This obviously conflicts with (9.69), such that we see there must exist a finite t_z such that $\|e_a\| = 0$, $t > t_e$. Further analysis shows that the larger the $k_{\min}$ is chosen the smaller t_e will be. Therefore, we could always be able to properly choose t_f and $k_{\min}$ to guarantee that $t_e \ll t_f$. This completes the proof. □

9.5 Adaptive Generator of Implicit Control Trajectory

The overall structure of the NN controller and reference trajectory generator is show in Fig. 9.1.

As above discussed, after finite time, $\bar{\zeta}_a$ will exactly track $\bar{\zeta}_{ad}$ such that the forward velocity dynamics (9.12) becomes a double integrator system as follows:

$$\Sigma_b: \quad \begin{aligned}\dot{\bar{\zeta}}_{b1} &= \bar{\zeta}_{b2} \\ \dot{\bar{\zeta}}_{b2} &= h_b(\bar{\zeta}_b, \bar{\zeta}_{ad}, \pi)\end{aligned} \tag{9.71}$$

with $\frac{\partial h_b(\cdot)}{\partial \pi} = b(\zeta_{3d}) < 0$. Consider the desired forward position and forward velocity of the vehicle as $\bar{\zeta}_{bd} = [\bar{\zeta}_{b1d}, \bar{\zeta}_{b2d}]^T = [\zeta_{bd}, \dot{\zeta}_{bd}]^T$. Then, our design objective is

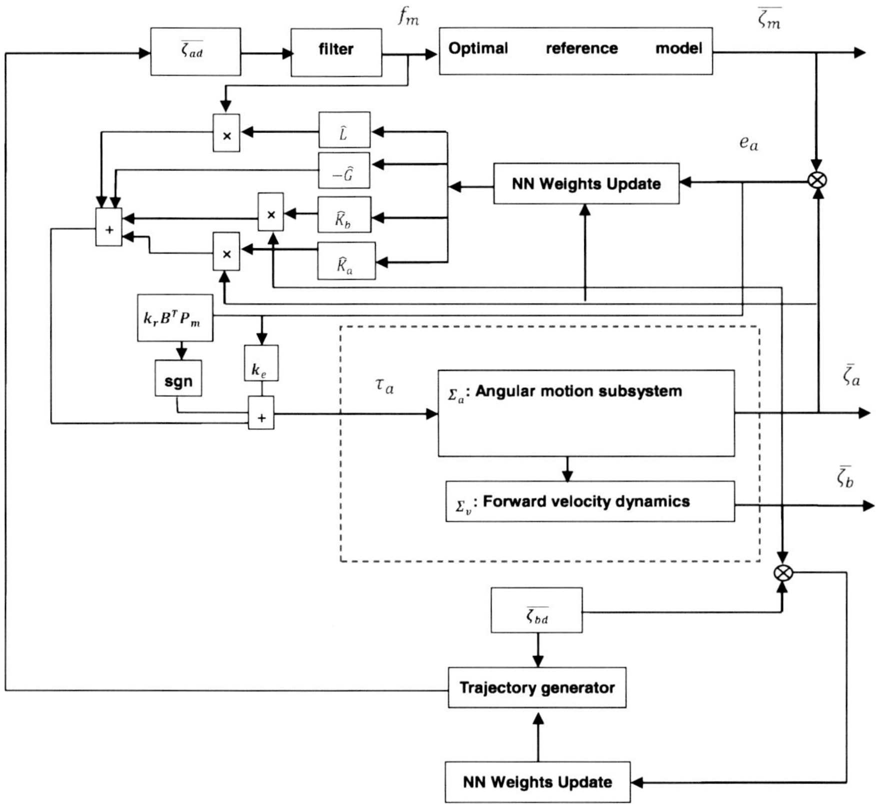

Fig. 9.1 Scheme of overall system: controller and trajectory generator

to construct a π (subsequently $\bar{\zeta}_3$) such that $\bar{\zeta}_b$ of system (9.71) follow the reference trajectory $\bar{\zeta}_{br} = [\bar{\zeta}_{b1r}, \bar{\zeta}_{b2r}]^T = [\zeta_{br}, \dot{\zeta}_{br}]^T$ generated from the following reference model

$$\begin{aligned} \dot{\bar{\zeta}}_{b1r} &= \bar{\zeta}_{b2r} \\ \dot{\bar{\zeta}}_{b2r} &= h_r(\bar{\zeta}_{bd}, \bar{\zeta}_{br}) \end{aligned} \tag{9.72}$$

where

$$h_r(\bar{\zeta}_{bd}, \bar{\zeta}_{br}) \triangleq -k_1(\zeta_{br} - \zeta_{bd}) - k_2(\dot{\zeta}_{br} - \dot{\zeta}_{bd}) + \ddot{\zeta}_{bd} \tag{9.73}$$

It can be easily checked that the reference model (9.72) ensures that $\bar{\zeta}_{br} - \bar{\zeta}_{bd} \to 0$. According to implicit function theorem (refer to Lemma 7.28) based neural network design [132], there must exist a function h_π

$$\pi^* = h_\pi(\bar{\zeta}_b, \bar{\zeta}_{ad}, h_r) \tag{9.74}$$

such that

$$h_b\big(\bar{\zeta}_b, \bar{\zeta}_{ad}, \pi^*\big) = h_r(\bar{\zeta}_{bd}, \bar{\zeta}_{br}) \tag{9.75}$$

Refer to Sect. 9.2.1, we can see that there exists an ideal RBFNN weight such that

$$\pi^* = W_\pi^{*T} S_\pi(z) + \epsilon_\pi, \quad z = [\bar{\zeta}_b, \bar{\zeta}_{ad}, h_r]^T \tag{9.76}$$

where ϵ_π is the neural network approximation error. Let us employ a RBFNN to approximate π^* as follows:

$$\hat{\pi} = \hat{W}_\pi^T(k) S_\pi(z) \tag{9.77}$$

Substituting $\hat{\pi}$ into (9.71) and using $h_b(\bar{\zeta}_b, \bar{\zeta}_{ad}, \pi^*) = h_r(\bar{\zeta}_{bd}, \bar{\zeta}_{br})$, we have

$$\begin{aligned} \dot{\bar{\zeta}}_{b1} &= \bar{\zeta}_{b2} \\ \dot{\bar{\zeta}}_{b2} &= h_r(\bar{\zeta}_{bd}, \bar{\zeta}_{br}) + \frac{\partial h_r(\cdot)}{\partial \pi}\big(\tilde{W}_\pi^T S_\pi(z) - \epsilon_\pi\big) \end{aligned} \tag{9.78}$$

where Mean Value theorem is used to obtain $h_b(\bar{\zeta}_b, \bar{\zeta}_{ad}, \hat{\pi}) - h_b(\bar{\zeta}_b, \bar{\zeta}_{ad}, \pi^*) = \frac{\partial h_r(\cdot)}{\partial \pi}(\hat{\pi} - \pi^*) = b(\zeta_3)(\tilde{W}_\pi^T S_\pi(z) - \epsilon_\pi)$ where $\tilde{W}_\pi = \hat{W} - W^*$.

Define $\tilde{\bar{\zeta}}_{b1} = \bar{\zeta}_{b1} - \bar{\zeta}_{b1r}$ and $\tilde{\bar{\zeta}}_{b2} = \bar{\zeta}_{b2} - \bar{\zeta}_{b2r}$ such that $\tilde{\bar{\zeta}}_b = \hat{\bar{\zeta}}_b - \bar{\zeta}_b$. Then, the comparison between (9.72) and (9.78) yields

$$\begin{aligned} \dot{\tilde{\bar{\zeta}}}_{b1} &= \tilde{\bar{\zeta}}_{b2} \\ \dot{\tilde{\bar{\zeta}}}_{b2} &= b(\zeta_{3d})\big(\tilde{W}_\pi^T S(z) - \epsilon\big) \\ &= -k_1\tilde{\bar{\zeta}}_{b1} - k_2\tilde{\bar{\zeta}}_{b2} + b(\zeta_{3d})\big(\tilde{W}_\pi^T S(z) - \epsilon\big) \end{aligned} \tag{9.79}$$

Theorem 9.6 *Consider the following weight adaptation law for RBFNN employed in* (9.77)

$$\dot{\hat{W}}_\pi = \Gamma_{W_\pi} S_\pi \tilde{\bar{\zeta}}^T P_{W_\pi} [0,\ 1]^T - \sigma \Gamma_{W_\pi} \hat{W}_\pi \tag{9.80}$$

where Γ_W and σ are suitably chosen positive definite matrix and positive scalar. Then, the tracking errors $\tilde{\bar{\zeta}}_{b1}$ and $\tilde{\bar{\zeta}}_{b2}$ in (9.79) *will be eventually bounded into a small neighborhood around zero.*

Proof Let us rewrite the error dynamics (9.79) as

$$\dot{\tilde{\bar{\zeta}}}_b = A_r \tilde{\bar{\zeta}}_b + b(v')[0,\ 1]^T\big(\tilde{W}_\pi^T S(z) - \epsilon_\pi\big) \tag{9.81}$$

where $A_r = \begin{bmatrix} 0 & 1 \\ -k_1 & -k_2 \end{bmatrix}$ satisfies the Lyapunov equation

$$A_r^T P_W + P_W A_r = -Q_W$$

i.e., for any symmetric positive definite matrix Q_W, there exists a symmetric positive definite P_W satisfying the above equation.

Considering the following Lyapunov function

$$V_2(t) = \tilde{\bar{\zeta}}_b^T P_W \tilde{\bar{\zeta}}_b + \big|b(\xi_{3d})\big| \tilde{W}^T \Gamma_W^{-1} \tilde{W}_\pi \tag{9.82}$$

and the closed-loop dynamics (9.81) with the update law (9.80), and notice $b(\xi_{3d}) < 0$, we obtain

$$
\begin{aligned}
\dot{V}_2(t) &= 2\bar{\tilde{\zeta}}_b^T P_W \dot{\bar{\tilde{\zeta}}}_b + 2\big|b(\xi_{3d})\big| \tilde{W}_\pi^T \Gamma_W^{-1} \dot{\tilde{W}}_\pi \\
&= 2\bar{\tilde{\zeta}}_b^T P_W A_r \bar{\tilde{\zeta}}_b + 2\bar{\tilde{\zeta}}_b^T P_W b(\xi_{3d})[0,\ 1]^T \big(\tilde{W}_\pi^T S(z) - \epsilon_\pi\big) \\
&\quad + 2\big|b(\xi_{3d})\big| \tilde{W}_\pi^T \Gamma_W^{-1} \Gamma_W S \bar{\tilde{\zeta}}_b^T P[0,\ 1]^T - 2\big|b(\xi_{3d})\big| \tilde{W}_\pi^T \Gamma_W^{-1} \Gamma_W \sigma \hat{W} \\
&= \bar{\tilde{\zeta}}_b^T \big(A_r^T P_W + P_W A_r\big) \bar{\tilde{\zeta}}_b + 2\big|b(\xi_{3d})\big| \bar{\tilde{\zeta}}_b^T P_W [0,\ 1]^T \epsilon_\pi \\
&\quad - 2\sigma \big|b(\xi_{3d})\big| \tilde{W}_\pi^T \big(\tilde{W}_\pi + W^*\big) \\
&\le -\bar{\tilde{\zeta}}_b^T Q_W \bar{\tilde{\zeta}}_b - 2\sigma \big|b(\xi_{3d})\big| \|\tilde{W}_\pi\|^2 + 2\big|b(\xi_{3d})\big| \big\|\bar{\tilde{\zeta}}_b^T P_W [0,\ 1]^T\big\| \epsilon_\pi \\
&\quad - 2\sigma \big|b(\xi_{3d})\big| \tilde{W}_\pi^T W^* \\
&\le -\lambda_{Q_W} \|\bar{\tilde{\zeta}}_b\|^2 - 2\sigma \big|b(\xi_{3d})\big| \|\tilde{W}_\pi\|^2 + \varepsilon^2 \|\bar{\tilde{\zeta}}_b\|^2 + \varepsilon^2 \|\tilde{W}_\pi\|^2 \\
&\quad + \frac{1}{\varepsilon^2} \epsilon_\pi^2 \big|b(\xi_{3d})\big|^2 \big\|P_W [0,\ 1]^T\big\|^2 + \frac{1}{\varepsilon^2} \sigma^2 \big|b(\xi_{3d})\big|^2 \big\|W^*\big\|^2
\end{aligned}
$$

where $|\epsilon_\pi| \le \epsilon_\pi$, λ_{Q_W} is the minimum eigenvalue of Q_W, ε is any given positive constant and we can choose it sufficiently small. Furthermore, we can choose the suitable Q_W and σ making $\lambda_{Q_W} \ge \varepsilon^2$, $\sigma|b(\xi_{3d})| \ge \varepsilon^2$, and it follows that $\dot{V}_2(t) \le 0$ in the complementary set of a set S_b defined as

$$
S_b \triangleq \left\{ (\bar{\tilde{\zeta}}_b, \tilde{W}_\pi) \,\middle|\, \frac{\|\tilde{W}_\pi\|^2}{\bar{a}^2} + \frac{\|\tilde{\varphi}\|^2}{\bar{b}^2} - 1 \le 0 \right\}
$$

with

$$
\begin{aligned}
\bar{a} &= \frac{1}{\varepsilon} \big|b(\xi_{3d})\big| \sqrt{\epsilon_0^2 \big\|P_W[0,\ 1]^T\big\|^2 + \sigma^2 \big\|W^*\big\|^2} \Big/ \sqrt{\lambda_{Q_W} - \varepsilon^2} \\
\bar{b} &= \frac{1}{\varepsilon} \big|b(\xi_{3d})\big| \sqrt{\epsilon_0^2 \big\|P_W[0,\ 1]^T\big\|^2 + \sigma^2 \big\|W^*\big\|^2} \Big/ \sqrt{2\sigma \big|b(\xi_{3d})\big| - \varepsilon^2}
\end{aligned}
$$

Obviously, the set S_b defined above is compact. Hence, by LaSalle's theorem, it follows that all the solutions of (9.81) are bounded. Thus, the proof is completed. □

9.6 Simulation Studies

The simulation study is carried out under the same condition as in Chap. 8. For RBFNN employed to control the angular motion subsystem Σ_a, each basis function $S_{K_{a}i,j}$, $S_{K_{b}i,j}$, $S_{Li,j}$ and $S_{\mathcal{G}i}$ for $\hat{K}_a(\bar{\zeta})$, $\hat{K}_b(\bar{\zeta})$, $\hat{L}(\bar{\zeta})$ and $\hat{\mathcal{G}}(\bar{\zeta})$, respectively contains 45 nodes. The RBFNN used to generate trajectory ζ_{3d} contains 65 nodes. The initial values of NN weights are chosen in such a way that each element is selected as an random variable with magnitude 0.01. The tuning gain matrix as well as forgetting factor of NN weights estimator are chosen as $\Gamma_{K_a i} = \mathrm{diag}[0.5, \ldots, 0.5]$, $\Gamma_{K_b i} = \mathrm{diag}[0.5, \ldots, 0.5]$, $\Gamma_{Li} = \mathrm{diag}[0.2, \ldots, 0.2]$, $\Gamma_{\mathcal{G}i} = \mathrm{diag}[0.6, \ldots, 0.6]$, $\Gamma_{W_\pi} = \mathrm{diag}[0.8, \ldots, 0.8]$ and $\sigma = 0.01$.

Similarly as in Chap. 8, we compare the developed angular motion subsystem with the same model based controller under the same operation condition and using same NN reference trajectory generator for ζ_{3d}. The tracking performance of

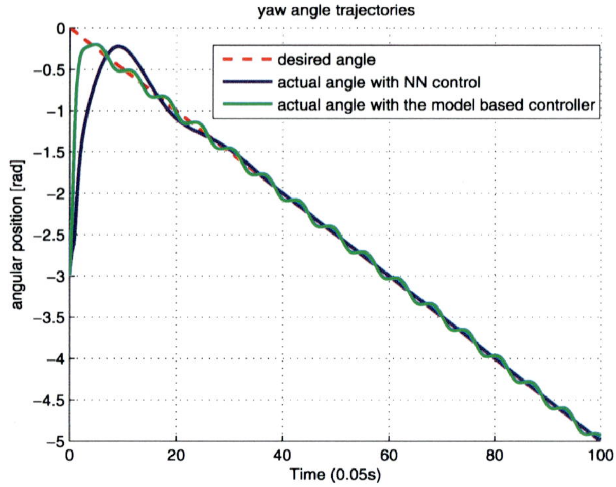

Fig. 9.2 Trajectories of yaw angle

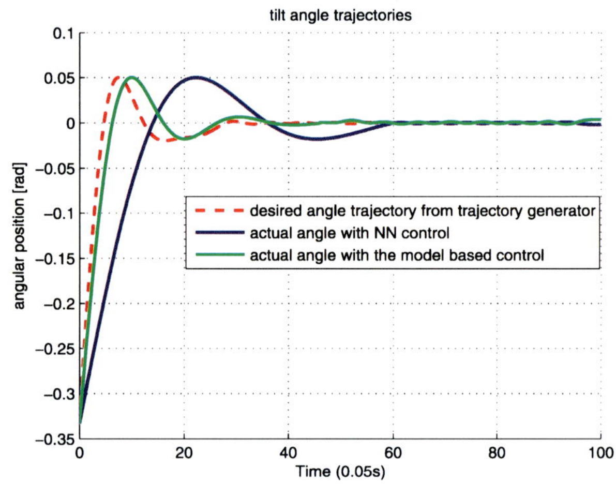

Fig. 9.3 Trajectories of tilt angle

the yaw angle and tilt angles are shown in Figs. 9.2 and 9.3, respectively. Forward velocity is show in Fig. 9.4 and the boundedness of NN weights adaptation for both controller and trajectory generator are shown in Fig. 9.5. The control torques inputs τ_1 and τ_2 are shown in Fig. 9.6 and the ground floor trajectory of the center of WIP is shown in Fig. 9.7.

9.7 Conclusion

In this chapter, NN adaptive controller and trajectory generator have been developed for WIP models which are widely employed for a class of transportation vehicles. Model reference control is designed using NN and variable structure for the yaw

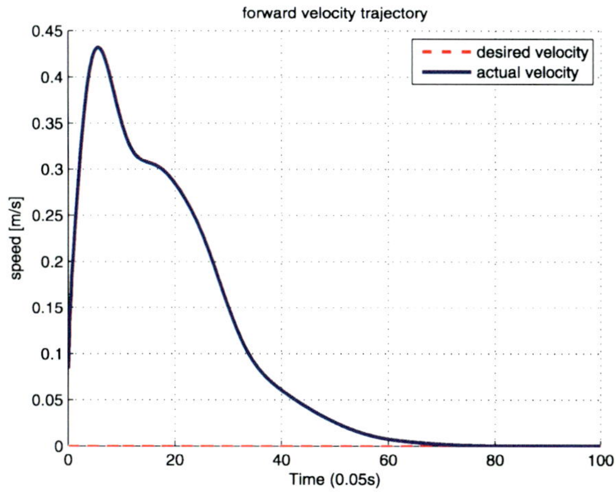

Fig. 9.4 Trajectories of forward velocity

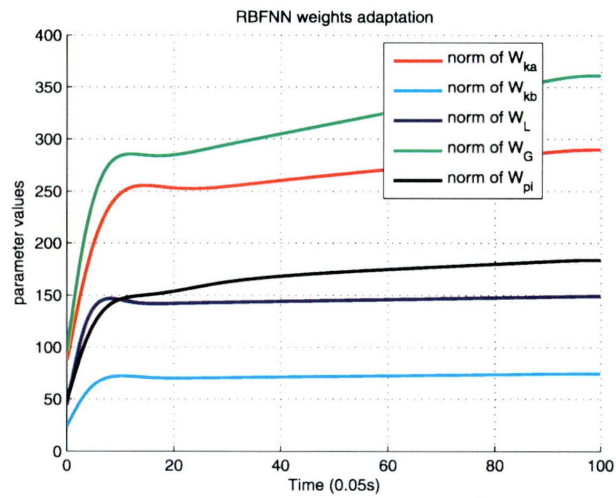

Fig. 9.5 Estimates of parameters

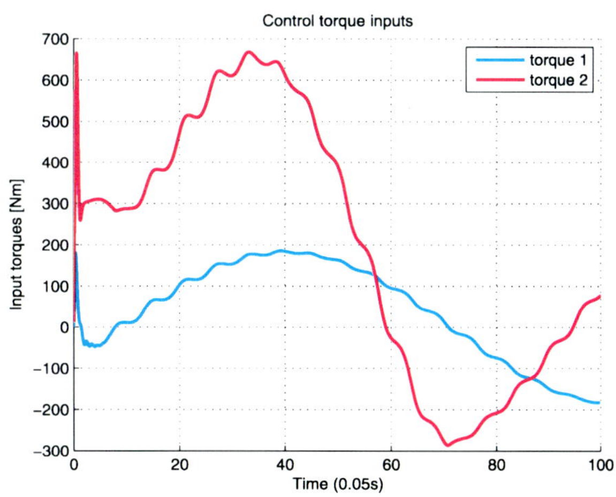

Fig. 9.6 Control input torques

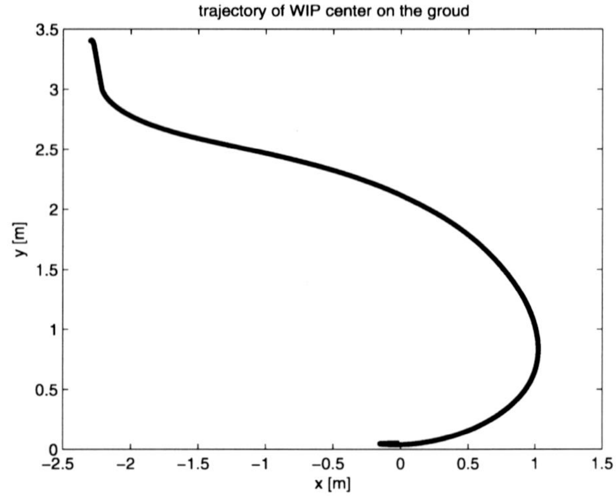

Fig. 9.7 Trajectory of center of WIP on the ground

and tilt angular motion subsystem of the WIP model such that the angular motion dynamics can match the optimal model in prescribed finite time, while the trajectory generated from the NN based adaptive generator of implicit control trajectory of tilt angle indirectly "control" the forward velocity such that it tracks the desired velocity asymptotically. The developed scheme achieves dynamic balance and desired motion tracking. The performance has been rigorously established with theoretic analysis and has also been verified by the simulation studies.

References

1. http://coecsl.ece.illinois.edu/ge423/spring04/group9/index.htm
2. http://en.wikipedia.org/wiki/AIBO
3. http://en.wikipedia.org/wiki/ASIMO
4. http://en.wikipedia.org/wiki/BigDog
5. http://en.wikipedia.org/wiki/Nao_robot
6. http://en.wikipedia.org/wiki/TOPIO
7. http://homepage.mac.com/sigfpe/Robotics/equibot.html
8. http://search.japantimes.co.jp/member/member.html?nn20011231a9.htm
9. http://www.geology.smu.edu/~dpa-www/robo/nbot/
10. http://www.rezero.ethz.ch/project_en.html
11. http://www-robotics.cs.umass.edu/Robots/UBot-5
12. Anderson, B.D.O., Moore, J.B.: Optimal Control. Prentice Hall, London (1989)
13. Anupoju, C.M., Su, C.Y., Oya, M.: Adaptive motion tracking control of uncertain nonholonomic mechanical systems including actuator dynamics. IEE Proc., Control Theory Appl. **152**(5), 575–580 (2005)
14. Arai, H., Tanie, K.: Nonholonomic control of a three-DOF planar underactuated manipulator. IEEE Trans. Robot. Autom. **14**(5), 681–694 (1998)
15. Astol, A.: Discontinuous control of nonholonomic systems. Syst. Control Lett. **27**, 37–45 (1996)
16. Barron, A.R.: Approximation and estimation bounds for artificial neural networks. In: Proc. 4th Ann. Workshop on Computational Learning Theory, pp. 243–249 (1991)
17. Barron, A.R.: Universal approximation bounds for superposition for a sigmoidal function. IEEE Trans. Inf. Theory **39**(3), 930–945 (1993)
18. Bi, D., Li, Y.F., Tso, S.K., Wang, G.L.: Friction modeling and compensation for haptic display based on support vector machine. IEEE Trans. Ind. Electron. **51**(2), 491–500 (2004)
19. Blankespoor, A., Roemer, R.: Experimental verification of the dynamic model for a quarter size self-balancing wheelchair. In: American Control Conference, Boston, MA, June 2004, pp. 488–492 (2004)
20. Brockett, R.W.: Asymptotic stability and feedback stabilization. In: Brockett, R.W., Millman, R.S., Sussmann, H.J. (eds.) Differential Geometric Control Theory, pp. 181–191. Birkhäuser, Boston (1983)
21. Brooks, R., Aryanada, L., Edsinger, A., Fitzpatrick, P., Kemp, C.C., O'Reilly, U., Torres-jara, E., Varshavskaya, P., Weber, J.: Sensing and manipulating built-for-human environments. Int. J. Humanoid Robot. **1**(1), 1–28 (2004)

Z. Li et al., *Advanced Control of Wheeled Inverted Pendulum Systems*,
DOI 10.1007/978-1-4471-2963-9, © Springer-Verlag London 2013

22. Campion, G., Bastin, G., d'Andréa-Novel, B.: Structural properties and classification of kinematic and dynamic models of wheeled mobile robots. IEEE Trans. Robot. Autom. **12**(1), 47–62 (1996)
23. Chang, Y.C., Chen, B.S.: Robust tracking designs for both holonomic and nonholonomic constrained mechanical systems: adaptive fuzzy approach. IEEE Trans. Fuzzy Syst. **8**, 46–66 (2000)
24. Chen, T.P., Chen, H.: Approximation capability to functions of several variables, nonlinear functionals, and operators by radial basis function neural networks. IEEE Trans. Neural Netw. **6**(4), 904–910 (1995)
25. Chen, W., Ballance, D.J., Gawthrop, P.J., O'Reilly, J.: A nonlinear disturbance observer for robotic manipulators. IEEE Trans. Ind. Electron. **47**(4), 932–938 (2000)
26. Chen, X., Komada, S., Fukuda, T.: Design of a nonlinear disturbance observer. IEEE Trans. Ind. Electron. **47**(2), 429–437 (2000)
27. Cherkassky, V., Ghering, D., Mulier, F.: Comparison of adaptive methods for function estimation from samples. IEEE Trans. Neural Netw. **7**(4), 969–984 (1996)
28. Colbaugh, R., Seraji, H., Glass, K.: Direct adaptive impedance control of manipulators. In: Proceedings of the 30th Conference on Decision and Control, pp. 2410–2415 (1991)
29. d'Andréa-Novel, B., Bastin, G., Campion, G.: Modelling and control of non holonomic wheeled mobile robots. In: Proceedings of 1991 International Conference on Robotics and Automation, Sacramento, CA, April 1991, pp. 1130–1135 (1991)
30. De Luca, A., Oriolo, G.: Trajectory planning and control for planar robots with passive last joint. Int. J. Robot. Res. **21**(5–6), 575–590 (2002)
31. de Wit, C.C., Berghuis, H., Nijmeijer, H.: Practical stabilization of nonlinear systems in chained form. In: Proceedings of the 33rd IEEE Conference on Decision & Control, Lake Buena, Vista, FL, USA, pp. 3475–3480
32. Desoer, C.A., Vidyasagar, M.: Feedback Systems: Input–Output Properties. Academic Press, New York (1975)
33. Dixon, W.E., Dawson, D.M., Zergeroglu, E., Behal, A.: Nonlinear Control of Wheeled Mobile Robots. LNCIS, vol. 262. Springer, London (2001)
34. Do, K., Pan, J.: Adaptive global stabilization of nonholonomic systems with strong nonlinear drifts. Syst. Control Lett. **46**, 195–205 (2002)
35. Do, K.D., Seet, G.: Motion control of a two-wheeled mobile vehicle with an inverted pendulum. J. Intell. Robot. Syst. **60**, 577–605 (2010)
36. Dong, W.: On trajectory and force tracking control of constrained mobile manipulators with parameter uncertainty. Automatica **38**(9), 1475–1484 (2002)
37. Dong, W., Xu, Y., Huo, W.: Trajectory tracking control of dynamics nonholonomic systems with unknown dynamics. Int. J. Robust Nonlinear Control **9**, 905–922 (1999)
38. Fahimi, F.: Sliding-mode formation control for underactuated surface vessels. IEEE Trans. Robot. Autom. **23**(3), 617–622 (2007)
39. Fierro, R., Lewis, F.L.: Control of a nonholonomic mobile robot using neural networks. IEEE Trans. Neural Netw. **9**(4), 589–600 (1998)
40. Gans, N.R., Hutchinson, S.A.: Visual servo velocity and pose control of a wheeled inverted pendulum through partial-feedback linearization. In: Proc. IEEE/RSJ Int. Conf. Intelligent Robots and Systems, pp. 3823–3828 (2006)
41. Ge, S.S., Wang, C.: Adaptive neural control of uncertain MIMO nonlinear systems. IEEE Trans. Neural Netw. **15**(3), 674–692 (2004)
42. Ge, S.S., Zhang, J.: Neural network control of nonaffine nonlinear system with zero dynamics by state and output feedback. IEEE Trans. Neural Netw. **14**(4), 900–918 (2003)
43. Ge, S.S., Hang, C.C., Woon, L.C.: Adaptive neural network control of robot manipulators in task space. IEEE Trans. Ind. Electron. **44**(6), 746–752 (1997)
44. Ge, S.S., Lee, T.H., Harris, C.J.: Adaptive Neural Network Control of Robot Manipulators. World Scientific, London (1998)
45. Ge, S.S., Hang, C.C., Zhang, T.: Adaptive neural network control of nonlinear systems by state and output feedback. IEEE Trans. Syst. Man Cybern., Part B **29**(6), 818–828 (1999)

46. Ge, S.S., Hang, C.C., Zhang, T.: A direct adaptive controller for dynamic systems with a class of nonlinear parameterizations. Automatica **35**, 741–747 (1999)
47. Ge, S.S., Hang, C.C., Lee, T.H., Zhang, T.: Stable Adaptive Neural Network Control. Kluwer Academic, Norwell (2001)
48. Ge, S.S., Sun, Z., Lee, T.H., Spong, M.W.: Feedback linearization and stabilization of second-order nonholonomic chained systems. Int. J. Control **74**(14), 1383–1392 (2001)
49. Ge, S.S., Wang, J., Lee, T.H., Zhou, G.Y.: Adaptive robust stabilization of dynamic nonholonomic chained systems. J. Robot. Syst. **18**(3), 119–133 (2001)
50. Ge, S.S., Wang, Z., Lee, T.H.: Adaptive stabilization of uncertain nonholonomic systems by state and output feedback. Automatica **39**(8), 1451–1460 (2003)
51. Ge, S.S., Ren, B., Lee, T.H.: Hard disk drives control in mobile applications. J. Syst. Sci. Complex. **20**(2), 215–224 (2007)
52. Ge, S.S., Yang, C., Lee, T.H.: Adaptive predictive control using neural network for a class of pure-feedback systems in discrete-time. IEEE Trans. Neural Netw. **19**(9), 1599–1614 (2008)
53. Ge, S.S., Ren, B., Tee, K.P., Lee, T.H.: Approximation based control of uncertain helicopter dynamics. IET Control Theory Appl. **3**(7), 941–956 (2009)
54. Giles, C.L., Maxwell, T.: Learning, invariance, and generalization in high-order neural networks. Appl. Opt. **26**(23), 4972–4978 (1987)
55. Gracanin, D., Valavanis, K.P., Tsourveloudis, N.C., Matijasevic, M.: Virtual-environment-based navigation and control of underwater vehicles. IEEE Robot. Autom. Mag. **6**(2), 53–63 (1999)
56. Grasser, F., Arrigo, A., Colombi, S., Rufer, A.C.: JOE: a mobile, inverted pendulum. IEEE Trans. Ind. Electron. **49**(1), 107–114 (2002)
57. Gruszka, A., Malisoff, M., Mazenc, F.: Tracking control and robustness analysis for planar vertical takeoff and landing aircraft under bounded feedbacks. Int. J. Robust Nonlinear Control (2011). doi:10.1002/rnc.1794
58. Gupta, M.M., Rao, D.H.: Neuro-control Systems: Theory and Applications. IEEE Press, New York (1994)
59. Ha, Y.S., Yuta, S.: Trajectory tracking control for navigation of the inverse pendulum type self-contained mobile robot. Robot. Auton. Syst. **17**, 65–80 (1996)
60. Hahn, W.: Stability of Motion. Springer, Berlin (1967)
61. Han, H., Su, C., Stepanenko, Y.: Adaptive control of a class of nonlinear systems with nonlinearly parameterized fuzzy approximators. IEEE Trans. Fuzzy Syst. **9**(2), 315–323 (2001)
62. Hespanha, J.P., Liberzon, S., Morse, A.S.: Towards the supervisory control of uncertain nonholonomic systems. In: Proceedings of the 1999 American Control Conference, San Diego, CA, USA, 1999, pp. 3520–3524 (1999)
63. Hong, F., Ge, S.S., Lewis, F.L., Lee, T.H.: Adaptive neural-fuzzy control of nonholonomic mobile robots. In: Ge, S.S., Lewis, F.L. (eds.) Autonomous Mobile Robots: Sensing, Control, Decision Making and Application, pp. 229–265. Taylor & Francis, Boca Raton (2006)
64. Hovakimyan, N., Nardi, F., Calise, A.J.: A novel error observer-based adaptive output feedback approach for control of uncertain systems. IEEE Trans. Autom. Control **47**(8), 1310–1314 (2002)
65. Huo, W., Ge, S.S.: Exponential stabilization of nonholonomic systems: an ENI approach. Int. J. Control **74**(15), 1492–1500 (2001)
66. Ibanez, C.A., Frias, O.G., Castanon, M.S.: Lyapunov-based controller for the inverted pendulum cart system. Nonlinear Dyn. **40**(4), 367–374 (2005)
67. Isidori, A.: Nonlinear Control Systems, 3rd edn. Springer, London (1995)
68. Isidori, A., Marconi, L., Serrani, A.: Robust Autonomous Guidance: An Internal Model Approach. Springer, New York (2003)
69. Ito, K., Matsuno, F., Takahashi, R.: Underactuated crawling robot. In: Proceedings of the 2000 IEEE/RSJ International Conference on Intelligent Robots and Systems, pp. 1684–1689 (2000)
70. Jiang, Z.P.: Robust exponential regulation of nonholonomic systems with uncertainties. Automatica **36**(2), 189–209 (2000)

71. Jiang, Z.P., Nijmeijer, H.: A recursive technique for tracking control of nonholonomic systems in chained form. IEEE Trans. Autom. Control **44**(2), 265–279 (1999)
72. Kaashoek, M.A., van Schuppen, J.H., Ran, A.C.M.: Robust Control of Linear Systems and Nonlinear Control. Birkhäuser, Boston (1990)
73. Karnopp, D.: Computer simulation of stick-slip friction in mechanical dynamic systems. ASME J. Dyn. Syst. Meas. Control **107**, 100–103 (1985)
74. Khalil, H.K.: Nonlinear Systems, 3rd edn. Prentice Hall, Upper Saddle River (2002)
75. Kim, Y., Kim, S.H., Kwak, Y.K.: Dynamic analysis of a nonholonomic two-wheeled inverted pendulum robot. J. Intell. Robot. Syst. **44**, 25–46 (2005)
76. Kim, Y.H., Lewis, F.L.: Neural network output feedback control of robot manipulators. IEEE Trans. Robot. Autom. **15**(2), 301–309 (1999)
77. Kolmanovsky, I., McClamroch, N.H.: Developments in nonholonomic control problems. IEEE Control Syst. Mag. **15**(6), 20–36 (1995)
78. Kosmatopoulos, E.B., Polycarpou, M.M., Christodoulou, M.A., Ioannou, P.A.: High-order neural network structures for identification of dynamical systems. IEEE Trans. Neural Netw. **6**(2), 422–431 (1995)
79. Krstic, M., Kanellakopoulos, I., Kokotovic, P.: Nonlinear and Adaptive Control Design. Wiley, New York (1995)
80. Kwan, C., Lewis, F.L., Dawson, D.M.: Robust neural-network control of rigid-link electrically driven robots. IEEE Trans. Neural Netw. **9**(4), 581–588 (1998)
81. Lasalle, J.P., Lefschetz, S.: Stability by Liapunov's Direct Method. Academic Press, New York (1961)
82. Lauwers, T., Kantor, G., Hollis, R.: A dynamically stable single-wheeled mobile robot with inverse mouse-ball drive. In: IEEE International Conference on Robotics and Automation, pp. 2889–2895 (2006)
83. Lewis, F.L., Yesildirek, A., Liu, K.: Multilayer neural network robot controller with guaranteed tracking performance. IEEE Trans. Neural Netw. **7**(2), 388–399 (1996)
84. Lewis, F.L., Jagannathan, S., Yesildirek, A.: Neural Network Control of Robot Manipulators and Nonlinear Systems. Taylor & Francis, London (1999)
85. Li, Z., Yang, C.: Neural-adaptive output feedback control of a class of transportation vehicles based on wheeled inverted pendulum models. IEEE Trans. Control Syst. Technol. (2011). doi:10.1109/TCST.2011.216822
86. Li, Z., Gu, J., Ming, A., Xu, C., Shimojo, M.: Intelligent complaint force/motion control of nonholonomic mobile manipulator working on the non-rigid surface. Neural Comput. Appl. **15**(3–4), 204–216 (2006)
87. Li, Z., Yang, C., Gu, J.: Neuro-adaptive compliant force/motion control for uncertain constrained wheeled mobile manipulator. Int. J. Robot. Autom. **22**(3), 206–214 (2007)
88. Li, Z., Chen, W., Luo, J.: Adaptive compliant force–motion control of coordinated nonholonomic mobile manipulators interacting with unknown non-rigid environments. Neurocomputing **71**(7–9), 1330–1344 (2008)
89. Li, Z., Yang, C., Ding, N.: Robust adaptive motion control for remotely operated vehicles with velocity constraints. Int. J. Control Autom. Syst. (accepted)
90. Lin, W., Qian, C.: Adding one power integrator: a tool for global stabilization of high-order cascade nonlinear systems. Syst. Control Lett. **39**, 339–351 (2000)
91. Liu, Z., Svoboda, J.: A new control scheme for nonlinear systems with disturbances. IEEE Trans. Control Syst. Technol. **14**(1), 176–181 (2006)
92. Loreto, G., Garrido, R.: Stable neurovisual servoing for robot manipulators. IEEE Trans. Neural Netw. **17**(4), 953–965 (2006)
93. Lu, G., Song, J., Hua, L., Sun, C.: Inverse system control of nonlinear systems using LS-SVM. In: Proceedings of the 26th Chinese Control Conference, China, 2007, pp. 233–236 (2007)
94. Lu, W.S., Meng, Q.H.: Impedance control with adaptation for robotic manipulations. IEEE Trans. Robot. Autom. **7**(3), 408–415 (1991)
95. Lyapunov, A.M.: Stability of Motion. Academic Press, New York (1966)

96. M'Closkey, R.T., Murray, R.M.: Convergence rate for nonholonomic systems in power form. In: Proceedings of the American Control Conference, Chicago, USA, pp. 2489–2493 (1992)
97. Miyashitaa, T., Ishiguroa, H.: Human-like natural behavior generation based on involuntary motions for humanoid robots. Robot. Auton. Syst. **48**, 203–212 (2004)
98. Munkres, J.R.: Analysis on Manifolds. Addison-Wesley, Reading (1991)
99. Murray, R., Sastry, S.: Nonholonomic motion planning: steering using sinusoids. IEEE Trans. Autom. Control **38**, 700–716 (1993)
100. Ogata, K.: Modern Control Engineering. Prentice Hall, New York (1997)
101. Oryschuk, P., Salerno, A., Al-Husseini, A.M., Angeles, J.: Experimental validation of an underactuated two-wheeled mobile robot. IEEE/ASME Trans. Mechatron. **14**(2), 252–257 (2009)
102. Paretto, P., Niez, J.J.: Long term memory storage capacity of multiconnected neural networks. Biol. Cybern. **54**(1), 53–63 (1986)
103. Pathak, K., Franch, J., Agrawal, S.K.: Velocity and position control of a wheeled inverted pendulum by partial feedback linearization. IEEE Trans. Robot. **21**(3), 505–513 (2005)
104. Petrov, P., Parent, M.: Dynamic modeling and adaptive motion control of a two-wheeled self-balancing vehicle for personal transport. In: 13th International IEEE Annual Conference on Intelligent Transportation Systems, Madeira Island, Portugal, September 19–22, 2010, pp. 1013–1018 (2010)
105. Polycarpou, M.M., Ioannou, P.: Learning and convergence analysis of neural-type structured networks. IEEE Trans. Neural Netw. **3**(1), 39–50 (1992)
106. Salerno, A., Angeles, J.: On the nonlinear controllability of a quasiholonomic mobile robot. In: Proc. IEEE Int. Conf. Robotics and Automation, pp. 3379–3384 (2003)
107. Salerno, A., Angeles, J.: The control of semi-autonomous two-wheeled robots undergoing large payload-variations. In: Proc. IEEE Int. Conf. Robotics and Automation, pp. 1740–1745 (2004)
108. Samson, C.: Time-varying feedback stabilization of a nonholonomic wheeled mobile robot. Int. J. Robot. Res. **12**, 55–66 (1993)
109. Segway Human Transporter. http://www.segway.com (2004)
110. Siciliano, B., Sciavicco, L., Villani, L., Oriolo, G.: Robotics: Modelling, Planning and Control. Springer, Berlin (2008)
111. Simaan, N., Taylor, R., Flint, P.: High dexterity snake-like robotic slaves for minimally invasive telesurgery of the upper airway. In: Medical Image Computing and Computer-Assisted Intervention—MICCAI 2004. Lecture Notes in Computer Science, vol. 3217, pp. 17–24 (2004)
112. Slotine, J.J.E., Li, W.: On the adaptive control of robotic manipulators. Int. J. Robot. Res. **6**(3) (1987)
113. Slotine, J.J.E., Li, W.: Applied Nonlinear Control. Prentice Hall, Englewood Cliffs (1991)
114. Sordalen, O.J., de Wit, C.C.: Exponential stabilization of nonholonomic chained systems. IEEE Trans. Autom. Control **40**(1), 35–49 (1995)
115. Sotirov, Z.M., Botev, R.G.: A model reference approach to adaptive impedance control of robot manipulators. In: Proceedings of the 1993 IEEE/RSJ International Conference on Intelligent Robots and Systems, pp. 727–733 (1993)
116. Suykens, J.A.K., Vandewalle, J., Moor, B.D.: Optimal control by least squares support vector machines. Neural Netw. **14**(1), 23–35 (2001)
117. Tinos, R., Terra, M.H., Ishihara, J.Y.: Motion and force control of cooperative robotic manipulators with passive joints. IEEE Trans. Control Syst. Technol. **14**(4), 725–734 (2006)
118. Vapnik, V.: An overview of statistical learning theory. IEEE Trans. Neural Netw. **10**(5), 955–999 (1999)
119. Vapnik, V.N.: Statistical Learning Theory. Springer, New York (1998)
120. Vilchis, J.C.A., Brogliato, B., Dzul, A., Lozano, R.: Nonlinear modeling and control of helicopters. Automatica **39**(9), 1583–1596 (2003)
121. Walsh, G.C., Bushnell, L.G.: Stabilization of multiple input chained form control systems. Syst. Control Lett. **25**, 227–234 (1995)

122. Wang, G.L., Li, Y.F., Bi, D.X.: Support vector machine networks for friction modeling. IEEE/ASME Trans. Mechatron. **9**(3), 601–606 (2004)
123. Wang, J., Chen, Q., Chen, Y.: RBF kernel based support vector machine with universal approximation and its application. In: Support Vector Machines, Part III. Lecture Notes in Computer Science, vol. 3173, pp. 512–517 (2004)
124. Wang, L.: Adaptive Fuzzy Systems and Control, Design, and Stability Analysis. Prentice Hall, Englewood Cliffs (1994)
125. Wang, S., Lai, L., Wu, C., Shiue, Y.: Kinematic control of omni-directional robots for time-optimal movement between two configurations. J. Intell. Robot. Syst. **49**(4), 397–410 (2007)
126. Wang, Z., Ge, S.S., Lee, T.H.: Robust motion/force control of uncertain holonomic/nonholonomic mechanical systems. IEEE/ASME Trans. Mechatron. **9**(1), 118–123 (2004)
127. Wang, Z., Su, C., Ge, S.S.: Adaptive control of mobile robots including actuator dynamics. In: Ge, S.S., Lewis, F.L. (eds.) Autonomous Mobile Robots: Sensing, Control, Decision Making and Application, pp. 267–293. Taylor & Francis, Boca Raton (2006)
128. Wedeward, K., Colbaugh, R.: New stability results for direct adaptive impedance control. In: Proceedings of the 1995 IEEE International Symposium on Intelligent Control, pp. 281–287 (1995)
129. Williams, R.L. II, Carter, B.E., Gallina, P., Rosati, G.: Dynamic model with slip for wheeled omni-directional robots. IEEE Trans. Robot. Autom. **18**(3), 285–293 (2002)
130. Xin, X., Kaneda, M.: Swing-up control for a 3-DOF gymnastic robot with passive first joint: design and analysis. IEEE Trans. Robot. **23**(6), 1277–1285 (2007)
131. Xu, J., Chen, S.: Adaptive control of a class of nonlinear discrete-time systems using support vector machine. In: Proceedings of the 5th World Congress on Intelligent Control and Automation, China, 2004, pp. 440–443 (2004)
132. Yang, C., Ge, S.S., Xiang, C., Chai, T.Y., Lee, T.H.: Output feedback NN control for two classes of discrete-time systems with unknown control directions in a unified approach. IEEE Trans. Neural Netw. **19**(11), 873–1886 (2008)
133. Yang, Z., Hara, S., Kanae, S., Wada, K., Su, C.Y.: An adaptive robust nonlinear motion controller combined with disturbance observer. IEEE Trans. Control Syst. Technol. **18**(2), 454–462 (2010)
134. Zhang, H.R., Wang, X.D., Zhang, C.J., Cai, X.S.: Robust identification of non-linear dynamic systems using support vector machine. IEE Proc. Sci. Meas. Technol. **153**(3), 125–129 (2006)
135. Zhang, M., Tarn, T.: Hybrid control of the Pendubot. IEEE/ASME Trans. Mechatron. **7**(1), 79–86 (2002)

Index

Z. Li et al., *Advanced Control of Wheeled Inverted Pendulum Systems*,
DOI 10.1007/978-1-4471-2963-9, © Springer-Verlag London 2013

Printed by Publishers' Graphics LLC
MO20120918-2010-133